Essential Mathematics for Electronics Technicians

FRED R. MONACO, PH.D.
Los Angeles Trade Technical College

Merrill, an Imprint of
Macmillan Publishing Company
New York

Collier Macmillan Canada, Inc.
Toronto

Maxwell Macmillan International Publishing Group
New York   Oxford   Singapore   Sydney

Administrative Editor: David Garza
Developmental Editor: Carol Hinklin Thomas
Production Editor: Peg Gluntz
Art Coordinator: Mark Garrett
Text Designer: Anne Daly
Cover Designer: Brian Deep
Cover Photo: Larry Hamill

This book was set in Times Roman.

Copyright © 1991 by Macmillan Publishing Company.
Merrill is an imprint of Macmillan Publishing Company.

Printed in the United States of America

All rights reserved. No part of this book may be reproduced or transmitted in any form or by any means, electronic or mechanical, including photocopy, recording, or any information storage and retrieval system, without permission in writing from the Publisher.

Macmillan Publishing Company
866 Third Avenue, New York, NY 10022

Collier Macmillan Canada, Inc.

Library of Congress Catalog Card Number: 90-60648
International Standard Book Number: 0-675-21172-7

Printing: 1 2 3 4 5 6 7 8 9   Year: 1 2 3 4

**In memory of my parents,
Charles and Bertha**

Merrill's International Series in Engineering Technology

ADAMSON	*Applied Pascal for Technology,* 20771–1 *Structured BASIC Applied to Technology,* 20772–X *Structured C for Technology,* 20993–5 *Structured C for Technology* (w/ disks), 21289–8
ANTONAKOS	*The 68000 Microprocessor: Hardware and Software, Principles and Applications,* 21043–7
ASSER/STIGLIANO/ BAHRENBURG	*Microcomputer Servicing: Practical Systems and Troubleshooting,* 20907–2 *Microcomputer Theory and Servicing,* 20659–6 *Lab Manual to accompany Microcomputer Theory and Servicing,* 21109–3
ASTON	*Principles of Biomedical Instrumentation and Measurement,* 20943–9
BATESON	*Introduction to Control System Technology, Third Edition,* 21010–0
BEACH/JUSTICE	*DC/AC Circuit Essentials,* 20193–4
BERLIN	*Experiments in Electronic Devices, Second Edition,* 20881–5 *The Illustrated Electronics Dictionary,* 20451–8
BERLIN/GETZ	*Experiments in Instrumentation and Measurement,* 20450–X *Fundamentals of Operational Amplifiers and Linear Integrated Circuits,* 21002–X *Principles of Electronic Instrumentation and Measurement,* 20449–6
BOGART	*Electronic Devices and Circuits, Second Edition,* 21150–6
BOGART/BROWN	*Experiments in Electronic Devices and Circuits, Second Edition,* 21151–4
BOYLESTAD	*DC/AC: The Basics,* 20918–8 *Introductory Circuit Analysis, Sixth Edition,* 21181–6
BOYLESTAD/ KOUSOUROU	*Experiments in Circuit Analysis, Sixth Edition,* 21182–4 *Experiments in DC/AC Basics,* 21131–X
BREY	*Microprocessors and Peripherals: Hardware, Software, Interfacing, and Applications, Second Edition,* 20884–X *The 8086/8088 Microprocessor Family, Second Edition,* 21309–6
BROBERG	*Lab Manual to accompany Electronic Communication Techniques, Second Edition,* 21257–X
BUCHLA	*Digital Experiments: Emphasizing Systems and Design, Second Edition,* 21180–8 *Experiments in Electric Circuits Fundamentals, Second Edition,* 21409–2 *Experiments in Electronics Fundamentals: Circuits, Devices and Applications, Second Edition,* 21407–6
COOPER	*Introduction to VersaCAD,* 21164–6
COX	*Digital Experiments: Emphasizing Troubleshooting, Second Edition,* 21196–4
CROFT	*Getting a Job: Resume Writing, Job Application Letters, and Interview Strategies,* 20917–X
DAVIS	*Technical Mathematics,* 20338–4 *Technical Mathematics with Calculus,* 20965–X *Study guide to Accompany Technical Mathematics,* 20966–8 *Study Guide to Accompany Technical Mathematics with Calculus,* 20964–1
DELKER	*Experiments in 8085 Microprocessor Programming and Interfacing,* 20663–4
FLOYD	*Digital Fundamentals, Fourth Edition,* 21217–0 *Electric Circuits Fundamentals, Second Edition,* 21408–4 *Electronic Devices, Second Edition,* 20883–1 *Electronics Fundamentals: Circuits, Devices, and Applications, Second Edition,* 21310–X *Essentials of Electronic Devices,* 20062–8

FLOYD, *continued*	*Principles of Electric Circuits, Electron Flow Version, Second Edition,* 21292–8
	Principles of Electric Circuits, Third Edition, 21062–3
FULLER	*Robotics: Introduction, Programming, and Projects,* 21078–X
GAONKAR	*Microprocessor Architecture, Programming, and Applications with the 8085/8080A, Second Edition,* 20675–8
	The Z80 Microprocessor: Architecture, Interfacing, Programming, and Design, 20540–9
GILLIES	*Instrumentation and Measurements for Electronic Technicians,* 20432–1
GOETSCH/RICKMAN	*Computer-Aided Drafting with AutoCAD,* 20915–3
HUBERT	*Electric Machines: Theory, Operation, Applications, Adjustment, and Control,* 21136–0
HUMPHRIES	*Motors and Controls,* 20235–3
HUTCHINS	*Introduction to Quality: Management, Assurance and Control,* 20896–3
KEOWN	*PSpice and Circuit Analysis,* 22135–8
KEYSER	*Materials Science in Engineering, Fourth Edition,* 20401–1
KIRKPATRICK	*The AutoCAD Textbook,* 20882–3
	Industrial Blueprint Reading and Sketching, 20617–0
KULATHINAL	*Transform Analysis and Electronic Networks with Applications,* 20765–7
LAMIT/LLOYD	*Drafting for Electronics,* 20200–0
LAMIT/WAHLER/ HIGGINS	*Workbook in Drafting for Electronics,* 20417–8
LAMIT/PAIGE	*Computer-Aided Design and Drafting,* 20475–5
LAVIANA	*Basic Computer Numerical Control Programming, Second Edition,* 21298–7
MARUGGI	*Technical Graphics: Electronics Worktext, Second Edition,* 21378–9
	The Technology of Drafting, 20762–2
	Workbook for the Technology of Drafting, 21234–0
McCALLA	*Digital Logic and Computer Design,* 21170–0
McINTYRE	*Study Guide to accompany Electronic Devices, Second Edition,* 21145–X
	Study Guide to accompany Electronics Fundamentals, Second Edition, 21406–8
MILLER	*The 68000 Microprocessor: Architecture, Programming, and Applications,* 20522–0
MONACO	*Essential Mathematics for Electronics Technicians,* 21172–7
	Introduction to Microwave Technology, 21030–5
	Laboratory Activities in Microwave Technology, 21031–3
	Preparation for the FCC General Radio Operators License Exam, 21313–4
	Student Workbook to accompany Essential Mathematics for Electronics Technicians, 21173–5
MOTT	*Applied Fluid Mechanics, Third Edition,* 21026–7
	Machine Elements in Mechanical Design, 20326–0
NASHELSKY/ BOYLESTAD	*BASIC Applied to Circuit Analysis,* 20161–6
PFEIFFER	*Proposal Writing: The Art of Friendly Persuasion,* 20988–9
	Technical Writing: A Practical Approach, 21221–9
POND	*Introduction to Engineering Technology,* 21003–8
QUINN	*The 6800 Microprocessor,* 20515–8
REIS	*Digital Electronics Through Project Analysis,* 21141–7
	Electronic Project Design and Fabrication, 20791–6

REIS, *continued*	*Laboratory Exercises to accompany Digital Electronics Through Project Analysis,* 21254–5
ROLLE	*Thermodynamics and Heat Power, Third Edition,* 21016–X
ROSENBLATT/ FRIEDMAN	*Direct and Alternating Current Machinery, Second Edition,* 20160–8
ROZE	*Technical Communication: The Practical Craft,* 20641–3
SCHOENBECK	*Electronic Communications: Modulation and Transmission,* 20473–9
SCHWARTZ	*Survey of Electronics, Third Edition,* 20162–4
SELL	*Basic Technical Drawing,* 21001–1
SMITH	*Statistical Process Control and Quality Improvement,* 21160–3
SORAK	*Linear Integrated Circuits: Laboratory Experiments,* 20661–8
SPIEGEL/LIMBRUNNER	*Applied Statics and Strength of Materials,* 21123–9
STANLEY, B. H.	*Experiments in Electric Circuits, Third Edition,* 21088–7
STANLEY, W. D.	*Operational Amplifiers with Linear Integrated Circuits, Second Edition,* 20660–X
SUBBARAO	*16/32-Bit Microprocessors: 68000/68010/68020 Software, Hardware, and Design Applications,* 21119–0 *Applications Manual to accompany 16/32-Bit Microprocessors: 68000/68010/ 68020 Software, Hardware, and Design Applications,* 21118–2
TOCCI	*Electronic Devices: Conventional Flow Version, Third Edition,* 20063–6 *Fundamentals of Pulse and Digital Circuits, Third Edition,* 20033–4 *Introduction to Electric Circuit Analysis, Second Edition,* 20002–4
TOCCI/OLIVER	*Fundamentals of Electronic Devices, Fourth Edition,* 21259–6 *Laboratory Manual to accompany Fundamentals of Electronic Devices, Fourth Edition,* 21260–X
TOMASI	*Fundamental Concepts of Operational Amplifiers and Linear Integrated Circuits,* 26637–X *Laboratory Exercises to accompany Fundamental Concepts of Operational Amplifiers and Linear Integrated Circuits,* 20637–5
WATERMAN	*Study Guide to accompany Digital Fundamentals, Fourth Edition,* 22172–2
WEBB	*Programmable Controllers: Principles and Applications,* 20452–6
WEBB/GRESHOCK	*Industrial Control Electronics,* 20897–1
WEISMAN	*Basic Technical Writing, Fifth Edition,* 20288–4
WOLANSKY/AKERS	*Modern Hydraulics: The Basics at Work,* 20987–0
WOLF	*Statics and Strength of Materials: A Parallel Approach,* 20622–7

Preface

As the title of this book suggests, we are going to be dealing primarily with mathematics. Of course, we also will be talking a good deal of the time about electronics and how mathematics can be applied to it. But our real emphasis here is on numerical methods, not circuit theory. The text is divided into seven parts and organized to enhance the study of mathematics through the coverage of electronics topics:

- Fundamentals of Arithmetic
- Fundamentals of Algebra
- Elementary Applications of Algebra to Electronics
- Advanced Topics in Algebra
- Applications of Algebra to Electronics
- Advanced Topics in Electronic Mathematics
- Advanced Applications of Mathematics to Electronics

When the first digital calculators began to replace the longtime friend and companion, the analog slide rule, their acceptance was slow. Now, little red LEDs (or black LCDs) blink out their answers more quickly and with a greater degree of precision than ever dreamed of before. With this increased use, students must know how to use their calculators to be more efficient and effective in their careers, so *Essential Mathematics for Electronics Technicians* includes calculator examples—complete with keypad sequences—and calculator exercises to test students' skills.

More recently, the personal computer has become equally widespread and is far more intimidating than the digital calculator. We must deal with the presence of a technology whose vocabulary is often strange and forbidding; yet, the use of computers is fast becoming an important ingredient in the electronics technician's arsenal of salable skills. Many instructors are understandably concerned about the risk of letting the computer become the student's major "thinking machine." Thus, the approach to computer usage taken in this book is one of learning through discovery; for example, the student *discovers* BASIC programming techniques by experimenting with prototype programs while dealing with the mathematics of circuit problems. In this way, more thought is placed on the problem rather than the program.

Essential Mathematics for Electronics Technicians also includes the following pedagogical features:

- Learning Objectives
- Rules and Axioms highlighted in boxes
- Examples in a step-by-step format
- Questions and Problems following every section in the text
- Summaries of Key Terms and Important Rules, Axioms, and Guidelines for every chapter

The entire package—textbook and supplementary materials—has been designed to give instructors all the support they need to teach electronics mathematics effectively:

Preface

- A Floppy Disk containing BASIC programs and commands allows students further exploration with little or no guidance from the instructor.
- The Instructor's Manual includes complete solutions to all the problems in the text.
- The Student Resource manual provides additional practice problems and their solutions.

    I would like to acknowledge the contributions of the following colleagues in the development of this textbook: A. F. Adkins, Amarillo College, Amarillo, Texas; Christine Barber, DeVry Institute of Technology, Chicago, Illinois; Leonard Bundra, Lincoln Technical Institute, Allentown, Pennsylvania; Laura Clarke, Milwaukee Area Technical College, Milwaukee, Wisconsin; Charles Corkhill, ITT Technical Institute, Dayton, Ohio; Algernon Daly, College of San Mateo, San Mateo, California; Robert Derby, DeVry Institute of Technology, Los Angeles, California; Melvin Duvall, Sacramento City College, Sacramento, California; Harry Edison, North Idaho College, Coeur d'Alene, Idaho; Robert Effland, Wabash Valley College, Mt. Carmel, Illinois; John Fitzen, Idaho State University, Pocatello, Idaho; Ed Geckler, Hocking Technical College, Nelsonville, Ohio; John Jellema, Eastern Michigan University, Ypsilanti, Michigan; Judy Jones, Madison Area Technical College, Madison, Wisconsin; Donald King, ITT Technical Institute, Dayton, Ohio; Harvey Laabs, North Dakota State School of Science, Wahpeton, North Dakota; Ron Moody, Pima Community College, Tucson, Arizona; James Murdoch, Bismark State College, Bismark, North Dakota; Roland Nelson, Waukesha Technical College, Palmyra, Wisconsin; Randy Ratliff, National Education Center, Cross Lanes, West Virginia; John Schira, Clover Park Vocational Technical Institute, Tacoma, Washington; Stan Vittetoe, Indian Hills Community College, Ottumwa, Iowa; and Davis Wilson, Arizona Western College, Yuma, Arizona.

F. R. Monaco

Contents

PART ONE
Fundamentals of Arithmetic — 1

1   Review of Arithmetic Operations — 3

- 1–1   Introduction   4
- 1–2   The Real Numbers   4
- 1–3   The Operations of Arithmetic   5
- 1–4   Addition   6
- 1–5   Subtraction   9
- 1–6   Multiplication   11
- 1–7   The Square Root   15
- 1–8   Division   16
- 1–9   Fractions   18
- 1–10   Operations with Fractions   18
  - Addition   18
  - Subtraction   19
  - Multiplication   20
  - Division   20
- 1–11   Special Problems with Zero   21
- 1–12   Ratio and Proportion   23
- 1–13   Percentage   25

PART TWO
Fundamentals of Algebra — 29

2   Algebraic Addition and Subtraction — 31

- 2–1   Introduction   32
- 2–2   Addition of Signed Numbers   34
- 2–3   Subtraction of Signed Numbers   36
- 2–4   Literal Numbers   38
- 2–5   Terms   39
- 2–6   Factors   39
- 2–7   Coefficients   40

Contents

    2–8   Evaluation of Expressions   40
    2–9   Symbols of Grouping   41

3   Algebraic Multiplication and Division   47

    3–1   Introduction   48
    3–2   Multiplication Involving Exponents   49
    3–3   Products of Monomials and Multinomials   51
            Multiplication of Monomials   51
            Multiplication of a Multinomial by a Monomial   53
            Multiplication of Multinomials   54
            Division Involving Exponents   56
    3–4   Quotients of Monomials and Multinomials   60
            Division of Monomials   60
            Division of a Multinomial by a Monomial   62
            Division of Multinomials   64

4   Fundamentals of Linear Equations   71

    4–1   Introduction   72
    4–2   Axioms   72
    4–3   Solving Equations   74
    4–4   Inequalities   78
    4–5   Word Problems   79

5   Factoring   85

    5–1   Introduction   86
    5–2   Factors of a Radicand   88
    5–3   The Common Factor   90
    5–4   The Perfect Trinomial Square   92
    5–5   The Difference of Squares   93
    5–6   Factoring Trinomials of the Type $x^2 + (a + b)x + ab$ 95
    5–7   Factoring Trinomials of the Type $acx^2 + (ad + bc)x + bd$ 98
    5–8   Further Topics in Factoring   101

6   Algebraic Fractions   107

    6–1   Introduction   108
    6–2   Properties of Fractions   108
    6–3   Addition and Subtraction   112
    6–4   Multiplication and Division   115
    6–5   Operations with Fractions Containing Radicals   119
            Rationalizing the Denominator   119
            Addition and Subtraction   120
            Multiplication and Division   121

Contents　　　　xi

PART THREE
Elementary Applications of Algebra to Electronics　　　　125

7　Series and Parallel Circuits　　　　127

- 7-1　Introduction　128
- 7-2　The Series Circuit　128
- 7-3　Power in Series Circuits　132
- 7-4　Summation Notation　133
- 7-5　Direct and Inverse Variation　134
- 7-6　The Voltage Divider　137
- 7-7　The Parallel Circuit　146
- 7-8　Resistances in Parallel　146
- 7-9　The Current Divider　150
- 7-10　Analysis of Units　154

PART FOUR
Advanced Topics in Algebra　　　　159

8　Graphs and Equations　　　　161

- 8-1　Introduction　162
- 8-2　Two-dimensional Reality　162
- 8-3　The Coordinate System　162
- 8-4　The Theorem of Pythagoras　167
- 8-5　The Distance Between Two Points　168
- 8-6　Special Triangles　172
  - The 45-45 Triangle　172
  - The 30-60 Triangle　173
- 8-7　The Equation of a Straight Line　176
  - The Slope of a Line　176
  - Equations of a Straight line　178
- 8-8　Graphing Linear Equations　180
- 8-9　Interpretation of Graphs　182

9　Simultaneous Linear Equations　　　　193

- 9-1　Introduction　194
- 9-2　The Simultaneous Solution　194
- 9-3　Solving Simultaneous Linear Equations　194
- 9-4　Solution by Addition and Subtraction　196
- 9-5　Solution by Substitution　202
- 9-6　Solution by Comparison　204
  - Second-Order Determinants　205
  - Determinants and Computer-Assisted Analysis　208

Third-Order Determinants   210
Third-Order Determinants Applied to Computer Analysis   213

10 Quadratic Equations   217

10-1 Introduction   218
10-2 Graphing Quadratic Relationships   218
10-3 The Quadratic Formula   221
10-4 The Nature of Quadratic Roots   226
10-5 Using Roots To Find an Equation   227

PART FIVE
Applications of Algebra to Electronics   231

11 Applications in Network Analysis   233

11-1 Introduction   234
11-2 Kirchhoff's Laws   234
11-3 Network Analysis   238

PART SIX
Advanced Topics in Electronics Mathematics   245

12 Elementary Trigonometry   247

12-1 Introduction   248
12-2 Angles   248
12-3 Measurement of Angles   249
12-4 The Trigonometric Relationships   251
12-5 Trigonometry and the Right Triangle   256
12-6 The Tangent of θ 257
12-7 The Inverse Trigonometric Relationships   259
12-8 Solutions of Right Triangles   260
12-9 Graphs of the Trigonometric Relationships   270
12-10 Graphs of $y = A \sin(Bx - C)$ and $A \cos(Bx - C)$ 274

13 Exponents and Logarithms   287

13-1 Introduction   288
13-2 Exponential Relationships   288
13-3 Logarithmic Relationships   290
13-4 Operations with Logarithms   292
13-5 Common and Natural Logarithms   295
The Common Log   296

Contents   **xiii**

	The Natural Log	296
	Change of Base	297
13–6	Exponential Equations	299
13–7	Logarithmic Equations	300
13–8	Decibels 305	
	Decibel Voltage Gain (A')	307
	Which Decibel?	308

14   Introduction to Vectors   312

- 14–1   Introduction   312
- 14–2   Vector Summation   313
-         Phasors   317

PART SEVEN
Advanced Applications of Mathematics to Electronics   323

15   Series AC Circuits   325

- 15–1   Introduction   326
- 15–2   The Resistive AC Circuit   326
  - Effective AC Value   327
  - Inductors in Series and Parallel   330
- 15–3   Transformers   333
- 15–4   Inductance in AC Circuits   337
- 15–5   Capacitors in Series and Parallel   340
- 15–6   Capacitance in AC Circuits   342
- 15–7   *RC* Time Constant   344
- 15–8   Impedance of Series Circuits   348
  - Phase Angle   352
- 15–9   Series Resonance   355
- 15–10   Power in AC Circuits   357

16   Complex Algebra and Parallel AC Circuits   365

- 16–1   Introduction   366
- 16–2   Imaginary Numbers   366
- 16–3   Complex Numbers   371
- 16–4   Complex Numbers Applied   373
- 16–5   Addition and Subtraction of Complex Numbers   375
- 16–6   Multiplication and Division of Complex Numbers   376
- 16–7   The Polar Form of a Complex Number   378
- 16–8   Calculator Conversions   380
- 16–9   Complex Parallel Impedances   381

APPENDICES 389

A	Conversion Factors	391
B	Resistor Color Codes	397
C	A Proof of the Theorem of Pythagoras	399
D	Powers of Ten	401
E	Mathematical Symbols	403
F	American Wire Gauge Conductor Sizes	405
G	The Greek Alphabet	407
H	Review of Boolean Algebra and Karnaugh Maps	409
I	Rules Review	417
J	Answers to Selected Problems	431

GLOSSARY 467
INDEX 479

PART ONE
Fundamentals of Arithmetic

1
Review of Arithmetic Operations

1 Review of Arithmetic Operations

Upon completion of this chapter, you will have:
1. Reviewed the real number system and the number line.
2. Reviewed the four fundamental operations of arithmetic:
    a. Addition
    b. Subtraction
    c. Multiplication
    d. Division
3. Reviewed arithmetic fractions and operations with them.
4. Reviewed special problems in dealing with zero.
5. Reviewed ratio and proportion.
6. Reviewed percentage.

1-1 INTRODUCTION

Before we begin our study of mathematics as it applies most particularly to electronics, we will take some time to review the fundamental operations of arithmetic. It cannot be emphasized too often that most students who fail to do well in algebra and higher mathematics are those who have a poor understanding of basic arithmetic concepts to begin with. Let us do away with the myth that those who do well in mathematics have a mathematical mind. There is no such thing. What these successful people do have, however, is a firm grasp of basic arithmetic skills.

There are people who do, regrettably, suffer from varying degrees of **mathophobia**. Mathophobia literally means "fear of mathematics." This fear is most commonly a product of bad social experiences encountered by very young children in front of their peers. For example, a child who is repeatedly laughed at or ridiculed while attempting an arithmetic problem at the chalkboard is apt to grow up hating math and everything associated with it. Most often, these people do not seek help for fear of further ridicule. Consequently, they tend to look for careers that have little if anything to do with mathematics. In fact, mathematics has often been referred to as a *career filter*, since it tends to filter out those who cannot, for whatever reason, deal with the subject. The only way to gain a deep understanding of electronics, however, is to approach it through the principles of mathematics. Otherwise, the subject is reduced to the show-and-tell level, and little of lasting value can come from such a superficial treatment of this fascinating field.

If you are a mathophobe or believe you might be, there are several courses of action open to you, including various forms of therapy and counseling. There are also several good books on the subject that might help:

Kogelman, S. and J. Warren. *Mind Over Math*. New York: Dial Press, 1978.
Tobias, S. *Overcoming Math Anxiety*. New York: W. W. Norton, 1978.

First, always talk with your instructor. Most instructors are a marvelous source of help in these matters and may even have experienced the problem themselves.

In the final analysis, proficiency is developed in mathematics the same way skills are acquired in any subject, namely, through practice. If you wish to *learn* mathematics, then you must *do* mathematics. For this reason, a variety of questions and problems appear at the end of each major section. You are encouraged to do *all* the problems, the odd-numbered as well as the even-numbered. Attempt to work through these problems without looking back at the text for help and without first looking at the answer. Only after you have thought and perhaps struggled a bit should you refer to the text.

Where personal computers are available, you will find it fun as well as instructive to work out the BASIC programming exercises presented in the problem section. The purpose of these exercises is not to make you a computer programmer but to give you—the novice technical student—an introduction to the computer as an engineering tool. Working through the programming examples and problems will also help fix certain essential mathematical and electronic ideas in mind. You are encouraged to play, explore, and refine the programming techniques presented.

1-2 THE REAL NUMBERS

If the title of this section suggests to you that perhaps there are other numbers that are *not* real (i.e., "imaginary"), then you are indeed an astute person. The idea may sound a little bizarre, but these imaginary numbers, as they are actually called, do exist and form a very real part of electronics. Their treatment is reserved for the last chapter in this book. In the meantime, we will concentrate exclusively on what are called the **real numbers**. These are all the various numbers that we have become accustomed to using in our ordinary calculations. For example, all the *whole numbers*—the counting numbers, or *digits*—are included in this set of real numbers. These include the Arabic numerals 0, 1, 2, 3, 4, and so on. Real numbers also include such quantities as ⅔, π (the Greek letter pi), 0.000259, 12.375, 1000, 100,000,000, and so on toward infinity. We say *toward* **infinity** (symbolized as ∞) rather than *to* infinity because infinity is not a number. It's not even a really big number. For no matter how big a number you think of, you can always find one considerably larger just by multiplying your number by 1 million, for example. But infinity *might* be defined simply as the *direction* one heads when looking for larger and larger numbers. Consider the straight line shown in Figure 1–1.

Figure 1–1   0 1 2 3 4 5 6 7 8 9 ⟶ ∞

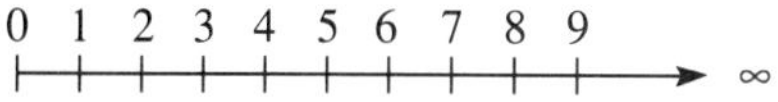

If we agree that this **number line** will start at zero on the left and will be marked off in arbitrary real-number increments extending toward the right, then the phrase "toward the right" implies direction. And as we move further and further toward the right, the numbers get bigger and bigger without end. In this limited sense, infinity means "toward the right and without end."

On the other hand, if we were to ask how many numbers there were between the digits 3 and 4, for example, what would be our answer? Let's make a list of some of the numbers, beginning at 4 and working backward toward 3: 4, 3.999999, 3.999998, 3.999997, 3.999996, 3.999995, 3.999994, 3.999993, 3.999992, 3.999991, 3.999990, 3.999989, 3.999988, ... Since we have been counting backward in intervals of one-millionth, it would take us quite some time just to get to 3.000000. But please note that between each of these intervals of one-millionth are a million still-smaller intervals. And between each of these, a million yet-smaller intervals, and so on. In other words, no matter how small the interval gets, we can continue to divide it into still smaller and smaller parts, as long as we please, and still never come to an end. We are forced to conclude, therefore, that there must be an unlimited number of numbers between 3 and 4. In fact, there is an infinitely large number of numbers between *any* two points on our number line. In this sense, infinity does not involve the idea of direction, but it certainly does involve the idea of bigness. As you progress in your studies of electronics, the concept of infinity becomes more and more important, especially when you begin to study that branch of mathematics known as calculus.

With this idea of boundlessness in mind, we may now define a real number as any of infinitely many numbers found on the number line. The number line itself, then, represents the picture of all real numbers, large or small, as points starting at zero and continuing toward the right without end. For each point on the line, there is one and only one real number corresponding to that point. Since we cannot possibly show all the numbers on the number line (remember, there are infinitely many such numbers), it has become customary to show only zero and the first few whole numbers, as we did in Figure 1–1.

1–3   THE OPERATIONS OF ARITHMETIC

Webster's Dictionary defines arithmetic as "The science or art of computation by positive, real numbers, the fundamental operations of which are performed by addition,

6   Fundamentals of Arithmetic

subtraction, multiplication, and division."* In keeping with this definition, we will limit our review to these fundamental operations and also introduce a few additional ideas such as powers of ten, ratio and proportion, percents, and the square root.

1–4   **ADDITION**

Recall that addition can be thought of as the operation of counting the total number of things and is symbolized by a + (plus) sign. For example, 3 + 5 = 8.

It is possible to give a graphical interpretation to the process of arithmetic addition by using a number line. For example, the problem just given, 3 + 5, may be represented as a line 3 units in length and another 5 units in length drawn along the number line, as shown in Figure 1–2. Note that the combined length is 8 units.

Figure 1–2   0 1 2 3 4 5 6 7 8 9
                ├─┼─┼─►┼─┼─┼─┼─►┼─┼──► +∞

In doing these arithmetic problems, you, as an electronics student, will probably want to perform the computation on a digital, hand-held, electronic calculator such as the one shown in Figure 1–3.

Figure 1–3

Webster's Unabridged Dictionary, 2d ed., 1979, s. v. "arithmetic."

Review of Arithmetic Operations

The calculator shown is called a **scientific calculator** because, in addition to performing the basic operations of arithmetic, it performs many other mathematical operations needed for advanced work in electronics and related fields of science and engineering. If you already own a calculator, check with your instructor as to its suitability.

A few words of caution about calculators. First, avoid the purchase of calculators beyond your needs. For example, programmable calculators, while often convenient, cost considerably more than their conventional counterparts. If you don't need the programming feature, why pay for it? Generally speaking, if you've paid more than about $30 for your calculator (at this point in your studies), you've either paid too much or bought too much calculator for your present needs.

Second, you should not allow your calculator to do your thinking for you. Many beginning students who do not completely understand certain concepts in a given problem simply start plugging numbers into the calculator in the hope of coming up with the right answer. The result, more often than not, is trash. The computer industry acronym GIGO (garbage in, garbage out) is very appropriate for this syndrome. In other words, the answer to a problem is no better than the thinking human brain behind its solution, and the best computer in the world can't do your thinking for you. Remember, too, if your batteries go dead, you will have to resort to your basic understanding of mathematical operations. It is because so many students learn to rely too heavily on the calculator that many schools prohibit their use in all but the most advanced courses in mathematics and science. Finally, a very common mistake among students is to discard the instruction manual that comes with the calculator. *Do not be so foolish!* Many calculator operations are performed using different keystrokes, and there are many variations, so don't expect your instructor to teach you how to use *your* calculator. Keep the instruction booklet with the calculator and refer to it often. The sample calculator problems worked in this book use keystrokes common to many manufacturers and will serve merely as a guide. Your instructor will assist you in understanding your calculator, but it is *your* responsibility to discover the peculiarities of your own digital machine.

Example 1-1 $24 + 152 = ?$

Keystroke	Display
2 4	24
+	24
1 5 2	152
=	176

Example 1-2 $0.0016 + 2.007 = ?$

Keystroke	Display
. 0 0 1 6	0.0016*
+	0.0016*
2 . 0 0 7	2.007
=	2.0086

---

*Some calculators might display this number as 1.6^{-03}, indicating that the number is being represented in *scientific notation,* which is discussed in the section on multiplication.

Example 1-3

126 + 12 + 8.5 = ?

Keystroke	Display
1 2 6	126
+	126
1 2	12
+	138
8 . 5	8.5
=	146.5

Note that any amount of numbers may be added, and the order in which they are summed does not determine the answer. In Example 1-3, we note that 126 + 12 + 8.5 = 8.5 + 12 + 126 = 12 + 8.5 + 126, and so forth. In mathematics, this is called the **commutative property** of addition. We may also pair up certain quantities before adding them to other numbers without affecting the outcome. Again in Example 1-3, (126 + 12) + 8.5 = 126 + (12 + 8.5). We call this property the **associative property** of addition.

Exercise Set 1-1

Questions

1. Why is an understanding of basic arithmetic skills important to the study of electronics?
2. What is a real number?
3. What does the picture of all real numbers look like?
4. What is mathophobia, and what can be done about it?
5. How many numbers are there between the numbers 3.5 and 5.5? (*Note:* We are *not* asking for the difference between these two numbers.)
6. What is the symbol for infinity, and what does it mean that some number is infinitely large? Infinitely small?
7. What are the fundamental operations of arithmetic?
8. What were the three fundamental precautions mentioned in the previous section concerning electronic calculators? Can you think of other precautions? (See Appendix A.)
9. Draw the keyboard for your particular calculator, and label the following keys: (a) the digits; (b) π; (c) the equal sign; (d) the keys used to perform the basic operations of arithmetic; (e) the decimal point; (f) the *inverse* key; (g) the *clear* and *all clear* keys; and (h) the *on-off* key (or switch).
10. In your own words, define the associative property of addition, and give an example.
11. What is meant by the commutative property of addition? Give an example.
12. Record your calculator make and model number on a separate piece of paper. Record the serial number of your calculator on the same sheet and store it somewhere safe. DO NOT engrave your calculator with any vibrating or impact engraving tool. Paint your name on the face of the calculator using, for example, a marker pen available from most stationery stores.

Problems

1. Draw a number line and label the following points: 0.1; 2.5; π (approximately 3.14); 9.9; 85.
2. Can the *point* infinity (∞) be labeled on the number line? Explain your answer.
3. Using a number line, show the graphical solution to the following sums:
    a. 2 + 9
    b. 5 + 3
    c. 250 + 125
    d. π + 4 + 10

Review of Arithmetic Operations

4. Perform the following additions on your calculator:
   a. $31 + 109$
   b. $11.01 + 1001.11$
   c. $2.09 + \pi$
   d. $1.35 + 0.0001006 + 256.065$
   e. $25.001 + (24.903 + 25.01)$
   f. $(15.97 + 22) + 13.5 + 86.892$

5. If you change the order of the numbers given in each part of Problem 4 and then add, will the answers be different? Why or why not?

6. The following **BASIC*** program computes the sum of two positive real numbers:

```
10 REM  This program computes the sum of
20 REM  two positive real numbers A and B.
30 LET S=0
40 PRINT"Enter the value of A"
50 INPUT A
60 PRINT"Enter the value of B"
70 INPUT B
80 LET S=A+B
90 PRINT"The sum of A and B is:";S
100 END
```

The REM statement in lines 10 and 20 means "remarks." These remarks have absolutely no affect on the program, and you may include as many of them as you wish. The statement LET $S = 0$ in line 30 is called an initialization statement. It ensures that the sum is, in fact, zero before the program begins. In lines 40, 60, and 90, you may enter the BASIC abbreviation ? for PRINT instead of spelling it out. When the program is listed, however, the word PRINT will appear, as shown.

Write a BASIC program that will compute the sum of three positive real numbers X, Y, and Z.

1-5 SUBTRACTION

Subtraction may be defined as the process of finding the difference between two numbers. Unlike addition, however, the subtraction process is neither associative nor commutative. For example, $12 - 9 = 3$, but what meaning do we attach to $9 - 12$? Moreover, $(12 - 9) - 1.5 = 1.5$; however, $12 - (9 - 1.5) = 4.5$. You must be very careful to maintain the proper order when doing subtraction.

In the previous section, we gave a graphical interpretation of addition using the number line. In that case, we interpreted the + sign as meaning "go right." In the case of subtraction, we will regard the − sign as meaning "go left." For example, in Figure 1–4, we have shown a graphical interpretation of the problem $8 - 3 = 5$. Note that a line 8 units in length is placed along the axis of the number line pointing toward the right. The number 3, since it is preceded by a − (minus) sign, is represented by a line 3 units long directed toward the left, where its end (the head of the arrow) comes to rest at 5.

Figure 1–4

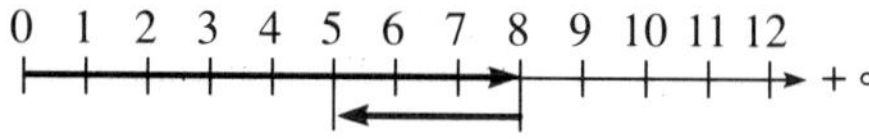

The calculator operation of subtraction is similar to that of addition.

---

*BASIC is an acronym for <u>B</u>eginners' <u>A</u>ll-purpose <u>S</u>ymbolic <u>I</u>nstruction <u>C</u>ode.

Fundamentals of Arithmetic

Example 1-4 $56 - 21 = ?$

Keystroke	Display
5 6	56
−	56
2 1	21
=	35

Example 1-5 A technician purchases fifteen 68 ohm (Ω) ±1% precision resistors for $8.25. If he gives the salesperson $10, how much change will he receive?

Solution Note that some of the numbers given in the problem are completely irrelevant to our purpose. For example, the total number of resistors (15) has no bearing on the amount of change received. Furthermore, the ohmic value of the resistors (68) and their tolerance (±1%) does not have the least influence on the answer. You must become accustomed to disregarding extraneous numbers while, at the same time, paying close attention to those that have a critical role in the problem. In the example, the amount of change received is $10 − $8.25 = $1.75. In many real-life situations, there may be much extraneous information associated with a particular problem, and the technician must be able to separate the important from the unimportant.

Example 1-6 Twelve 0.047 microfarad (μF) capacitors are required to complete a circuit assembly. The circuit also requires 8 resistors, 6 2N2222 transistors, and 3 1N914 diodes. If 4 of the resistors are found to be defective, how many more must be ordered?

Solution It should be apparent that the only usable information given in the problem is that 4 of the 8 resistors are defective. Therefore, $8 - 4 = 4$ represents the number of new resistors that must be requisitioned.

Exercise Set 1-2

Questions

1. What is meant by the statement that subtraction is neither commutative nor associative? Give examples.
2. What is meant by the phrase *extraneous information* in connection with practical problems?
3. What graphical interpretation may be applied to the − sign associated with arithmetic subtraction?

Problems

1. Perform the following subtraction problems graphically using a number line.
   a. $100 - 50$
   b. $0.015 - 0.012$
   c. $21.5 - 17.3$
   d. $3000 - 1850$
2. Perform the following subtractions on your calculator.
   a. $\pi - 2.99$
   b. $2008 - 15.01$
   c. $27 - 0.09$
   d. 1.8 kilohms (kΩ) $-$ 0.5 kΩ
   e. $0.00693 - 0.00239$
   f. $0.500 \, \mu F - 0.047 \, \mu F$

Review of Arithmetic Operations

3. A technician receives 134 diodes in one box, 214 resistors in another box, 408 transistors in a plastic bag, and 186 RF chokes in another box. If the technician lost the box of RF chokes (worth about $139.50) how many containers of parts does he still have?
4. A technician makes $8.50 an hour and time-and-a-half for each hour overtime. If she is supposed to quit work at 4:30 P.M. but works until 6 P.M., how many hours of overtime pay will she receive?
5. The following BASIC program finds the difference between two real positive numbers: $D = P - Q$.

```
10 REM   This program computes the difference (D) between
16 REM   two real, positive numbers P and Q, where the
17 REM   number P is larger than the number Q.
20 LET D=0
30 PRINT"Enter the values P and Q in the proper order"
33 INPUT P, Q
46 LET D=P-Q
67 PRINT"The difference P-Q is:";D
104 END
```

Note that the number statements are not given in increments of 10 as was the case with the program presented in Exercise Set 1–1. Although this program will still run, you may wish to modify it by adding other statements at some future time. It would probably be a good idea, then, to renumber the statements in decades—that is, 10, 20, 30, etc. After you have keyed the program exactly as just shown, type RENUM and then press the enter key. Now list the program again. What do you observe?

Write a BASIC program that will find the difference between any two numbers less than or equal to 25.

1–6 MULTIPLICATION

Multiplication may be interpreted as a form of continued addition. For example, the **product** $2 \times 4 = 8$ (where $\times$ means to multiply) may be taken to mean 2 added 4 times, or 4 added twice. Verify this assertion graphically using the number line.

Any amount of numbers can be multiplied together in any order and in any grouping without affecting the result. For example, $2 \times 5 \times 4 = 2 \times 4 \times 5 = 4 \times 2 \times 5 = 40$. Moreover, $(2 \times 5) \times 4 = (4 \times 5) \times 2 = (4 \times 2) \times 5$, etc. All this demonstrates that multiplication is both commutative and associative, as was the case with addition.

Multiplication problems are keyed into the calculator as shown in the following example.

Example 1–7 $2 \times 3 \times 18 = ?$

Keystroke	Display
2	2
×	2
3	3
×	6
1 8	18
=	108

In multiplication, each of the numbers is called a **factor**. In Example 1–7, there were three factors: 2, 3, and 18.

Fundamentals of Arithmetic

Very often in electronics, it is desirable to multiply the same number by itself over and over again. For example, one may wish to find the product $2 \times 2 \times 2 \times 2 \times 2 \times 2 \times 2 \times 2 \times 2 = ?$. Here we see that there are nine factors, all identical, to be multiplied together. Although you could key each one of these factors individually, one after another, into the calculator as in Example 1–7, there is a much easier method. On your calculator, you should have a key (often called the *power key*) designated as x^y. The x on this key stands for the value of the factor (2 in the preceding example), and y stands for the number of times the factor is to be multiplied by itself (9 in this example). The keystrokes involved in solving this particular problem are as follows:

Keystroke	Display
2	2
x^y	2
9	9
=	512

The reason the x^y key is called the power key is that the process of repeatedly multiplying the same number over and over again by itself a certain number of times is called *raising a number to a power*. The number to be raised to a given power (x) is called the **base,** whereas the other number (y) is called the **exponent.** The result is called the **power.** Therefore, in the preceding example, the number 2 is the base, 9 is the exponent, and 512 is the power. We say, "2 is raised to the ninth power." In the usual notation of mathematics, we would write this as:

$$2^9 = 512$$

Example 1–8   Find the value of 5^{16}.

Keystroke	Display
5	5
x^y	5
1 6	16
=	1.525878906^{11}

The small 11 written at the upper right of the answer is another convenient way that engineers have of expressing very large (as well as very small) numbers. In a practical sense, the 11 means move the decimal point 11 places to the right. The answer is, therefore, approximately 152,587,890,600 (note that the calculator rounds to 10 places) or, in words, 152 billion, 587 million, 890 thousand, 6 hundred. Suppose we wanted to round this number to 150,000,000,000, for instance, and then do something with the result. A question that may come immediately to your mind is, Why carry all those zeros around? And you are, of course, absolutely correct. After all, this number itself is really nothing more than 1.5 multiplied by 100,000,000,000, or, in other words, 1 with 11 zeros after it. Ahha! There's that 11. And what do you suppose you would get as an answer if you raised 10 to the 11th power? Right! 100,000,000,000. All this has just been another way of saying that the 11 indicates a **power of ten**—a shorthand way of writing very big (or very small) numbers in a way with which most of us are familiar, since we are accustomed to using the decimal system in counting: 10; 100; 1,000; 10,000; . . .; 100,000,000,000; and so on. Therefore, when an engineer wants to express a number with a lot of decimal places in it, he or she simply writes the quantity as some number between 1 and 10 times the appropriate power of 10. This form of a number is called **scientific notation.** A positive exponent on the 10 indicates the number of places the decimal point must be moved to the left when converting from standard to scientific notation. A few examples should make the idea a little clearer.

Example 1-9

a. $126{,}804 = 1.26804 \times 10^5$ (decimal point moved 5 places)
b. $124 = 1.24 \times 10^2$ (decimal point moved 2 places)
c. One ampere (A) of electrical current is defined as the movement of 6,250,000,000,000,000,000 electrons past a given point in a circuit in 1 second (s). This number may be written much more simply as 6.25×10^{18} using scientific notation, where the decimal point is moved 18 places.
d. The mean distance to the sun is 93,000,000 miles (mi). That is, the distance is 9.3×10^7 mi. The decimal point has been moved 7 places.
e. The speed of a radio wave in a vacuum is 30,000,000,000 centimeters per second (cm/s)—that is, 3×10^{10} cm/s.

As a check, note in Part (a) that $1.26804 \times 10^5 = 1.26804 \times 100{,}000 = 126{,}804$.

Unfortunately, we have not progressed far enough in our studies to make the idea of a negative exponent especially meaningful at this point. What meaning, for example, should be given to 10^{-4}? We will delay answering this question until Chapter 3, where we develop the various laws of exponents. Until then, the following chart will be of great help in using the concept of powers of ten.

$$1{,}000{,}000{,}000{,}000 = 1 \times 10^{12} \quad \text{(tera-)} \quad \text{T}$$
$$1{,}000{,}000{,}000 = 1 \times 10^{9} \quad \text{(giga-)} \quad \text{G}$$
$$100{,}000{,}000 = 1 \times 10^{8}$$
$$10{,}000{,}000 = 1 \times 10^{7}$$
$$1{,}000{,}000 = 1 \times 10^{6} \quad \text{(mega-)} \quad \text{M}$$
$$100{,}000 = 1 \times 10^{5}$$
$$10{,}000 = 1 \times 10^{4}$$
$$1{,}000 = 1 \times 10^{3} \quad \text{(kilo-)} \quad \text{k}$$
$$100 = 1 \times 10^{2} \quad \text{(hecto-)} \quad \text{h}$$
$$10 = 1 \times 10^{1} \quad \text{(deka-)} \quad \text{da}$$
$$1 = 1 \times 10^{0}$$

Note the pattern. The power of 10 is simply the number of zeros contained in the number. Note, too, that many powers of 10 occur so often that they are given special prefix names: for example, a 3 megohm (3×10^6 MΩ) resistor or an 8 gigahertz (8×10^9 GHz) microwave radio signal.

Fundamentals of Arithmetic

The calculator has an EXP (exponent) key that allows you to express numbers as powers of ten and operate on these numbers. For example, $256{,}000 \times 1047 = 268{,}032{,}000$. This problem is keyed into the calculator as follows:

Keystroke	Display
[2] [.] [5] [6]	2.56
[EXP]	2.56^{00}
[5]	2.56^{05}
[×]	256000
[1] [.] [0] [4] [7]	1.047
[EXP]	1.047^{00}
[3]	1.047^{03}
[=]	268032000

The answer may be written in scientific notation as 2.68032×10^8. Note that if the number 268032000 is keyed into the calculator and then the ENG (**engineering notation**) key is pressed, the answer is displayed as 268.032^{06}, where the exponent 06 is the power of ten. This display differs from the ordinary powers of 10 we have been using thus far in that the power of ten in engineering notation is always a multiple of 3. For example, this problem may be keyed into the calculator as follows:

$$2.56\ [EXP]\ 5\ [\times]\ 1.047\ [EXP]\ 3\ [=]\ [ENG]$$

The answer is now displayed as 268.032^{06}.

Alternatively, the problem may be keyed as follows:

$$256000\ [\times]\ 1047\ [=]\ [ENG]\ 268.032^{06}$$

The principal advantage of this specialized use of powers of 10 is that the answer is always displayed as one of the prefixes shown in the previous table. For example,

$$372{,}000\ [ENG] = 372^{03} \quad \text{(kilo-)}$$
$$12{,}304{,}000\ [ENG] = 12.304^{06} \quad \text{(mega-)}$$
$$8{,}000{,}000{,}000\ [ENG] = 8^{09} \quad \text{(giga-)}$$

Displaying answers in this way is common in electrical engineering computations.

Powers of 10 are carried over to the **color code** commonly used in electronics work. For example, the third band on a composition resistor specifies the power of ten used as the resistance multiplier as follows:

$$\text{Black} = 1 \times 10^0 \quad \text{(no zeros)}$$
$$\text{Brown} = 1 \times 10^1 \quad \text{(1 zero)}$$
$$\text{Red} = 1 \times 10^2 \quad \text{(2 zeros)}$$
$$\text{Orange} = 1 \times 10^3 \quad \text{(3 zeros)}$$
$$\text{Yellow} = 1 \times 10^4 \quad \text{(4 zeros)}$$
$$\text{Green} = 1 \times 10^5 \quad \text{(5 zeros)}$$
$$\text{Blue} = 1 \times 10^6 \quad \text{(6 zeros)}$$
$$\text{Violet} = 1 \times 10^7 \quad \text{(7 zeros)}$$
$$\text{Gray} = 1 \times 10^8 \quad \text{(8 zeros)}$$
$$\text{White} = 1 \times 10^9 \quad \text{(9 zeros)}$$

A more complete table of multipliers and color codes is given in Appendix B.

Review of Arithmetic Operations

1-7 THE SQUARE ROOT

A topic closely related to raising a number to a power is that of extracting the square root of a number. The square root operation is indicated by the symbol $\sqrt{}$, which is often referred to as a **radical sign.** Finding the square root of N ($\sqrt{N}$), where N, called the **radicand,** is any positive real number, means finding a number such that the product of the number with itself equals N. For example, $\sqrt{49} = 7$, since 7×7, or 7^2, is 49.

On the calculator, finding the square root is accomplished as shown by the following example: Find $\sqrt{7.84}$.

Keystroke	Display
7 . 8 4	7.84
$\sqrt{}$	2.8

Example 1-10

The power in a circuit is proportional to the square of the current, which, in this case, is 25 A. What is the actual current in the circuit?

Solution: The current (I) is given as $I = \sqrt{25} = 5$ A.

Exercise Set 1-3

Questions

1. In your own words, define the following terms and give an example of each:
   a. Product
   b. Factor
   c. Base
   d. Exponent
   e. Power
   f. Square root
2. Why is multiplication said to be both associative and commutative?
3. What are powers of 10 and why are they used? Give an example.
4. What does a positive exponent on a power of 10 tell you to do with the decimal point?
5. Explain the differences and similarities between the EXP and ENG keys of the calculator. Give examples.
6. List the powers of 10 associated with the color code multiplier values.

Problems

1. Perform the following operations on your calculator. Make appropriate use of the EXP, ENG, x^y, and $\sqrt{}$ keys. Express all answers both in scientific notation and in engineering notation.
   a. 32.2×8.02
   b. $0.075 \times 10,000$
   c. $\pi \times 18.5 \times 24,008$
   d. $10,000,000 \times 186,278.569$
   e. $3.449 \times 10^8 \times 668.223 \times 4^{12}$
   f. $6 \times 6 \times 6 \times 6 \times 6 \times 6$
   g. 2 raised to the tenth power
   h. 33.47 raised to the sixth power
   i. 9.2 GHz $\times$ 3.3 kHz
   j. 47 MΩ $\times$ 1.5 kΩ
   k. $0.55 \times 15,862$
   l. $1 \times 10^6 \times 4^7 \times 258,442$
   m. Twice the distance between the earth and the sun

16 Fundamentals of Arithmetic

n. Four times the speed of a radio wave in a vacuum
o. The number of electrons flowing past a point in 5 s when the current is 1 A
p. The product of the numerical values obtained in the answers to parts (m), (n), and (o)
q. $\sqrt{105,000,000}$
r. $\sqrt{27.016}$

2. A technician purchases one thousand thirty-seven 50 μF capacitors at a 22% discount. The discount price is 3.5¢ each. What is the total price she paid?
3. The speed of light (c) in a vacuum is the product of the frequency (f) and the wavelength (λ). Confirm this fact if the frequency is 12.4 GHz and the wavelength is 2.419355 cm.
4. If one 2N3904 transistor costs $0.118 and this represents a discount given on lots of 1000 pieces, find the cost of 5608 transistors bought at the same discount rate, excluding a 6.5% sales tax and a 5% cash discount. (Beware of extraneous information!)
5. A certain drawing of the real number line has its whole numbers spaced 0.1 inch (in.) apart. How many inches is it between the real numbers 3 and 72?
6. The following BASIC computer program multiplies two real numbers A and B together. In BASIC, the symbol ' means multiply.

```
10 REM This program finds the product of two real numbers.
20 LET P=0
30 PRINT"Enter the values of A and B"
40 INPUT A
50 INPUT B
60 LET P=A*B
70 PRINT"The product of A and B is";P
80 END
```

Write a BASIC program for computing the product of three real numbers.

7. The following program extracts the square root of the positive real number N. In BASIC, the letter string SQR means "square root."

```
10 REM  This program extracts the square root
20 REM  of the positive, real number N.
30 LET Z=0
40 PRINT"Enter the value of N"
50 INPUT N
60 LET Z=SQR(N)
70 PRINT"The square root of N is";Z
80 END
```

Write a BASIC program that will extract the square root of the sum of R^2 and X^2, that is, $Z = \sqrt{R^2 + X^2}$.
(*Note:* In BASIC, the symbol ∧ may be used to mean "raise to the power of." For example, R∧2 means R raised to the second power.)

1–8 DIVISION

As younger students in elementary school, most of us played a division game called "gozinta," as in "eight gozinta sixteen two times." This was often written as

$$8 \overline{\smash{)}16}^{\,2}$$

where 16 was called the **dividend**, 8, the **divisor**, and 2, the **quotient**. This division process may be viewed as a form of continued subtraction. In this example, we may regard the division as taking 8 away from 16 two times; that is, $16 - 8 = 8$ and $8 - 8 = 0$, for a total of two subtractions. This interpretation is fairly obvious from the fact that $8 \times 2 = 16$. You can demonstrate this fact by using the number line.

The calculator operations for performing division are similar to those used in doing multiplication, except that the ÷ key is used. For example, what is 54 ÷ 9?

Keystroke	Display
5 4	54
÷	54
9	9
=	6

Example 1–11 Find 106,000,000 ÷ 83,500.

Keystroke	Display
1 0 6	106
EXP	106^{00}
6	106^{06}
÷	106000000
8 3 . 5	83.5
EXP	83.5^{00}
3	83.5^{03}
=	1269.461078
ENG	1.269461078^{03}

Exercise Set 1–4

Questions

1. Using your own words, define the following terms and give examples:
   a. dividend
   b. divisor
   c. quotient
2. Explain how division may be regarded as continued subtraction.

Problems

1. Using your calculator, perform the following divisions:
   a. 965,000,000,000 divided by 3,000,899
   b. 5×10^8 divided by 4500
   c. 64.09 divided by 18.007
   d. 0.5 divided by 0.01
   e. 2^{10} divided by 16
   f. 1 MΩ divided by 500 Ω
2. The following BASIC program calculates the quotient A/B.

```
10 REM   This program divides the real number A
20 REM   by the real (nonzero) number B.
30 LET Q=0
40 PRINT"Enter the value of A"
50 INPUT A
60 PRINT"Enter the value of B"
70 INPUT B
80 LET Q=A/B
90 PRINT"The quotient is:";Q
100 END
```

Write a BASIC program that will add 25 to the quotient obtained by dividing a real number P by another real number Q.

18 Fundamentals of Arithmetic

1-9 FRACTIONS

Fractions are often referred to as **indicated divisions,** since we do not necessarily wish to actually carry out the division process as defined in the previous section. For example, ¾ is a fraction *indicating* the division of 3 by 4, which would be 0.75 if we actually carried out the division.

Like most other things, the parts of a fraction have names. For example, in the fraction ¾, the 4 is called the **denominator** and refers to the *total* number of parts involved. The 3 is called the **numerator** and refers to the *number of parts* being considered. The fraction line is technically known as a **vinculum** but is sometimes referred to as a slash, or division bar. For example, suppose we are considering 39 out of a total of 142 batteries. We could indicate this by the fraction ³⁹⁄₁₄₂. Note that no actual division was performed, although it is *indicated*.

1-10 OPERATIONS WITH FRACTIONS

Fractions may be added, subtracted, multiplied, and divided.

Addition

In your early experiences with fractions, you were probably told something to the effect that in order to add two (or more) fractions, they first had to be expressed with the lowest common denominator. *This is not true!* All that is required is a *common* denominator, *any* common denominator, so that the numerators can be added together and placed over the common denominator.

Using the lowest common denominator came about as a result of performing fractional computations by longhand in elementary school. For example, consider the sum of the fractions ⁴⁹⁄₁₁₂ and ²⁵⁄₈₀. Since it is a fact that the value of a fraction remains unchanged if *both* numerator and denominator are multiplied by the same *nonzero* number, we may proceed as follows:

$$\frac{49}{112} \times \frac{80}{80} = \frac{3920}{8960}$$

and

$$\frac{25}{80} \times \frac{112}{112} = \frac{2800}{8960}$$

Since we now have a common denominator, we simply add the numerators, obtaining

$$\frac{3920}{8960} + \frac{2800}{8960} = \frac{6720}{8960}$$

If we carry out the actual division, we obtain 0.75, or ¾. If we had first reduced ⁴⁹⁄₁₁₂ to ⁷⁄₁₆ and ²⁵⁄₈₀ to ⁵⁄₁₆, we could have computed the sum mentally: ⁷⁄₁₆ + ⁵⁄₁₆ = ¹²⁄₁₆, or ¾. However, it is not necessary to find the lowest common denominator first. In our operations with algebraic fractions in Chapter 6, however, it will greatly simplify things in most cases if fractions are first written in simplest form. We discuss this matter in more depth in Chapter 6.

Review of Arithmetic Operations

Example 1-12 $5/8 + 3/16 = ?$

Solution   Note that if we first multiply both the numerator and denominator of $5/8$ by 2, we obtain the fraction $10/16$. Since this now has the same denominator as $3/16$, we may simply add their numerators:

$$\frac{10}{16} + \frac{3}{16} = \frac{13}{16}$$

Example 1-13 $9/192 + 11/64 = ?$

Solution   Note that 3 divides evenly into both 9 and 192: $9/3 = 3$ and $192/3 = 64$. Therefore, we may write the problem as $3/64 + 11/64$. Since we now have a common denominator, we may simply add the numerators and write the answer as $14/64$, which reduces to $7/32$.

Example 1-14 $8/11 + 5/9 = ?$

Solution   Note that neither of these denominators is a multiple of the other, as was the case with the previous two examples. Such problems are approached as follows:

$$\frac{8}{11} \times \frac{9}{9} = \frac{72}{99}$$

and

$$\frac{5}{9} \times \frac{11}{11} = \frac{55}{99}$$

Note that multiplying both fractions by the denominator of the other always produces a common denominator. We may now write:

$$\frac{72}{99} + \frac{55}{99} = \frac{127}{99}, \text{ or } 1\,28/99$$

Subtraction

Subtraction of one fraction from another is carried out in a similar manner as addition. For example, $17/32 - 3/8$ can be found as follows:

$$\frac{17}{32} \times \frac{1}{1} = \frac{17}{32}$$

and

$$\frac{3}{8} \times \frac{4}{4} = \frac{12}{32}$$

Therefore,

$$\frac{17}{32} - \frac{12}{32} = \frac{5}{32}$$

Example 1-15

$8 - 5/16 = ?$

Solution

$$\frac{8}{1} \times \frac{16}{16} = \frac{128}{16}$$

Therefore,

$$\frac{128}{16} - \frac{5}{16} = \frac{123}{16}, \text{ or } 7\tfrac{11}{16}$$

Multiplication

Multiplication is perhaps the easiest arithmetic operation to perform on fractions. To multiply two or more fractions together, simply write the product of all the numerators over the product of all the denominators. For example,

$$\frac{2}{3} \times \frac{5}{16} = \frac{10}{48}$$

which may be reduced to $5/24$ by dividing both numerator and denominator by 2.

Example 1-16

$12/5 \times 1/8 = 12/40$, which can be reduced to $3/10$ by dividing both numerator and denominator by 4.

Example 1-17

$8 \times 11/32 = 8/1 \times 11/32 = 88/32$, which can be reduced to $11/4$, or $2\tfrac{3}{4}$. Note that 8 is considered to be a fraction with an "understood" denominator of 1. This is true of all whole numbers.

Example 1-18

$0.3 \times 5/8 \times 4/9$

Solution Note that the decimal 0.3 may be written as $3/10$. Therefore, we proceed as follows: $3/10 \times 5/8 \times 4/9 = 60/720$. This may be reduced to $1/12$, since 60 divides evenly into both the numerator and denominator.

Division

In your earlier courses in arithmetic, you were probably told that in order to divide one fraction by another, you have to invert the divisor and then multiply. For example,

$$\frac{2}{3} \div \frac{5}{8} = \frac{2}{3} \times \frac{8}{5} = \frac{16}{15}$$

However, you may not have been told why this particular strategy works as it does. We will now explain. Let us begin by writing the problem in fractional form.

$$\frac{\frac{2}{3}}{\frac{5}{8}}$$

Review of Arithmetic Operations

As stated earlier, the value of a fraction is not changed if both numerator and denominator are multiplied by the same nonzero number. Therefore, let us multiply both terms of this fraction by 24, which is the product of the denominators of the given fractions. We now have

$$\frac{\frac{2}{3} \times 24}{\frac{5}{8} \times 24}$$

In the numerator, the 3 divides evenly into 24 eight times; in the denominator, the 8 divides evenly into 24 three times. Our fraction now looks like this:

$$\frac{2 \times 8}{5 \times 3}$$

Or, in other words, since multiplication is commutative,

$$\frac{2}{3} \times \frac{8}{5}$$

which shows that to divide $\frac{2}{3}$ by $\frac{5}{8}$, you invert the divisor and multiply. Therefore,

$$\frac{2}{3} \div \frac{5}{8} = \frac{2}{3} \times \frac{8}{5} = \frac{16}{15}, \text{ or } 1\frac{1}{15}$$

Example 1-19

$\frac{3}{7} \div \frac{1}{3} = \frac{3}{7} \times \frac{3}{1} = \frac{9}{7}$, or $1\frac{2}{7}$

Example 1-20

$\frac{3}{100} \div \frac{3}{4} = \frac{3}{100} \times \frac{4}{3} = \frac{4}{100} = \frac{1}{25}$

Note that the associative and commutative properties do *not* apply to the division of any numbers whatsoever. That is, 3 divided by 4 (which equals 0.75) is not the same as 4 divided by 3 (which is 1.333).

1-11 SPECIAL PROBLEMS WITH ZERO

There are three special cases concerning division with zero that deserve attention.

1. $0/N$, where N is some whole number not equal to zero. The answer to this problem is a number n, if it exists, such that $n \times N = 0$. Clearly, the only value of n that makes $n \times N = 0$ is 0. We conclude, therefore, that division of zero by a *nonzero* number is, indeed, a permissible operation, and the answer is always 0. As examples, $0/3 = 0$, $0/105 = 0$, and $0/\pi = 0$.
2. $0/0$. Here, we are looking for a quotient n, if it exists, such that $n \times 0 = 0$. After a little thought, you are forced to conclude that n may be any number, since any number multiplied by zero is zero. Therefore, this case is called the indeterminate case. Division of 0 by 0 is not a permissible operation in mathematics.

Fundamentals of Arithmetic

3. *N*/0, where *N* is any nonzero number. In this case, we seek a quotient *n*, if it exists, such that $n \times 0 = N$. It should be apparent, however, that no number exists with this property in the real number system. Therefore, division of a nonzero number by 0 is also a prohibited operation in mathematics.

These three cases involving division with zero cause a great deal of trouble for the beginning student of mathematics, and a great deal of caution should be exercised in working with zero. You are strongly urged to commit these three cases to memory at once.

Example 1–21

$0/1058 = 0$ (case 1)
$0/0$ cannot be determined   (case 2)
$27/0$ is not a permissible operation in mathematics   (case 3)

Exercise Set 1–5

Questions

1. What is meant by the statement that a fraction is an *indicated* division?
2. In your own words, define the following terms:
   a. Fraction
   b. Numerator
   c. Denominator
   d. Vinculum
3. Explain why division by zero is not a permissible operation in mathematics.
4. Use an example to explain why a common denominator must be used when adding or subtracting fractions.
5. Why is a common denominator *not* required in the multiplication or division of fractions?

Problems

Perform the indicated operations. Express your answer as a fraction.

1. $\dfrac{3}{5} + \dfrac{1}{8}$
2. $0.5 + \dfrac{3}{10}$
3. $\dfrac{1}{2} - \dfrac{1}{3}$
4. $\dfrac{5}{8} - \dfrac{11}{16}$
5. $\dfrac{37}{64} - \dfrac{19}{256}$
6. $\dfrac{7}{12} \times \dfrac{4}{6}$
7. $\dfrac{3}{32} \times \dfrac{1}{8} \times \dfrac{5}{16}$
8. $\dfrac{16}{3} \times \dfrac{1}{9}$
9. $\dfrac{4}{5} \div \dfrac{2}{3}$
10. $\dfrac{12}{0}$
11. $\dfrac{0}{9}$
12. $\dfrac{4}{5} + \dfrac{2}{10}$
13. $3 - \dfrac{11}{32}$
14. $0.2 - \dfrac{2}{10}$
15. $\dfrac{19}{32} \div \dfrac{1}{2}$
16. $\dfrac{0}{5} \div \dfrac{2}{6}$
17. $\dfrac{3}{16} \div \dfrac{1}{16}$
18. $9 \times \dfrac{5}{8} \times \dfrac{11}{32}$
19. $1 + \dfrac{1}{2} + \dfrac{1}{3} + \dfrac{1}{4} + \dfrac{1}{5}$
20. $\dfrac{2 - 1/3}{4}$

1-12 RATIO AND PROPORTION

Now that we have developed an understanding of fractions, we are ready to proceed with a related subject. By definition, a **ratio** is simply a fraction comparing or relating similar quantities. For example, a particular circuit contains 22 transistors (Q) and 110 resistors (R). The ratio of transistors to resistors, then, may be written as the fraction $22/110$, or $1/5$. We say that the ratio of Q to R is 1 to 5, which is sometimes written as 1:5, where the colon notation has the same meaning as the vinculum of a fraction.

Example 1-22 A certain company has 108 female technicians and 312 male technicians. What is the ratio of male to female technicians?

Solution $312/108 \approx 2.89:1$ (i.e., 2.89 to 1), where the symbol $\approx$ means "approximately equals."

Example 1-23 The drive shaft of an electric motor is attached to a spur gear having 214 teeth. The load gear (or driven gear), having 904 teeth, meshes with the drive gear. If the motor speed is 1670 revolutions per minute (rev/min), how fast is the load gear turning? A picture of the problem is shown in Figure 1-5, where the dashed line indicates a mechanical linkage.

Figure 1-5

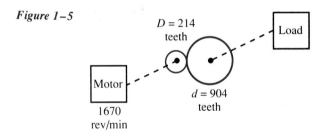

Solution Note that the ratio of the drive gear teeth (D) to driven gear teeth (d) is given by the ratio $214:904 \approx 0.237:1$. Another way of looking at this is that for every 1000 revolutions of the driven gear (d), the drive gear (D) must make 2370 revolutions. Obviously, then, the driven gear (d) rotates more slowly than the drive gear. Therefore, the speed of d is $0.237 \times 1670 \approx 396$ rev/min.

A **proportion** is a statement that two ratios are equal. For example,

$$\frac{4}{7} = \frac{8}{14}$$

is a proportion. We can readily see that the two ratios are equal, since the ratio on the right has been formed by multiplying the numerator and denominator of the ratio on the left by 2. In fact, if we did not know the value of the denominator of the right ratio, for example, we could have figured it out simply by observing the other three quantities. For example, in

$$\frac{3}{4} = \frac{?}{10}$$

24 Fundamentals of Arithmetic

we can see that since 4 (the denominator on the left) × 2.5 = 10 (the denominator on the right), then the numerator on the right must also be 2.5 times the numerator on the left. Thus

$$\frac{3}{4} = \frac{7.5}{10}$$

We often express this using another form of notation, and write

$$3:4::7.5:10$$

which is read, "3 is to 4 as 7.5 is to 10."

These examples of proportion were intentionally chosen for their simplicity. Before we tackle more difficult problems, however, we must wait until the chapter on algebraic equations.

Example 1–24 The ratio of resistors (R) to capacitors (C) is 2:4. This is the same as the ratio of transistors (Q) to inductors (L). If there are 10 transistors, how many inductors are there?

Solution We may set up the problem initially using the symbols for the various parts.

$$\frac{R}{C} = \frac{Q}{L}$$

$$\frac{2}{4} = \frac{10}{?}$$

Since the numerator of the ratio on the right is 5 times the numerator on the left, we infer that the denominator of the right ratio must be 5 times the denominator on the left, or 20. Therefore, there are 20 inductors.

Exercise Set 1–6

Questions

1. Using your own words, define the following terms:
   a. Ratio
   b. Proportion

Problems

1. If the ratio of voltage to current is 1 to 3 and the voltage measures 7 volts (V), what is the value of the current?
2. The ratio of diodes to resistors is 2:8. This is the same as the ratio of transistors to capacitors. If there are 62 diodes and 124 transistors, how many resistors and capacitors are there?
3. In each of the following proportions, determine the value of x:
   a. $\dfrac{1}{3} = \dfrac{x}{27}$
   b. $\dfrac{5}{20} = \dfrac{90}{x}$
   c. $\dfrac{x}{3.5} = \dfrac{5}{17.5}$
   d. $\dfrac{122}{x} = \dfrac{1952}{64}$

1-13 PERCENTAGE

In an earlier section, we discovered how convenient it was to express numbers as powers of 10 because of our familiarity with the decimal system. No doubt this system evolved as a result of our having 10 fingers with which to count. In fact, the word *digit* is derived from the Latin word *digitus*, meaning finger.

It is not uncommon to find other numerical systems or familiar ideas based entirely on 10 or its multiples. For example, the metric system is based entirely on 10s; our system of currency, too, is based on 10; an average IQ is often specified as 100; and a perfect test score is frequently taken as 100. It seems almost a natural thing to think and work in 10s. It is not surprising, then, that a system of fractions based on a denominator of 100 appears in the study of mathematics. The word for this fractional system, known as **percentage,** is derived from the Latin words *per centum,* meaning by the hundreds. Here's how the system works.

Suppose that the Acme Resistor Company delivered a certain quantity of resistors, and it was discovered that there were 25 defective resistors out of a box of 154. Further suppose that the Ace Resistor Company delivered the identical resistors with 53 defective out of 308. Which company has provided more dependable resistors? By merely looking at the numbers, it is difficult to decide. Moreover, arranging the numbers as fractions doesn't give us any better idea ($25/154$ and $53/308$), since the denominators are different. If, however, we use the common denominator 47,432 (154 × 308 = 47,432), we see instantly that Ace has a higher failure rate than Acme.

$$\begin{array}{cc} \textit{Acme} & \textit{Ace} \\ \dfrac{25}{154} = \dfrac{7700}{47{,}432} & \dfrac{53}{308} = \dfrac{8162}{47{,}432} \end{array}$$

Even though these numbers reveal the information we were seeking, their sheer size makes them difficult to deal with. There is, however, an easier way.

Suppose Acme and Ace both delivered 10,000 resistors with the same rate of defects just shown (i.e., $25/154 \approx 0.1623$ for Acme and $53/308 \approx 0.1721$ for Ace). Since the denominator in each case is 10,000, it is instantly apparent which failure rate is larger:

$$\begin{array}{cc} \textit{Acme} & \textit{Ace} \\ \dfrac{1623}{10{,}000} & \dfrac{1721}{10{,}000} \end{array}$$

Of course we could have expressed that same rate in hundreds as:

$$\begin{array}{cc} \textit{Acme} & \textit{Ace} \\ \dfrac{16.23}{100} & \dfrac{17.21}{100} \end{array}$$

Since these rates have been expressed by the hundreds, we call each a percent, and write:

$$\begin{array}{cc} \textit{Acme} & \textit{Ace} \\ 16.23\% & 17.21\% \end{array}$$

Not only are these numbers much more compact, but the needed information is immediately apparent.

As a general rule, we may find the percent by expressing the fraction as a decimal and then moving the decimal point two places to the right and attaching a percent sign. For example, to find what percent 12 is of 84, the fraction is first formed: $12/84$. Then, the division is done, obtaining 0.142857, which may be rounded to 0.143. Finally, the decimal point is moved two places to the right and the percent sign is attached, giving 14.3%.

Example 1–25

A total of 104 technicians applied for a job. Only three were hired. What percent of the total applicant pool was actually hired?

Solution $3/104 = 0.029$, which corresponds to 2.9%.

Example 1–26

A certain resistor has a nominal (color-coded) value of 500 with a tolerance of 5%. Find the highest and lowest permissible values of this resistor.

Solution In order to find 5% of the resistance value, we first express the percent as a decimal, moving the decimal point two places to the left, and drop the percent sign. Therefore, 5% corresponds to the decimal 0.05.

We may now multiply 500 by 0.05, obtaining 25. Therefore, the actual measured value of the resistance may vary between 475 and 525 and still be within tolerance.

Example 1–27

The output of a 120 V, 400 Hz alternator varies between 118.5 V and 121.2 V. What percent of the nominal output is represented by the range of variance?

Solution $121.2 - 118.5 = 2.7$ V
Therefore,

$$\frac{2.7}{120} = 0.0225$$

which corresponds to 2.25%.

Example 1–28

A technician purchases 128 capacitors at 2.5¢ each. If the tax on the purchase is 6.5%, what is the total amount paid?

Solution The cost of the resistors, before taxes, is $128 \times 0.025 = \$3.20$. Therefore, $3.2 \times 0.065 = 0.208$, or about 21¢. The total cost, then, is $\$3.20 + 0.21 = \$3.41$.

Example 1–29

A technician is paid an hourly rate of $7.50 for a 40-hour (h) workweek. Every two weeks, 8.5% of his gross earnings is deducted and placed in a stock investment program. How many dollars does this represent?

Solution In 2 weeks (wk), the technician's gross pay is $40 \times 2 \times 7.5 = \$600$. So, 8.5% of $\$600 = 0.085 \times 600 = \$51$.

Exercise Set 1–7

Questions

1. Using your own words, define the meaning of percent.
2. What symbol is used to indicate percent?

Review of Arithmetic Operations

Problems

1. Express each of the following fractions as a percent:
   a. $\frac{3}{5}$
   b. $\frac{1}{3}$
   c. $\frac{27}{89}$
   d. $\frac{100}{25}$

2. Seventy-five resistors were purchased at a cost of 3¢ each. If the sales tax was 5%, what was the total amount paid?

3. A 47 kΩ resistor has a tolerance of 10%. The measured value is 42.4 kΩ. Is the resistor within tolerance?

4. A technician received a 1.8% cost-of-living wage increase on an hourly wage of $8.04. What is her new pay rate?

5. A total of 216 relays were received at incoming inspection, where it was discovered that 56 were defective. What defect rate does this represent, written as a percent?

6. The power output from an audio amplifier changed from 15 watts (W) to 18 W. What percent increase does this represent?

7. Approximately 11% of the 1086 employees at a certain company were involved in minor job-related accidents during 1987. How many employees does this represent?

8. A 1 kΩ ± 5% resistor and a 500 Ω ± 10% resistor are connected in series. What are the possible extreme resistance limits for this circuit? (*Note:* Resistances in series are additive.)

9. The annual absentee rate for one electronics company with 1100 employees dropped 5.5% per year for the last 3 years (y). If 22 employees were absent the first year, how many were absent during the third (last) year? Round your answer to the smallest whole number of people. (*Hint:* The number of employees absent the first year is 22. The number absent the second year equals 22 − 5.5% of 22 = 20.79.)

10. Write a BASIC program that computes the sales tax in your state on any purchase.

Summary

The emphasis of the chapter has been a review of arithmetic concepts. The four basic operations of arithmetic (addition, subtraction, multiplication, and division) have been applied to simple review problems with whole numbers, decimals, and fractions. Ratio, proportion, and percentage problems were reviewed briefly. Introductory exercises in BASIC computer programming were introduced in the problems.

Key Terms

mathophobia
real number
infinity
number line
scientific calculator
commutative property
associative property
BASIC
product
factor

base
exponent
power
power of ten
scientific notation
engineering notation
color code
radical sign
radicand
dividend

divisor
quotient
indicated divisions
denominator
numerator
vinculum
ratio
proportion
percentage

PART TWO
Fundamentals of Algebra

2
Algebraic Addition and Subtraction

3
Algebraic Multiplication and Division

4
Fundamentals of Linear Equations

5
Factoring

6
Algebraic Fractions

2 Algebraic Addition and Subtraction

Upon completion of this chapter, you will have:
1. Learned the meaning of signed numbers and how they are applied.
2. Learned how to add and subtract signed numbers.
3. Learned how signed numbers are grouped together.
4. Learned the meaning of important terms associated with algebraic expressions.

32   Fundamentals of Algebra

2-1 INTRODUCTION

In the previous chapter, we were concerned with positive numbers and the operations of arithmetic. The number line shown in Figure 1-1 was used to represent all such positive numbers—that is, *all the real numbers to the right of zero*. We are now at a point, however, where it is necessary to consider **negative numbers** so that we may deal with concepts of algebra. Negative numbers are those that lie to the left of zero and are, therefore, said to be less than zero. The first question you might reasonably ask is, "How can a number be less than nothing? If you have zero dollars in your bank account, you're broke!" Although the answer may surprise you, *zero does not mean the same thing as nothing*. Going back to the banking example, if you have no money remaining in your account but accidently write a check for $10, for instance, then you owe that amount to the bank. In other words, you are using money you don't have. Obviously, the amount of money you have is less than zero. Theoretically, of course, there is no limit to how far below zero (in the red) you can get. The confusion between the terms *zero* and *nothing* comes about as a result of our conventional thinking that once you've reached zero, there's no further you can go. Obviously, this is not the case.

As another example, consider temperatures below 0 degrees Celsius (°C). Here, zero is chosen to coincide with the freezing point of water (since water is abundant) and, therefore, provides a familiar reference point from which to measure other temperatures. Nearly everyone has an intuitive idea of how cold an ice cube is.

Note, however, that it is possible to get much colder than the freezing point of water, as is the case when nitrogen boils at about 196°C *below zero*.* In order to indicate that a temperature is below (rather than above) zero, we use negative signs. We write the boiling point of nitrogen as $-196°C$ and read this as "negative 196 degrees Celsius." Note that this is often incorrectly read as "minus 196 degrees Celsius." You should note that the term *minus* refers to the *operation* of subtraction, whereas the term *negative* refers to **polarity**— that is, whether a number is above or below zero. For example, it is generally the *negative* terminal of an automobile battery that is attached to the frame of the car. It would be technically incorrect to say that the *minus* terminal was attached to the frame. The point may seem trivial, but many beginning students of mathematics frequently get into trouble by not using the nomenclature properly.

Consider, now, the example of the voltage divider shown in Figure 2-1. For the time being, do not concern yourself with the particular values of the resistances or the manner in which the voltages shown might be obtained. In Chapter 7, we will have ample opportunity to explore the series circuit. For now, we are interested only in the placement of the circuit ground shown by the ⏚ symbol. This ground point in the actual circuit may be the metal chassis, a conductive trace on a printed wiring board, or simply the place chosen to connect the common (black) meter lead, as is the case here.

Using a galvanometer whose scale is calibrated directly in volts, we place the red (positive) lead at points D, C, B, and A in succession and note that the voltage readings obtained are $+20$ V, $+10$ V, 0 V, and -10 V, respectively. We note that as long as our red test lead stays to the right of the circuit ground ⏚, all our voltages are positive or zero. However, when the lead is moved to the left of ground, the voltage reads negative. Is this some magical property of the ground? Before answering that question, look at Figure 2-2, which is the same as Figure 2-1, except that the ground has been relocated to position C. We now see that as the red galvanometer lead is moved to D, C, B, and then A, the voltages read are $+10$ V, 0 V, -10 V, and -20 V, respectively. Again, we see

---

*Gray, H. and A. Issacs, eds. *A New Dictionary of Physics*. London, England: Longman, 1975, p. 593.

Algebraic Addition and Subtraction

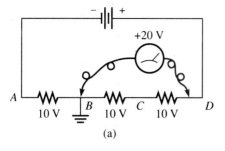
(a)

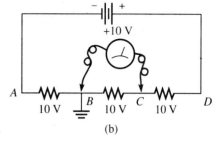
(b)

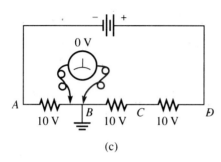
(c)

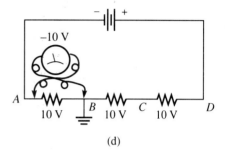
(d)

Figure 2–1

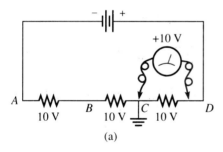
(a)

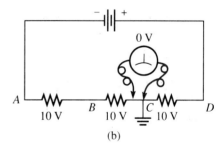
(b)

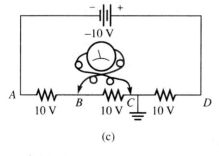
(c)

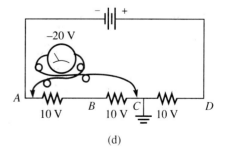
(d)

Figure 2–2

that as long as our red lead remains to the right of ground, our voltage reading is positive. As we move to the left of ground, however, we obtain negative values. It is clear, then, that the ground has two important properties. First, it serves as an arbitrary *point of reference* from which all other voltage measurements are taken. Second, the ground is considered the arbitrary *point of zero voltage*. In this way the voltage readings are positive or negative depending simply on where the ground (reference) point is located. There is certainly nothing magical about that. If point A were grounded, then *all* the voltage readings would be positive. Conversely, if point D were selected as the ground, then *all* the voltage readings would be negative. In the same manner, we can regard zero in mathematics as the reference point that separates all positive numbers from all negative numbers. It is, so to speak, the mathematical ground point, as shown in Figure 2–3, where every tenth integer is numbered in a manner similar to the voltage readings obtained in Figures 2–1 and 2–2.

Figure 2–3

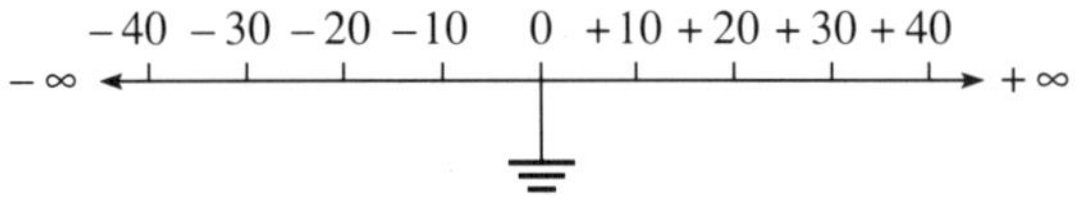

$-\infty \quad -40 \quad -30 \quad -20 \quad -10 \quad 0 \quad +10 \quad +20 \quad +30 \quad +40 \quad +\infty$

From the foregoing discussion, the following properties of zero should be evident: (1) Zero is in the exact middle of the number line and (2) zero is greater than any negative number. The fact that negative numbers are less than zero should not lead you into believing that negative -1000 V, for example, is harmless! You can be electrocuted just as quickly with -1000 V as you can with $+1000$ V.

Exercise Set 2–1

Questions

1. If the positive terminal of a 500 V DC power supply is grounded, explain why this point is considered as being at *zero volts*. Why isn't it $+500$ V?
2. Why are negative numbers considered to be less than zero?

Problems

1. Express the following quantities using positive or negative numbers.
   a. The boiling point of water (degrees Celsius)
   b. The temperature at which solder melts (degrees Fahrenheit)
   c. The voltage of an automobile battery relative to the positive terminal
   d. In Figure 2–2, the voltage of point B if point D is grounded
   e. Your new balance if you have $12.83 in your checking account and write a check for $20
2. Draw the number line (to scale), and label the following points: -21, 4.5, $-\pi$, 0, 15.09, -3.00.

2–2 ADDITION OF SIGNED NUMBERS

Now that we have justified the existence of the negative numbers, we will establish rules by which positive and negative numbers (called **signed numbers**) can be algebraically manipulated and operated upon. In Figure 1–2, a graphical interpretation was given to the operation of adding $+3$ and $+5$ to obtain $+8$. In a similar manner, we can add -3 to -5 to obtain -8. This graphical interpretation is shown in Figure 2–4 using the left half of the number line.

Algebraic Addition and Subtraction

Figure 2–4

In Figure 2–4, we see that a line of length 3 units and directed toward the left is added physically to a line of length 5 units and also directed toward the left, resulting in a total length of 8 units to the left. In other words, the total length of the resultant line is simply the sum of the individual lengths—that is, 3 units and 5 units—and directed toward the left. Note that in adding these lengths, we disregarded the negative signs and merely paid attention to the magnitude of their lengths. Whenever we consider magnitude without regard to sign, we say that we are dealing with the **absolute value** of a quantity—that is, its size without regard to its polarity. To indicate the absolute value of -3, for example, we place the -3 between two vertical bars: $|-3|$. In other words, $|-3| = 3$—not $+3$, just 3. There are no signs ($+$ or $-$) associated with the absolute value of a number. This leads us to the following rule for adding two or more numbers having the same sign.

> **Rule 2–1**
>
> To add two or more numbers with the same sign, find the sum of their absolute values and affix the common sign.

Example 2–1 Add the following numbers: -3, -15.2, and -8.

Solution By the preceding rule, we add their absolute values: $3 + 15.2 + 8 = 26.2$. We then affix the common sign to obtain the final answer of -26.2.

Although Example 2–1 is fairly simple, what interpretation should we give to the addition of two numbers that differ in sign? Consider the addition of -3 and $+5$, as shown in Figure 2–5.

Figure 2–5

(a) $(-3) + (+5) = +2$ (b) $(+5) - (+3) = +2$

In obtaining the answer $+2$, we made use of the "go-left" interpretation of the $-$ sign, and the "go-right" interpretation of the $+$ sign, as we did previously in connection with Figure 1–4.

From Figure 2–5, we note that *adding* $+5$ to -3 is the same as *subtracting* $+3$ from $+5$. These two interpretations lead us to the following rule for addition of two numbers that differ in sign.

> **Rule 2–2**
>
> To add two numbers differing in sign, find the difference between their absolute values and affix the sign of the number having the larger absolute value.

Example 2-2   Add -22 and $+7.4$.

Solution   The difference between their absolute values is $22 - 7.4 = 14.6$. Because -22 has the larger absolute value, we affix a negative sign to the answer, obtaining -14.6.

If we wish to obtain the sum of more than two numbers, all differing in sign, we can apply Rules 2-1 and 2-2 separately and obtain the following general rule.

> **Rule 2-3**
>
> To add more than two numbers differing in sign, first find the sum of all the positive numbers using Rule 2-1 and then find the sum of all the negative numbers by applying Rule 2-1 again. Finally, obtain the sum of the two resulting numbers using Rule 2-2.

Example 2-3   Find the sum of -14, $+8.8$, -9.95, $+12$, and -2.2.

Solution   By Rule 2-1, the sum of the positive numbers is $8.8 + 12 = 20.8$. Again by Rule 2-1, the sum of the negative numbers is $(-14) + (-9.95) + (-2.2) = -26.15$. Finally, by Rule 2-2, $|-26.15| - |20.8| = 5.35$. Therefore, the sum is -5.35.

Exercise Set 2-2

Questions

1. How is the number line used to show addition of signed numbers?
2. What directional interpretation on the number line can be applied to the positive (+) and negative (−) signs of real numbers?

Problems

1. Find 34 added to 22.8.
2. Find the sum of 56.25 and 88.
3. Add -56 and 21.
4. Find $7/16$ added to $-1/4$.
5. Find the sum of 0.0045 and -0.00018.
6. Add 12.7, -8, 53.6, and $-7/32$.
7. Find $3\frac{7}{8} + -2\frac{1}{4}$.
8. Find -12.5×10^3 added to 10,362.
9. Find the sum of -3×10^2 and 300.
10. Add: $-15 + 15$.
11. Add -8.2, 17, $4\frac{1}{2}$, -3.6, 14.01, -1×10^1, and -2.
12. Find -44.03 added to the sum of 21 and -2.09.

2-3   SUBTRACTION OF SIGNED NUMBERS

In ordinary subtraction, we have always encountered the situation where the smaller number was subtracted from the larger number. For example, the difference between 3 and 5 always implied that 3 was to be subtracted from 5—that is, $5 - 3 = 2$. What

Algebraic Addition and Subtraction

interpretation, however, can we give to the problem 3 − 5? From the directional properties established earlier for signed numbers, we can derive two interpretations for this problem. First, as shown in Figure 2–6a, we may take the problem to mean, "Go right 3 units and then go left 5 units, stopping at −2." Alternately, as shown in Figure 2–6b, the problem can be stated as, "Go left 5 units and then go right 3 units, stopping at −2." We see that the answer is the same in either case. These results force us to conclude that subtracting a positive 5 (3 − 5) is identical to adding a negative 5 (3 + (−5)). This last statement gives us the basis for the rule used in subtracting signed numbers.

Figure 2–6

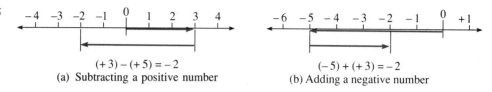

(a) Subtracting a positive number $(+3) - (+5) = -2$

(b) Adding a negative number $(-5) + (+3) = -2$

Rule 2–4

To subtract one signed number (the subtrahend) from another (the minuend), change the sign of the subtrahend and apply the rules of addition of signed numbers.

(*Note:* Minuend − subtrahend = difference.)

Example 2–4 $4 - 7 = ?$

Solution   The minuend is $+4$; the subtrahend is $+7$. We change the $+7$ to a -7 and then apply the rules of addition. In this case, since the two numbers differ in sign, we make use of Rule 2–2 of addition to obtain $4 - 7 = 4 + (-7) = -3$.

Example 2–5 $-3 - 8 = ?$

Solution   The sign of the subtrahend is changed to -8, giving $-3 - 8 = -3 + (-8) = -11$.

Example 2–6 $4 - (-2) = ?$

Solution   Since the subtrahend is -2, we change this to $+2$ and then add: $4 - (-2) = 4 + 2 = 6$.

Exercise Set 2–3

Questions

1. Define the following terms using your own words:
   a. Minuend
   b. Subtrahend
   c. Difference
2. Using the number line, show that *subtracting a negative* 3 is equivalent to *adding a positive* 3.
3. Is subtraction commutative? Explain your answer.
4. Is subtraction associative? Explain your answer.

Problems

Subtract the first number from the second number.
1. 5, 2
2. −5, −2
3. −100, 14
4. 22, 12
5. −3, −7
6. −π, −3
7. 0.138, 0.002
8. 1×10^3, 1×10^4
9. 0.00157, −0.707
10. 14.0072, −15.7889

2-4 LITERAL NUMBERS

Thus far in our study of signed numbers we have dealt only with concrete numbers. A concrete number is one having a definite, fixed value, such as 5, −56.3, or 1×10^3. Although such quantities are needed to reduce a particular problem to a final numerical answer, their very exactness limits their usefulness. For example, suppose you want to purchase 100 2N3904 transistors but have only $5 to spend. What would be the highest price (including tax) that you could pay for each transistor? It may be fairly obvious that you could not possibly pay more than 5¢ per transistor. To arrive at that answer, you simply divide 500¢ by 100 transistors, obtaining 5¢ per device. Can you think of a way to express the problem in more general terms—that is, the unit price for *any* number of *any* type of device for *any* fixed budget? If we let D stand for the number of *dollars* we have to spend, N be the *number* of devices we have to buy, and C be the *cost* for each, we might write the answer to the problem in a more general way as:

$$\frac{D}{N} = C$$

In other words, D (the *dollars* we have) divided by N (the *number* of devices needed) results in C (the unit *cost*). In this example, we allowed letters to take the place of concrete numbers. Whenever letters are used in this way to stand for numbers, the letters are called **literal numbers.** The word literal means letter, and any mathematical idea written using literal numbers, along with the signs and symbols of algebra, is referred to as an **algebraic expression.**

The principal advantage of algebraic expressions is that they provide a general solution for all other situations of the same type. We are no longer limited to particular concrete numbers in given instances. This simple idea represents one of the major strengths and advantages of algebra.

As a further example of the manner in which literal numbers may be used, suppose the current in a particular circuit is found to be the sum of three other individual currents. In this case, we might label the three currents as I_1, I_2, I_3 and the total current as I_t. We then write:

$$I_t = I_1 + I_2 + I_3 \quad \text{amperes}$$

For example, if $I_1 = 30$ A, $I_2 = 15.4$ A, and $I_3 = 2$ A, then we can write the total current as

$$I_t = I_1 + I_2 + I_3$$
$$= 30 \text{ A} + 15.4 \text{ A} + 2 \text{ A}$$
$$= 47.4 \text{ A}$$

Algebraic Addition and Subtraction

Here, we let the letter I (along with the proper subscript) stand for the individual currents, whereas A represents the *unit* of electrical current, the ampere.

Suppose that I_2 was found to be running through the circuit in a direction opposite to that of I_1 and I_3. We can indicate this fact by writing it as a negative number and then performing the addition according to the rules for adding signed numbers developed earlier. For example,

$$I_t = I_1 + (-I_2) + I_3$$
$$= 30 \text{ A} + (-15.4 \text{ A}) + 2 \text{ A}$$
$$= 30 \text{ A} - 15.4 \text{ A} + 2 \text{ A}$$
$$= 16.6 \text{ A}$$

Example 2-7

Write an algebraic expression for the sum of three voltages in a particular series circuit.

Solution We can write $V_t = V_1 + V_2 + V_3$, where V_t stands for total voltage and the subscripts indicate individual voltages.

2-5 TERMS

In algebra, only those elements of an expression having the same literal parts may be combined. We refer to the individual parts of an expression, such as I_1, I_2, and I_3 in our earlier example, as the **terms** of the expression. More formally, a term is that part of an expression whose individual pieces are *not* separated by either a plus or minus sign. Terms that have the same literal pieces (e.g., $12P$ and $0.2P$) are called **like terms,** or similar terms. Terms that do not have the same literal parts (e.g., $0.75V$, $3W$, and $-j18$) are called **unlike,** or dissimilar, **terms.** Note that like as well as unlike terms may appear in a single expression, but only the like terms can be combined. For example, in the expression $-8G + 3LC + 50G - 0.1LC$, the numerical parts of the two like terms containing G can be combined to give $42G$; similarly, the two LC terms can be combined to form $2.9LC$, but the resulting two unlike terms cannot be combined any further. We must stop at $42G + 2.9LC$.

Example 2-8

Combine the similar terms in the expression $-8at + 3IR + IR + 4at - at$.

Solution We write $-8at + 4at - at = -5at$ and $3IR + IR = 4IR$, which can be written as $-5at + 4IR$.

2-6 FACTORS

Note that the literal parts of the previous expression, $42G + 2.9LC$, are not mere labels. In fact, the literal parts G and LC represent numerical quantities themselves, whose values may or may not be known. Moreover, these literal parts actually *form a product along with their numerical part*. For example, the term $2.9LC$ actually means the product formed in multiplying $2.9 \times L \times C$. Each of the individual pieces (or any combination of them) forming the product of the term $2.9LC$ is called a **factor**. The factors of the term $2.9LC$ are 2.9, L, C, $2.9L$, $2.9C$, and LC. More technically, if two or more quantities are multiplied together, each of them or any combination of them is referred to as a factor.

| **Example 2-9** | List the factors of the term $-xyz$.

Solution The factors are $-1, x, y, z, -x, -y, -z, xy, xz, yz, -xy, -xz, -yz,$ and xyz. |

2-7 COEFFICIENTS

Most often, we refer to the numerical factor as the **numerical coefficient.** In the foregoing example, 2.9 is said to be the numerical coefficient of LC. Likewise, any literal part of the term can be called the **literal coefficient** of the remaining factors. Factors containing numbers and letters, such as $2.9C$, are simply called coefficients without further qualification. If a term contains no explicit numerical coefficient, then the coefficient 1 is implied. For example, the term IR has an implied numerical coefficient of $+1$.

| **Example 2-10** | What is the numerical coefficient of the second term in the expression $5RC + 12LC - 8TC$?

Solution The second term is $12LC$, and the numerical coefficient is 12. |

2-8 EVALUATION OF EXPRESSIONS

Since the literal parts of a term represent specific numerical values and since expressions are made up of terms, it seems reasonable that if we knew the values of the literal parts, we could evaluate a given expression by simply replacing each literal number with its numerical equivalent. For example, in the expression $4AF - 6k + 2Ak$, suppose that $A = 2$, $F = 5$, and $k = \frac{1}{2}$. Upon replacing the literal numbers by their numerical equivalents, we can evaluate the given expression, obtaining:

$$4AF - 6k + 2Ak = 4 \times 2 \times 5 - 6 \times \frac{1}{2} + 2 \times 2 \times \frac{1}{2}$$
$$= 40 - 3 + 2 = 39$$

| **Example 2-11** | Evaluate the expression $-3IR + 14X + 12RX$ if $I = 1.5$, $R = 4$, and $X = -2$.

Solution We write

$$-3IR + 14X + 12RX = -3(1.5)(4) + 14(-2) + 12(4)(-2)$$
$$= -18 - 28 - 96 = -142$$ |

Exercise Set 2-4

Questions

1. For each pair of terms shown, (a) define each term; (b) compare and contrast the two terms; and (c) give an example of each.
   a. Literal number; concrete number
   b. Algebraic expression; arithmetic expression
   c. Term; factor
   d. Like terms; unlike terms
   e. Numerical coefficient; literal coefficient
2. What are the principal advantages of using literal numbers in algebra?

Algebraic Addition and Subtraction

Problems

1. In the algebraic expression $3VA - 12I^2R + 4ZLC^2$, give (a) the terms of the expression and (b) the factors in each term.
2. Simplify the following expressions by combining like terms where possible:
   a. $3V + i - 2V - 8i$
   b. $-C + L - C - L$
   c. $2R + 3r + 4R^2$
   d. $5 + 3Z + X^2 + 3X^2$
   e. $\frac{Z}{3} + 2Z + R$
   f. $30Q - 0.5Q + Q^2$
   g. $3v^2 - 8av^2 + 4a + a^2v$
   h. $K + \frac{1}{2}K - 0.02K - K$
   i. $2000V + (2 \times 10^3)V + (0.02 \times 10^5)V$
   j. $* + \# - 2* + 4\#$
3. Evaluate the following expressions if $x = 2$, $k = 3$, and $c = -1$.
   a. $x^2 - k + c$
   b. $-x - k + c$
   c. $-2c + k^2 + 4x$
   d. $-c^2 - k^2 + x^2$
   e. $-3c + k - 3x$
   f. $15 - x - k + c$
   g. $1000x + 500k - 250c$
   h. $-0.005c - 0.02x + k$
   i. $\frac{1}{2}x + \frac{1}{3}k - \frac{3}{5}c$
   j. $x^3 + k^2 + c - 16$
4. The following BASIC program evaluates the expression in Problem 3i for any real values of x, k, and c:

```
10 LET N=0
20 PRINT"Enter the values of x, k, c"
30 INPUT X, K, C
40 LET N=1/2*(X)+1/3*(K)-3/5*(C)
50 PRINT"The value of x/2 + k/3 - 3c/5 is:";N
60 END
```

Write a BASIC program that evaluates the expression in Problem 3j for all real values of x, k, and c.

2-9 SYMBOLS OF GROUPING

In performing certain algebraic operations, it is often required that two or more quantities be grouped together and treated as a single expression. In order to do this, certain **symbols of grouping*** are commonly employed. These symbols are *parentheses*, ( ), *brackets*, [ ], *braces*, { }, and the *vinculum*, ⎯⎯⎯. Any terms affected by any one of these symbols are to be treated as if they were a single quantity. For example, the expression

---

*Symbols of grouping are often called *signs* of grouping. However, since we have reserved the word sign to mean polarity (+ or −), we will use the word *symbol* here to avoid confusion.

$P - (W + VI)$ means that the sum of W and VI is to be found before being subtracted from P. This condition could have been written using any of the other symbols: $P - [W + VI]$, $P - \{W + VI\}$, or $P - \overline{W + VI}$, although the latter symbol, the vinculum, is generally reserved for use with fractions or the extractions of roots. For example, in the fraction

$$\frac{3}{Z + R}$$

the vinculum is used to indicate that the sum of $Z + R$ is to be divided into 3. In a similar manner, the expression

$$5 - \sqrt{R + K}$$

means that the square root of the sum of $R + K$ is to be extracted before the result is subtracted from 5.

In working with symbols of grouping, it often happens that some symbols are *nested* within others—that is, symbols may occur one within the other. For example, in the expression

$$B - (120 + K\{3 + M\})$$

the sum of $3 + M$ is to be multiplied by K and the result is to be added to 120; finally, this sum is to be subtracted from B. In order to maintain consistency, it has been established by general agreement that the operations within the innermost symbols be performed first, followed by the next-innermost grouping, and so forth. In other words, whenever nested symbols occur, the operations are performed from the inside out. For example,

$$\begin{aligned}
100 - [6 + \{-5 + 3(2 - 11) + 7.5\} - 8] &= 100 - [6 + \{-5 + 3(-9) + 7.5\} - 8] \\
&= 100 - [6 + \{-5 - 27 + 7.5\} - 8] \\
&= 100 - [6 + \{-24.5\} - 8] \\
&= 100 - [-26.5] \\
&= 100 + 26.5 = 126.5
\end{aligned}$$

In working with nested symbols of grouping, we apply the following rule.

Rule 2-5
Whenever symbols of grouping occur one within the other (i.e., are nested), remove the innermost symbols first.

Often it is possible to simplify an expression involving symbols of grouping by the application of a few simple rules. These rules are based on the rules of addition and subtraction developed earlier. For example, in the expression $8 + (2 + 4)$, we note that we arrive at the same answer whether we perform the parenthetical sum first and then add or simply perform the sum from left to right without any regard for the parentheses. In other words, $8 + (2 + 4) = 14 = 8 + 2 + 4$. This is always the case for simple addition, and leads us to the following rule.

Rule 2-6
To remove symbols of grouping preceded by a plus sign, simply remove the symbols without making any other changes.

Algebraic Addition and Subtraction

Example 2-12

Simplify the following expression by removing the brackets.
$$25 + [-3 + 16]$$
Solution Using Rule 2-6, we write $25 + [-3 + 16] = 25 - 3 + 16 = 38$.

Example 2-13

Simplify the expression $-3X^3 + (-X^3 + 5)$ by removing symbols of grouping and combining like terms.

Solution $-3X^3 + (-X^3 + 5) = -3X^3 - X^3 + 5 = -4X^3 + 5$

If parentheses or other symbols of grouping are preceded by a minus sign, indicating subtraction, we can apply the rule for subtraction developed previously. For example, $19 - (3 + 8)$ states that the sum $(3 + 8) = 11$ is to be subtracted from 19, giving 8 as a result. The rule for subtraction, however, states that we change the sign of the subtrahend and then add. Therefore, we change the signs of the numbers within the parentheses to -3 and -8, remove the parentheses, and then add as follows

$$19 - (3 + 8) = 19 - 3 - 8 = 8$$

This leads us to another rule for simplifying expressions containing symbols of grouping.

> **Rule 2-7**
> To remove symbols of grouping preceded by a minus sign, change the sign of each term within the symbols.

Example 2-14

Simplify the expression $28 - (3 - 12)$ by removing the parentheses.

Solution From Rule 2-7,
$$28 - (3 - 12) = 28 - 3 + 12 = 37$$

Example 2-15

Simplify the expression $4Z - \{-8X + Z\}$ by removing the symbols of grouping and combining like terms.

Solution $4Z - \{-8X + Z\} = 4Z + 8X - Z = 3Z + 8X$

If we remove the symbols of grouping from the expression $a + (b - c)$, we obtain $a + b - c$. Note that if we desired to reinsert the parentheses around the last two terms, we would merely reverse the process of removal and write $a + (b - c)$ without making any other change. This simple method of reversal is due to the fact that a plus sign precedes the grouping symbols. This example allows us to make the following rule.

> **Rule 2-8**
> To enclose terms within symbols of grouping preceded by a plus (+) sign, simply enclose the terms without making any other changes.

Example 2-16

Enclose the last two terms of the expression $3G + R - 4K$ within braces, preceded by a plus (+) sign.

Solution $3G + R - 4K = 3G + \{R - 4K\}$

In removing the parentheses from the expression $a - (b - c)$, we obtain $a - b + c$, by Rule 2-7. If we now wish to reverse the process and reinsert the last two terms in parentheses, the answer is *not* $a - (b + c)$, since removing these results in $a - b - c$, which is different than $a - b + c$. It is evident, however, that all we need do to accomplish the correct reversal is to change the sign of each grouped term prior to placing the terms within the grouping symbols. This last example permits us to state the following rule.

> **Rule 2-9**
> To enclose terms within symbols of grouping preceded by a minus (−) sign, change the sign of each term to be included within the group.

Example 2-17

Enclose the last three terms of the expression $8C - 2L + 0.5C - L$ in parentheses preceded by a minus sign.

Solution $8C - 2L + 0.5C - L = 8C - (2L - 0.5C + L)$

Note that the correctness of this result can be checked by removing the parentheses and observing that the original expression is obtained:

$$8C - (2L - 0.5C + L) = 8C - 2L + 0.5C - L$$

Exercise Set 2-5

Questions

1. Name the four most common symbols of grouping and show the symbol for each.
2. For what purpose are symbols of grouping used?
3. What is meant by nested symbols of grouping? Give an example.

Problems

1. Simplify each of the following expressions by removing symbols of grouping and combining like terms.
    a. $3a + (a - 5a)$
    b. $V^2 + 4(V^2 - R)$
    c. $3R^3 + (R^3 - \{2R^3 + 4\} + 3R^3)$
    d. $-a - (-a - a) - 2a$
    e. $-(Z - 3 + \{Z + [3Z - 3] + Z\} - 3)$
    f. $(R^2 - 3X^2) - (3R^2 + X^2)$
    g. $(3\phi - 2\lambda + \{-\phi + 4\lambda + 3\} + 4\lambda)$
    h. $-[V + 2I - \{2V - I + V\} + V]$
    i. $\overline{-4 - x} + (3 - \overline{2x + 5})$
    j. $-P - \{-2P - \overline{P - 2Q} + Q\} - P - 2Q$
2. Insert the last two terms of each expression within symbols of grouping preceded by a plus (+) sign. Check your answers by removing the grouping symbols.

Algebraic Addition and Subtraction

   a. $A + 4A - k$
   b. $Z + 7 + 2Z$
   c. $3C - 4D + 8$
   d. $-x - 5 - 2x$
   e. $V + 0.05V - V - 0.02V$
3. Insert the last two terms of each expression within symbols of grouping preceded by a minus $(-)$ sign. Check your answers by removing the grouping symbols.
   a. $4k + 3V - 4k$
   b. $2x - 4y + 5z$
   c. $3R + 2Xa - 4Xc$
   d. $-a - 2a - 3a + 4a$
   e. $2\lambda + 8f + \lambda - f$

Key Terms

negative numbers	difference	unlike terms
polarity	literal numbers	factor
signed numbers	algebraic expression	numerical coefficient
absolute value	terms	literal coefficient
minuend	like terms	symbols of grouping
subtrahend		

Important Rules

Rules of Addition

Rule 2–1 To add two or more numbers with the same sign, find the sum of their absolute values and affix the common sign.

Rule 2–2 To add two numbers differing in sign, find the difference between their absolute values and affix the sign of the number having the larger absolute value.

Rule 2–3 To add more than two numbers differing in sign, first find the sum of all the positive numbers using Rule 2–1 and then find the sum of all the negative numbers by applying Rule 2–1 again. Finally, obtain the sum of the two resulting numbers using Rule 2–2.

Rule for Algebraic Subtraction

Rule 2–4 To subtract one signed number (the subtrahend) from another (the minuend), change the sign of the subtrahend and apply the rules of addition.

Rules for Working with Symbols of Grouping

Rule 2–5 Whenever symbols of grouping occur one within the other (i.e., are nested), remove the innermost symbols first.

Rule 2–6 To remove symbols of grouping preceded by a plus sign, simply remove the symbols without making any other changes.

Rule 2–7 To remove symbols of grouping preceded by a minus sign, change the sign of each term within the symbols.

Rule 2–8 To enclose terms within symbols of grouping preceded by a plus $(+)$ sign, simply enclose the terms without making any other changes.

Rule 2–9 To enclose terms within symbols of grouping preceded by a minus $(-)$ sign, change the sign of each term to be included within the group.

3 Algebraic Multiplication and Division

Upon completion of this chapter, you will have:
1. Learned how to multiply and divide using exponents.
2. Learned how to multiply and divide multinomials.

48 Fundamentals of Algebra

3-1 INTRODUCTION

Multiplication might be looked upon as a time-saving tool for addition. For example, if you want to add the number 9 four times, you would write $9 + 9 + 9 + 9 = 36$. Multiplication, however, accomplishes the same task: $9 \times 4 = 36$. It is easy to see, then, why multiplication is often described as the shortcut process of repeated addition. This definition can even be applied to such fractional products as $\frac{3}{4} \times \frac{2}{5}$. For although it makes little sense to regard this as $\frac{3}{4}$ added two-fifths of a time, it is clear that the answer $\frac{6}{20}$ can be regarded as $\frac{1}{20}$ added six times.

There are a variety of ways of indicating algebraic multiplication. One way, of course, is to make use of the familiar "times" sign that we learned in arithmetic, as in $3 \times 2 \times 6$. In algebra, however, the letter x is often used to represent a literal number and so the times sign is seldom used to indicate multiplication in order to avoid confusion. Another means of indicating multiplication is to place a dot between the quantities to be multiplied. For example, $3 \cdot a \cdot y$ means 3 times a times y. Most often, the multiplication process is indicated by **juxtaposition,** which means placing the quantities next to one another without any other operational symbols, as in $3ay$. Finally, parentheses or other symbols of grouping can be used to group factors, as in $(3)(a)(y)$. We see that all the following expressions mean the same thing, namely, the product of the factors 3, a, and y:

$$3 \times a \times y = 3 \cdot a \cdot y = (3)(a)(y) = 3ay$$

Since algebraic multiplication involves the manipulation of signed ($+$ or $-$) numbers, it is important that we establish rules for the sign of the resulting product. However, these rules are not arbitrary but are based on the results obtained previously by algebraic addition, as will be shown. Whenever two signed numbers are multiplied together, there are four possible combinations of signs:

$$+a \cdot (+b)$$
$$+a \cdot (-b)$$
$$-a \cdot (+b)$$
$$-a \cdot (-b)$$

For example, suppose we let the absolute values of the two literal numbers a and b be 3 and 4. The following represent the four possible combinations of signs:

$$+3 \cdot (+4)$$
$$+3 \cdot (-4)$$
$$-3 \cdot (+4)$$
$$-3 \cdot (-4)$$

From our earlier discussion of addition, we can interpret the first product $(+3)(+4)$ graphically as $+3$ added four times or $+4$ added three times. In either case, the answer is $+12$, as shown in Figure 3–1.

Figure 3–1

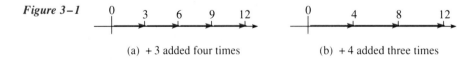

(a) $+3$ added four times

(b) $+4$ added three times

In the second and third cases, we can look upon this as -4 added three times or -3 added four times. In both cases, the answer is -12, as shown in Figure 3–2. In the last

Algebraic Multiplication and Division

case, $(-3)(-4)$, we can interpret this as -3 subtracted four times—that is, $-(-3) - (-3) - (-3) - (-3) = +12$.

Figure 3-2

(a) -4 added three times

(b) -3 added four times

From the foregoing, we write the following rules for multiplication of signed numbers.

> **Rule 3-1**
> The product of two numbers with the same sign is positive.
>
> **Rule 3-2**
> The product of two numbers with different signs is negative.
>
> **Rule 3-3**
> Rules 3-1 and 3-2 are to be applied in succession when there are more than two factors.
>
> **Corollary to Rule 3-3**
> The product of an *even* number of negative factors is positive. The product of an *odd* number of negative factors is negative.

Exercise Set 3-1

Questions

1. Define the term *juxtaposition* as it is used in algebraic multiplication.
2. What are four methods of indicating multiplication?
3. State the rules for determining the sign of a product.

Problems

1. If x is a negative real number, determine the sign of the product $-(x)(x)(x)$.
2. If -3 is used as a factor 25 times, determine the sign of the result without actually doing the problem.
3. Find the following products.
   a. -3×6
   b. $-(-3) \times 2$
   c. $½ \times (-¼)$
   d. $-5 \times [-(-2)]$
   e. $-0.0015 \times 2.07 \times (-0.0109)$
   f. $-1 \times a$

3-2 MULTIPLICATION INVOLVING EXPONENTS

By definition, an **exponent** is a number written to the upper right of another number, called the **base**, which tells how many times the base is to be taken as a factor. For example, in the expression $2^5 = 32$, the exponent 5 tells us to take the base 2 as a factor 5 times—that is, $2^5 = 2 \times 2 \times 2 \times 2 \times 2 = 32$.

Whenever the bases are the same, we can derive a simple law for finding the product of two or more exponential expressions. For example, to find the product of $b^3 \cdot b^4$, we note that b^3 means $b \cdot b \cdot b$ and b^4 means $b \cdot b \cdot b \cdot b$. Therefore, $b^3 \cdot b^4$ represents the product $b \cdot b \cdot b \cdot b \cdot b \cdot b \cdot b = b^7$, which can be written as b^{3+4}. This gives us the following law for the multiplication of exponential terms.

> **Rule 3–4 Exponential Law of Multiplication**
> To find the product of two or more exponential terms having the same base, add their exponents.
> $$b^p b^q = b^{p+q}$$

Caution: A common error made by students in working with exponents is to confuse $-b^2$ with $(-b)^2$. We read the first expression, $-b^2$, as "the negative of b squared." The latter expression, $(-b)^2$, is read as "the square of $-b$," which is a positive number.

In order to avoid this sort of confusion, remember that the exponent applies *only* to the factor to its immediate left; $-b^2$ is actually the product of (-1) and $(b)^2$. Therefore, $-b^2 = (-1)(b^2) = -b^2$.

There are two other situations concerning the multiplication of exponential terms that need to be investigated. First, we note that $(a^3)^2$ means $(a \cdot a \cdot a)(a \cdot a \cdot a)$, which is the same as a^6. Therefore, we see that a power such as a^3, for example, can itself be raised to another power. This gives us the law for raising a power to a power:

> **Rule 3–5 Law for the Power of a Power**
> $$(b^p)^q = b^{pq}$$

Secondly, exponential factors as a group can be affected by an exponent. This leads to the law for the power of a product.

> **Rule 3–6 Law for the Power of a Product**
> $$(bc)^p = b^p c^p$$

Example 3–1 $3^2 \times 3^4 = 3^{2+4} = 3^6 = 729$

Example 3–2 $a \times a^2 \times a^4 = a^{1+2+4} = a^7$

Example 3–3 $(-z^2)(z^3)(y^2) = -z^{2+3} y^2 = -z^5 y^2$

Algebraic Multiplication and Division

Example 3-4

$(3^2)^3 = 3^6 = 729$

Also,

$(3^2)^3 = (9)^3 = 729$

Example 3-5

$(a^2b^3)^2 = a^4b^6$

Example 3-6

$(3VI)^3 = 3^3V^3I^3 = 27V^3I^3$

Exercise Set 3-2

Questions

1. Define the following terms.
   a. Exponent
   b. Base
2. State the rule for the multiplication of exponential terms having the same base.
3. State the two power laws for exponential multiplication.

Problems

1. Find the following exponential products.
   a. x^3x^5
   b. $2^4 \times 2^5$
   c. $\epsilon^3(\epsilon^2\epsilon^4)$
   d. $-2^2 \times 2^3 \times 2^4$
   e. $(+a)^2(-a^2)$
   f. $(2^3)^2$
   g. $(½)^2$
   h. $(V^5)^2$
   i. $(I^2)^3 I$
   j. $\left(\dfrac{1}{x^2}\right)^5$
   k. $(3x)^2$
   l. $(-3k^2)^n$
   m. $\left(\dfrac{1}{2}LC\right)^2$
   n. $(2\pi f)^2$
   o. $-(2^m x^2)^r$

2. In BASIC, exponentiation is accomplished by using the carat symbol (∧) as in x∧3, which means "x raised to the third power." The following BASIC program will compute the product of $b^p b^q = b^{p+q}$ for all real numbers b, p, and q.

```
10 PRINT"Enter the values of b, p, and q."
20 INPUT B,P,Q
30 LET N=B^Q*B^P
40 PRINT"The product of the powers is:";N
50 END
```

Write a BASIC program for computing the power of a power, $(b^p)^q = b^{pq}$, and the power of a product, $(bc)^p = b^p c^p$, for all real numbers b, c, p, and q.

3-3 PRODUCTS OF MONOMIALS AND MULTINOMIALS

Multiplication of Monomials

The word *monomial* is derived from the Latin combining form *mono-*, meaning one, and the Latin word *nomen*, meaning name. A **monomial**, then, is an algebraic expression of one term. A binomial is an expression of two (bi-) terms. Similarly, a trinomial has three (tri-) terms. Although there are technical names for expressions having four or more terms, we ordinarily do not take the trouble to learn them, since the nomenclature is relatively unimportant from our point of view. In fact, any expression having two or more terms is generally referred to simply as a **multinomial,** since the combining form *multi-* means many. Some textbooks refer to a multinomial expression as a polynomial, since the Greek combining form *poly-* also means many. Technically speaking, however, the term polynomial as used in mathematics has a very precise meaning and is *not* the same as a multinomial.* The expression $-3IR$ is a monomial. An example of a multinomial having three terms is $2VI - IR + \sqrt{28W}$.

To multiply one monomial by another, we apply the following rule.

Rule 3-7
1. Multiply the numerical coefficients, paying careful attention to the sign of the result.
2. Multiply the literal factors using the rule for exponents.
3. Write the results of steps 1 and 2 as a single product.

For example, the product of the two monomials $(2X^2B)$ and $(-3X^3B^3)$ can be formed by applying the preceding steps:

1. $(2)(-3) = -6$
2. $(X^2)(X^3) = X^5$ and $(B)(B^3) = B^4$
3. $(2X^2B)(-3X^3B^3) = -6X^5B^4$

Note that in order to form the product of the literal factors, the bases must be identical. If, for example, we had the two terms $(11AX)$ and $(3RC)$, we could multiply their numerical coefficients, obtaining 33, but nothing further could be done with the literal parts because all four literal factors are dissimilar. The resulting product, $33AXRC$, cannot be simplified further.

Using the strategy just outlined, we can obtain the product of more than two monomials. For example, $(12kB)(-k^2B)(0.5kB) = -6k^4B^3$. Here, we multiplied all three numerical coefficients, observing the proper sign, and then combined the literal factors using the rule for the multiplication of exponents. We can find the product of any number of monomials in a similar manner.

Note that the result of any algebraic multiplication can be checked by substituting any convenient numbers for the literal factors and then comparing the results obtained. For instance, in the last example, if we let $k = 2$ and $B = 3$, we note that

$$(12kB)(-k^2B)(0.5kB) = (12 \times 2 \times 3)(-4 \times 3)(0.5 \times 2 \times 3)$$
$$= (72)(-12)(3) = -2592$$

We may now substitute these same numbers for the literal factors obtained in our earlier answer and note that

$$-6k^4B^3 = -6 \times 16 \times 27 = -2592$$

Since both answers agree, we conclude that our result must be correct.

---

*A **polynomial** is an expression of the form ax^n, where n is a *nonnegative integer*. A polynomial may have just one term; a multinomial may not.

Algebraic Multiplication and Division

Example 3-7 $(-3a)(-a) = 3a^2$

Example 3-8 $5ax^2 \times 4a^2x = 20a^3x^3$

Example 3-9 $(4V^2R)(V^2R)(-2V^2R) = -8V^6R^3$

Exercise Set 3-3

Questions

1. Define the following terms.
   a. Monomial
   b. Binomial
   c. Multinomial
2. What is the difference between a multinomial and a polynomial?
3. State the rule for the multiplication of monomials.

Problems

Find each product.

1. $(ax^2)(ax)$
2. $(VI^3)(IV)$
3. $(-3V^2)(-3V^2)$
4. $(-10k^3)(-10k)$
5. $(2\pi f^2)(3\pi fL)$
6. $-(2\pi f)(5\pi f^2)$
7. $-\pi r^2 \cdot 3\pi r^3$
8. $\pi r^3 \cdot (-3)^2 \pi r^3$
9. $P^2y \cdot Py^2$
10. $x^2i \cdot xi^5$
11. $(-c^2m^3)(cm^2)(cm)$
12. $-(I^2R)(IR^2)(IR)$
13. $\frac{1}{2}z^3 \cdot \frac{3}{4}z$
14. $\frac{1}{5}x^2 \cdot \frac{5}{2}x^3$
15. $(a^3)^3(a^2)$
16. $-(a^2)^2 a^3(-a)^3$
17. $(3bx^2)^3 bx$
18. $(5AM^2)^3 A^2 M^3$
19. $(a^2)^3(a^3)^5 a$
20. $(-a^3)^2(-a)^5(-a^3)^2$

Multiplication of a Multinomial by a Monomial

To multiply a multinomial by a monomial, we can apply the three steps given in Rule 1 to each term of the multinomial, then add the resulting terms. For example, suppose we were to multiply the multinomial $3sk - 8bt^3$ by the monomial $-2s^2t$. We begin by multiplying $3sk$ by $-2s^2t$, obtaining $-6s^3kt$. We then multiply the second term of the multinomial by $-2s^2t$ to obtain $16s^2bt^4$. Finally, we add the resulting terms forming the new multinomial $-6s^3kt + 16s^2bt^4$. This straightforward process leads us to the following rule:

> **Rule 3-8**
>
> To multiply a multinomial by a monomial, multiply each term of the multinomial by the monomial according to Rule 3-7, and write each partial product as a term of the new multinomial.

| Example 3–10 | $-2x(3x^2 + 8x) = -6x^3 - 16x^2$ |

| Example 3–11 | $I(3I + I - \frac{1}{2}I) = 3I^2 + I^2 - \frac{1}{2}I^2 = 3.5I^2$ |

| Example 3–12 | $0.05yz^2(0.001y - 1.09z) = 0.00005y^2z^2 - 0.0545yz^3$ |

Exercise Set 3–4

Problems

Simplify.
1. $-R(R - R^2)$
2. $k(k^2 + k)$
3. $a^3(-ax + a)$
4. $-g^2(gx - g)$
5. $-X^2(R^2 + X^2)$
6. $P^3(-Q^2 + P)$
7. $3^2b^3(-2b + b^2)$
8. $2^3x(-x + 2^5)$
9. $4ac(a^2c + 2ac^2 - 2c^3)$
10. $5x(2x^3y - xy + 5y^2)$
11. $2i^3\left(-3i^2 - i - \frac{1}{2}i^2\right)$
12. $5\phi^2\left(-5\phi^2 + \phi + \frac{1}{8}\phi^3\right)$
13. $ak(a^5k - a^4k^2 + a^3k^3 - a^2k^4 + ak^5)$
14. $vi(v^6i + v^5i^2 - v^4i^3 + v^3i^4 - v^2i^5 + vi^6)$
15. $-\frac{1}{3}x\left(3x + \frac{2}{3}x^2 - 9 + 12x\right)$
16. $\frac{1}{2}V\left(6 - \frac{4}{5}V^2 + 18 - 2V^3\right)$
17. $V^2(4V - V) + 2V(V^2 - 4V)$
18. $-a^3(2a^2 - a^2) + a(a - 3a^2)$
19. $-ax^2(ax - a^2x^2) - a(a^2x^4 - ax^3)$
20. $-MN^3(M - N^2) + M(M^2N + MN^3)$

Multiplication of Multinomials

To multiply two multinomials, we multiply each term of one multinomial by every term in the other, observing the proper signs, and then add the resulting partial products, combining like terms where necessary. This process is shown diagrammatically in Figure 3–3.

Algebraic Multiplication and Division

Figure 3–3 $(\overset{\frown}{a + b})\,\overset{\frown}{(c + d)} = ac + ad + bc + bd$

From the foregoing, we may state the following rule.

> **Rule 3–9**
> To multiply multinomials, multiply every term of one by every term of the other, and write each resulting product as a term of the new multinomial.

Example 3–13 $(a - x^2)(2a + x) = 2a^2 + ax - 2ax^2 - x^3$

Example 3–14 $(V - R)(V + R) = V^2 + VR - VR - R^2 = V^2 - R^2$

Example 3–15 $(-v + 2i - 4x)(vi - 3ix) = -iv^2 + 3ivx + 2i^2v - 6i^2x - 4ivx + 12ix^2$
$= -iv^2 - ivx + 2i^2v - 6i^2x + 12ix^2$

In this last example, note that we have arranged the factors of each term in alphabetical order. This does not affect the result in any way, since multiplication is commutative. However, it does make it easier to keep track of like terms so that they can be combined. In Example 3–15, the product of the two terms $-v$ and $-3ix$ is $3vix$ if we keep the original order. Moreover, the product of $-4x$ and vi is $-4xvi$ if the original order is maintained. On first observation, these two terms may not appear to be like terms. If we place their factors in alphabetical order, however, it becomes immediately apparent that these are like terms: $3ivx$ and $-4ivx$, which combine to give $-ivx$. For this reason, it is a good idea for you to get into the habit of arranging the factors alphabetically.

Exercise Set 3–5

Problems

Find the product of the following multinomials.

1. $(a - b)(a + b)$
2. $(m + n)(-n - m)$
3. $(V - 2)^2$
4. $(P + Q)^2$
5. $(P + W)(2P - W^2)$
6. $(I - R)(0.5I - R)$
7. $(M + k^2)(2k - M^3)$
8. $(a + b^3)(5a - b^2)$
9. $\left(x - \dfrac{1}{3}\right)(x + 9)$
10. $\left(Z + \dfrac{1}{8}\right)(Z - 4)$
11. $(R^2 - X^2)(2R + 3X)$

12. $(x^3 + y^3)(4x - 8y)$

13. $\left(1.03V - \dfrac{1}{3}v\right)(0.09v^2 + V^3)$

14. $(3 - ax)\left(\dfrac{ax}{9} + ax^3 - x^2\right)$

15. $(100 + VI)(0.01VI + VI - V)$

16. $-(20k + m)(0.05k + m/100 - mk)$

17. $-(Y + Z)(Y/20 - 0.3Z + YZ^3)$

18. $(\lambda^4 + 2f + \lambda^5 f)(\lambda^3 - \lambda^4 - 2\lambda f)$

19. $(a - 2b - a^2 b)(a + b^3 - 2ab)$

20. $(x + y)^3(x - 2xy + y)$

21. The following BASIC program evaluates the expression shown in Problem 20, $(x + y)^3(x - 2xy + y)$, for all real values of x and y.

```
10 PRINT"Enter the values of x and y."
20 INPUT X,Y
30 LET N=(X+Y)^3*(X-(2*X*Y)+Y)
40 PRINT"The product of the two multinomial terms is:";N
50 END
```

Write a BASIC program for computing the product in Problem 19, $(a - 2b - a^2 b)(a + b^3 - 2ab)$, for all real values of the literal numbers.

Division Involving Exponents

From our previous discussion of exponents, we know that we can write the numerator and denominator of the expression b^5/b^3 as

$$\frac{b \cdot b \cdot b \cdot b \cdot b}{b \cdot b \cdot b}$$

If we cancel (i.e., divide out) common factors, we obtain

$$\frac{b^5}{b^3} = \frac{\not{b} \cdot \not{b} \cdot \not{b} \cdot b \cdot b}{\not{b} \cdot \not{b} \cdot \not{b}} = b^2$$

The cancellation of common factors produces the same result as if we had subtracted the exponent of the denominator from the exponent of the numerator. In this example, $b^5/b^3 = b^{5-3} = b^2$. This procedure gives us the following law for dividing exponents.

Rule 3–10 Law of Exponential Division

To divide exponential terms having the same base, subtract the exponent in the denominator from the exponent in the numerator.

$$\frac{b^p}{b^q} = b^{p-q} \qquad (b \neq 0)$$

Example 3–16

$$\dfrac{Q^9}{Q^4} = Q^{9-4} = Q^5$$

Example 3-17
$$\frac{-V^4}{V^3} = -V^{4-3} = -V$$

If the exponent of the numerator is less than the denominator exponent, the cancellation process looks like this:

$$\frac{b^3}{b^5} = \frac{\cancel{b}\cdot\cancel{b}\cdot\cancel{b}}{\cancel{b}\cdot\cancel{b}\cdot\cancel{b}\cdot b \cdot b} = \frac{1}{b^2}$$

The application of the rule for dividing exponential terms results in $b^3/b^5 = b^{3-5} = b^{-2} = 1/b^2$. We must conclude, therefore, that if a base has a negative exponent, this expression is the same as 1 divided by the same base but with a positive exponent. This leads to the following mathematical statement for the law of negative exponents.

Rule 3-11 Law of Negative Exponents

$$b^{-p} = \frac{1}{b^p}, \quad (b \neq 0)$$

The foregoing gives us a convenient method of moving a term in the numerator of a fraction to the denominator or vice versa by merely changing the sign of the exponent.

Example 3-18
$$\frac{k^{-3}}{k} = \frac{1}{k(k^3)} = \frac{1}{k^4}, \text{ or } k^{-4}$$

Example 3-19
$$\frac{x^{-5}}{x^{-7}} = \frac{x^7}{x^5} = x^2$$

In Chapter 1, we developed the idea of positive **powers of ten** as a shorthand way to write very large numbers. For example, it was stated that 1,000,000 could be written as 1×10^6, where the exponent 6 indicated the number of zeros that were to be added to 1. From the previous discussion of exponential multiplication, it should be evident that this shorthand method was actually derived from the process of raising a base to a power. In this case, since the base is 10, the exponent represents the number of zeros. For example, $10^6 = 10 \times 10 \times 10 \times 10 \times 10 \times 10 = 1,000,000$, which has six zeros. If we were to divide 10^6 by 10^8, for example, we would obtain

$$\frac{10^6}{10^8} = 10^{-2} = \frac{1}{100}, \text{ or } 0.01$$

by the law of negative exponents. This can also be written as 1×10^{-2}. In this form, note that the exponent indicates the number of places the implied decimal point after the 1 is moved to the *left* in forming the answer 0.01.

Using this interpretation of the negative exponent of ten, we can now extend our table that we began in Chapter 1 to include numbers less than zero (see Table 3-1).

Table 3–1   *Powers of ten*

$1{,}000{,}000{,}000{,}000 = 1 \times 10^{12}$	(tera-)	T
$1{,}000{,}000{,}000 = 1 \times 10^{9}$	(giga-)	G
$100{,}000{,}000 = 1 \times 10^{8}$		
$10{,}000{,}000 = 1 \times 10^{7}$		
$1{,}000{,}000 = 1 \times 10^{6}$	(mega-)	M
$100{,}000 = 1 \times 10^{5}$		
$10{,}000 = 1 \times 10^{4}$		
$1{,}000 = 1 \times 10^{3}$	(kilo-)	k
$100 = 1 \times 10^{2}$	(hecto-)	h
$10 = 1 \times 10^{1}$	(deka-)	da
$1 = 1 \times 10^{0}$		
$0.1 = 1 \times 10^{-1}$	(deci-)	d
$0.01 = 1 \times 10^{-2}$	(centi-)	c
$0.001 = 1 \times 10^{-3}$	(milli-)	m
$0.0001 = 1 \times 10^{-4}$		
$0.00001 = 1 \times 10^{-5}$		
$0.000001 = 1 \times 10^{-6}$	(micro-)	μ
$0.0000001 = 1 \times 10^{-7}$		
$0.00000001 = 1 \times 10^{-8}$		
$0.000000001 = 1 \times 10^{-9}$	(nano-)	n
$0.000000000001 = 1 \times 10^{-12}$	(pico-)	p

As an example, a very small number such as 0.00000135 can be written using a power of ten as 1.35×10^{-6}. From what we have learned about negative exponents, we know that this means

$$\frac{1.35}{1{,}000{,}000} = 0.00000135, \quad \text{or} \quad 1.35 \times 10^{-6}$$

indicating that the decimal point is to be moved six places to the left when writing 1.35×10^{-6} in standard form.

Recall that powers of ten are often employed in what is called **scientific notation.** That is, a given quantity is written as a number between 1 and 10 times the appropriate power of 10. This practice allows you to estimate answers to problems involving very large or very small numbers in your head with a fair degree of accuracy. For example, suppose you want to multiply 149,910,411 by 0.0000198. By rounding these numbers and writing them in scientific notation, you get 1.5×10^{8} times 2×10^{-5}, which can be computed mentally as 3×10^{3}. Since the actual answer is 2968.226, the percent of error* is only about 1%—not bad for an estimate!

You might well ask why you need to estimate at all, since your calculator can give you an exact answer. First, you may not always have a calculator handy, and second, you should never allow your calculator to do your thinking for you. Your estimate should be used to check the calculator, not the other way around!

---

*Percent of error is calculated as:

$$\frac{|N - n|}{N} \times 100$$

where N = true value and n = estimated value.

Algebraic Multiplication and Division

It often happens that both terms of a fraction are identical. For example,

$$\frac{b^3}{b^3} = \frac{\cancel{b} \cdot \cancel{b} \cdot \cancel{b}}{\cancel{b} \cdot \cancel{b} \cdot \cancel{b}} = b^{3-3} = b^0 = 1$$

Therefore, we conclude that any base (except zero) raised to zero exponent is equal to 1. The mathematical statement for this law of zero exponents is written as follows.

Rule 3–12   The Law of Zero Exponents
$$b^0 = 1, \quad (b \neq 0)$$

For example, each of the following expressions is equal to 1:

$$x^0, \quad k^0, \quad 10^0, \quad 5^0$$

Exercise Set 3–6

Questions

1. State the three laws of exponential division.
2. Explain the concept behind powers of ten.
3. Define *scientific notation*, and explain how powers of ten are used with this method of notation.

Problems

Perform the following exponential divisions. Express your answers with positive exponents.

1. $\dfrac{a^7}{a^3}$

2. $\dfrac{3I}{-9I^4}$

3. $\dfrac{3W^2}{12W}$

4. $\dfrac{b^{-3}x^2}{bx}$

5. $\dfrac{-V^3}{3V^2}$

6. $\dfrac{(3^{-1})15x^2}{5x}$

7. $\dfrac{21X^4}{-3X^3}$

8. $\dfrac{1.6 \times 10^{-3}}{0.8 \times 10^2}$

9. $\dfrac{14M^x}{2M^k}$

10. $\dfrac{x^0}{-5}$

11. The following BASIC program evaluates the exponential expression $b^p/b^q = b^{p-q}$ for all real values of b, p, and q.

```
10 PRINT"Enter the values of b, p and q."
20 INPUT B,P,Q
30 LET N=B^P/B^Q
40 PRINT"The quotient of the two exponential terms is:";N
50 END
```

Write a BASIC program for evaluating the exponential quotient in Problem 9 for all real values of the literal exponents x and k.

3-4 QUOTIENTS OF MONOMIALS AND MULTINOMIALS

Division of Monomials

The division indicated by the fraction x/y can be regarded as the product of the two factors $x/1$ and $1/y$. In this manner, division can be viewed as a form of multiplication, and it should be evident that the same rules of signs developed earlier for products also hold for division. In other words, the quotient of two like-signed numbers is always positive, and the quotient of terms having unlike signs is always negative.

> **Rule 3-13**
>
> The quotient of two terms with the same sign is positive.
>
> **Rule 3-14**
>
> The quotient of two terms with different signs is negative.

In dividing one monomial by another, we begin by applying the rules of signs just stated to the numerical coefficients. For example, to divide $-21ax^3$ by $3a^2x$, we would begin by writing the quotient of the numerical coefficients as -7, since $-21/3 = -7$. We would then apply the rules of exponents stated earlier to the literal factors, obtaining $a^{-1}x^2$. Finally, we would combine these two steps to form our final answer, $-7a^{-1}x^2$, or $-7x^2/a$. The foregoing illustration allows us to state the following rule.

> **Rule 3-15**
>
> To divide one monomial by another:
>
> 1. Divide the numerical coefficients, paying careful attention to the sign of the result.
> 2. Divide the literal factors using the rule for the division of exponents.
> 3. Form a single quotient using the results of steps 1 and 2.

Just as it is possible to check the answer to a multiplication problem by numerical substitution, it is possible to check division in the same way. For example, we can substitute $a = 2$ and $x = 4$ (or any other nonzero numbers) and obtain

$$-21ax^3 = (-21)(2)(64) = -2688$$

Algebraic Multiplication and Division

and

$$3a^2x = (3)(4)(4) = 48$$

Therefore,

$$\frac{-2688}{48} = -56$$

Also,

$$\frac{-7x^2}{a} = \frac{(-7)(16)}{2} = -56$$

Therefore, we conclude that our division is correct.

Division can also be checked by multiplying the quotient by the divisor to obtain the dividend. In the preceding example, the quotient is $-7x^2/a$; the divisor is $3a^2x$. Therefore, if our division is correct, their product will equal the dividend, $-21ax^3$.

$$\left(\frac{-7x^2}{a}\right)(3a^2x) = -21ax^3$$

This result assures us that our answer is correct.

Example 3–20
$$\frac{3^{-1}a^{-2}V}{3RV^2} = \frac{V}{3(3)a^2RV^2} = \frac{1}{9a^2RV}$$

Example 3–21
$$\frac{3 \times 10^{-12}}{2 \times 10^3} = 1.5 \times 10^{-12-3} = 1.5 \times 10^{-15}$$

Example 3–22
$$\frac{12f\lambda^2}{2f^{-2}\lambda^{-1}} = 6f^{1+2}\lambda^{2+1} = 6f^3\lambda^3$$

Exercise Set 3–7

Questions

1. State the steps in the rule for dividing one monomial by another.
2. What are two common methods of checking that an algebraic division is correct?

Problems

Perform the following divisions. Express all answers using positive exponents.

1. $\dfrac{-37a^3b}{148ab}$

2. $\dfrac{16P^{-3}}{0.0625P^6}$

3. $\dfrac{0.01V^3I^3}{(-1 \times 10^{-2})VI}$

4. $\dfrac{x^3y^4}{(5 \times 10^{-6})xy^{-4}}$

5. $\dfrac{(1.64 \times 10^{-8})P^3}{(-2 \times 10^3)P^2}$

Fundamentals of Algebra

6. $\dfrac{1024Q^3}{2^8 Q^{12}}$

7. $\dfrac{33a^5 b^8 c^3}{11a^4 b^7 c^{-3}}$

8. $\dfrac{25^2 v i^2}{(5 \times 10^1) i^{-3} v^5}$

9. $\dfrac{-169 Q^5 P^{-1}}{13 Q P^0}$

10. $\dfrac{81 X^{-1}}{-3^2 X}$

11. $\dfrac{0.625(8c^2)}{5c}$

12. $\dfrac{(-1 \times 10^{-3}) M N^3}{(5 \times 10^2) M^{-2} N^{-1}}$

13. $\dfrac{0.01325(32k)}{-k^2}$

14. $\dfrac{27(50M)}{3^3(2M)}$

15. $\dfrac{(300 \times 10^6) f^3}{(9 \times 10^9) f}$

16. $\dfrac{(3 \times 10^{10}) \lambda}{(12 \times 10^3) \lambda^4}$

17. $\dfrac{-22 \lambda^5 m^{-2}}{-11 \lambda^3 m}$

18. $\dfrac{-3 \phi k^4 n^3}{10 \phi^2 n^5}$

19. $\dfrac{-5.5 v^2 M^{-1}}{110 v^5}$

20. $\dfrac{-196 Z^{-1} I^3}{14 Z^3 I^{-2}}$

21. The following BASIC program evaluates the expression shown in Problem 19 for all real values of the literal factors v and M. (*Note:* v and M cannot equal zero, since division by zero is not defined in mathematics. See Chapter 1.)

```
10 PRINT"Enter the values of v and M."
20 INPUT V,M
30 LET N=(-5.5*V^2*M^-1)/(110*V^5)
40 PRINT"The quotient is:";N
50 END
```

Write a BASIC program for evaluating the expression in Problem 17 for all real values of the literal numbers.

Division of a Multinomial by a Monomial

The division of 1296 by 6 can be looked upon as the process of dividing 6 into a multinomial made up of the four terms 1000 + 200 + 90 + 6. Expressed mathematically,

Algebraic Multiplication and Division

$$\frac{1000 + 200 + 90 + 6}{6} = \frac{1000}{6} + \frac{200}{6} + \frac{90}{6} + \frac{6}{6}$$
$$= 166\tfrac{2}{3} + 33\tfrac{1}{3} + 15 + 1$$
$$= 216$$

You should check the correctness of this result yourself by performing the division 1296/6 on your calculator.

In the same way, any algebraic multinomial that ultimately represents some numerical quantity can be divided by a monomial, and the partial quotients can be added to obtain a final answer. For example, consider the algebraic division problem

$$\frac{5ac^3x + 3ac^2x}{ac}$$

The quotient is formed term by term, as was done in the preceding division of 1296 by 6.

$$\frac{5ac^3x + 3ac^2x}{ac} = \frac{5ac^3x}{ac} + \frac{3ac^2x}{ac} = 5c^2x + 3cx$$

Note that each term of the multinomial was divided by the monomial according to the rules developed earlier for the division of monomials.

If we let $a = 2$, $c = 3$, and $x = 4$ in this example, we note that

$$\frac{5ac^3x + 3ac^2x}{ac} = \frac{5(2)(27)(4) + 3(2)(9)(4)}{(2)(3)}$$
$$= \frac{1296}{6} = 216$$

The substitution of these particular numerical values for the literal numbers demonstrates the equivalence of the arithmetical division process shown earlier and the algebraic division of a multinomial by a monomial.

To divide a multinomial by a monomial, we apply the following rule.

> **Rule 3-16**
>
> To divide a multinomial by a monomial, divide each term of the multinomial by the monomial according to Rule 3-15, and write each resulting partial quotient as a term of the new multinomial.

Example 3-23
$$\frac{-12PQ^2 + 6PQ^{-3} - 2Q}{2PQ} = -6Q + \frac{3}{Q^4} - \frac{1}{P}$$

Example 3-24
$$\frac{ax^5 - a^2x^4 + a^3x^3 - a^4x^2 + a^5x}{ax} = x^4 - ax^3 + a^2x^2 - a^3x + a^4$$

Exercise Set 3-8

Questions

1. State the rule for the division of a multinomial by a monomial.

2. What procedure may be used to check the correctness of the division in Question 1?

Problems

Divide.

1. $\dfrac{-81t^3 + 12at^2}{3at}$

2. $\dfrac{a^{-5}bc^3 - 3ab^3c}{2a^2bc^2}$

3. $\dfrac{-V^3 + 3V^2 - V}{2V}$

4. $\dfrac{gt^3 - g^2t^2 + g^3t}{gt^2}$

5. $\dfrac{-625P^3 + 175 - 225P^{-1}}{25P}$

6. $\dfrac{18x^5yz^2 - 36x^{-2}y^2z^{-1} + 12x}{4x^2y^{-3}z^4}$

7. $\dfrac{0.0015V^8R^4 + 0.024VR^5 - 0.21V^4R^4}{0.03V^2R^2}$

8. $\dfrac{1.35x^3y^{-3} - 1.05xy^{-4} + 1.5x^{-2}y^{-2}}{0.15x^{-2}y}$

9. $\dfrac{4(x - 2y) - 2(x + y)}{xy}$

10. $\dfrac{-4(Z + 4Y) + 2(2Z - 2Y)}{-10Y}$

Division of Multinomials

The process of dividing 1875 by 15 can be considered the same as dividing the multinomial $1000 + 800 + 70 + 5$ by the multinomial $10 + 5$. This division is accomplished as follows.

$$
\begin{array}{r}
100 + 30 - 8 \\
10 + 5 \overline{\smash{)}\,1000 + 800 + 70 + 5} \\
\underline{1000 + 500 } \\
300 + 70 \\
\underline{300 + 150 } \\
-80 + 5 \\
\underline{-80 - 40} \\
\dfrac{45}{15} = 3
\end{array}
$$

On combining the partial quotients with the remainder divided by the divisor, or 3, we obtain a final answer of $100 + 30 - 8 + 3 = 125$. You can see that this result is correct by dividing 1875 by 15 on your calculator.

    Note that only the 10 of the divisor was used to determine the partial quotients during each successive division—that is, the partial quotients are $1000/10$, $300/10$, $-80/10$, and $40/10$. This division process is faintly reminiscent of the long-division scheme we learned in elementary school. In a similar manner, we can divide algebraic multinomials by using a precise recipe. In mathematics, a recipe for solving a given type of problem is called an

Algebraic Multiplication and Division

algorithm. The following steps comprise the algorithm for the division of algebraic multinomials.

> **Rule 3–17 Rule for Dividing Multinomials**
> 1. Arrange the terms of both multinomials in either an increasing or decreasing order of some common literal factor.
> 2. Divide the first term of the dividend by the first term of the divisor. This is the first term of the quotient.
> 3. Multiply the first term of the quotient by the entire divisor, and subtract this new expression from the appropriate terms of the dividend.
> 4. The result of step 3 creates a new dividend, which is then divided by the divisor, and steps 2 and 3 are repeated until either (a) there is no longer a remainder or (b) the remainder can no longer be divided by the divisor.

The following examples clarify the algorithm.

Example 3–25

$$\begin{array}{r} 4h - 3 \\ h - 2 \overline{\smash{\big)}\, 4h^2 - 11h + 6} \\ \underline{4h^2 - 8h} \\ -3h + 6 \\ \underline{-3h + 6} \end{array}$$

Check $(h - 2)(4h - 3) = 4h^2 - 3h - 8h + 6$
$= 4h^2 - 11h + 6$

Example 3–26

Divide $-6 + 6V^2 + 5V$ by $-1 + 1.5V$.

Solution In order to perform this division, we apply step 1 of Rule 3–17, obtaining $6V^2 + 5V - 6$ for the dividend and $1.5V - 1$ for the divisor. We can now proceed as follows:

$$\begin{array}{r} 4V + 6 \\ 1.5V - 1 \overline{\smash{\big)}\, 6V^2 + 5V - 6} \\ \underline{6V^2 - 4V} \\ 9V - 6 \\ \underline{9V - 6} \end{array}$$

Check $(1.5V - 1)(4V + 6) = 6V^2 + 9V - 4V - 6$
$= 6V^2 + 5V - 6$

Example 3–27

$$\begin{array}{r} P^2 + 7P \\ P - 1 \overline{\smash{\big)}\, P^3 + 6P^2 + 8P} \\ \underline{P^3 - P^2} \\ 7P^2 + 8P \\ \underline{7P^2 - 7P} \\ 15P \\ \overline{P - 1} \end{array}$$

Note the remainder $15P$. Generally, we write this result as $P^2 + 7P + \dfrac{15P}{P - 1}$.

Check
$$(P - 1)\left(P^2 + 7P + \frac{15P}{P - 1}\right) = P^3 + 7P^2 + \frac{15P^2}{P - 1} - P^2 - 7P - \frac{15P}{P - 1}$$
$$= P^3 + 6P^2 - 7P + \frac{15P^2 - 15P}{P - 1}$$
$$= P^3 + 6P^2 - 7P + 15P\left(\frac{P - 1}{P - 1}\right)$$
$$= P^3 + 6P^2 - 7P + 15P = P^3 + 6P^2 + 8P$$

Note: Since we have not yet studied algebraic fractions and factoring, you may not have been able to follow all the steps in the checking procedure shown. This should not be a cause for concern, and you may simply wish to return to this section at a later time. The check has been provided here merely for those who may be more advanced.

Example 3–28

Divide $M^4 + M^2N^2 + N^4$ by $M^2 + N^2 - MN$.

Solution We must allow space in the dividend for the missing M^3N and MN^3 terms by placing zeros in the appropriate places, as follows:

$$\begin{array}{r} M^2 + MN + N^2 \\ M^2 - MN + N^2 \overline{\smash{)}M^4 + 0 + M^2N^2 + 0 + N^4} \\ \underline{M^4 - M^3N + M^2N^2} \\ M^3N \\ \underline{M^3N - M^2N^2 + MN^3} \\ M^2N^2 - MN^3 + N^4 \\ \underline{M^2N^2 - MN^3 + N^4} \end{array}$$

Check $(M^2 - MN + N^2)(M^2 + MN + N^2)$
$= M^4 + \cancel{M^3N} + \cancel{M^2N^2} - \cancel{M^3N} - \cancel{M^2N^2} - \cancel{MN^3} + M^2N^2 + \cancel{MN^3} + N^4$
$= M^4 + M^2N^2 + N^4$ Note how the same terms with opposite signs cancel.

Exercise Set 3–9

Questions

1. State the steps in the rule for the division of one multinomial by another.
2. Define the term *algorithm* in your own words.
3. What procedure may be used to check the division of multinomials?

Problems

Perform the following multinomial divisions. Check the answer to each problem.

1. $\dfrac{Z^2 + 3Z - 4}{Z - 1}$

2. $\dfrac{V^2 - 5V}{V + 2}$

3. $\dfrac{P^2 + 2R + 1}{P + 1}$

Algebraic Multiplication and Division

4. $\dfrac{k^3 + 3k^2 - 4k}{k - 1}$

5. $\dfrac{a^2 - 2ab + b^2}{a - b}$

6. $\dfrac{i^2 + 2it + t^2}{i + t}$

7. $\dfrac{R^2 - X^2}{R + X}$

8. $\dfrac{M^2 - N^2}{M - N}$

9. $\dfrac{2C^4 + C^3 - 8C^2 + 3C - 2}{C^2 - 5}$

10. $\dfrac{6z^3 + 12yz - 2yz^2 - 4y^2}{3z - y}$

11. $\dfrac{x^5 - y^5}{x - y}$

12. $\dfrac{12m^2 - 36n^2 - 11mn}{4m - 9n}$

13. $\dfrac{4W^3 - 8W^2 - 9W}{2W - 3}$

14. $\dfrac{3x^3 + 5x^2 + 7x + 9}{x^2 + 2}$

15. $\dfrac{x^4 - 2x^3 + 4x^2 - 8x + 1}{x^2 + 3x - 2}$

16. $\dfrac{I^3 - 9I^2 + 20I - 38}{I^2 - 3I + 5}$

17. $\dfrac{f^3 + 8f^2 - 20f - 4}{f - 2}$

18. $\dfrac{V^3 + 125}{V + 5}$

19. $\dfrac{x^4 + 3x^2 + 6}{x^2 - 2}$

20. $\dfrac{Q^4 + 3Q - 7}{Q^2 + 2Q - 3}$

21. The following BASIC program evaluates the expression shown in Problem 19 for all real values of the literal number x except $x = \sqrt{2}$ (why?).

```
10 PRINT"Enter the value of x."
20 INPUT X
30 LET N=(X^4+3*X^2+6)/(X^2-2)
40 PRINT"The quotient is";N
50 END
```

Write a BASIC program to evaluate Problem 20 for all permissible values of the literal numbers.

Key Terms

exponent	multinomial	juxtaposition
base	polynomial	powers of ten
monomial	algorithm	scientific notation

Important Algebraic Rules

The Sign of a Product

Rule 3–1 The product of two numbers with the same sign is positive.

Rule 3–2 The product of two numbers with different signs is negative.

Rule 3–3 Rules 3–1 and 3–2 are to be applied in succession when there are more than two factors.

Corollary to Rule 3–3 The product of an *even* number of negative factors is positive. The product of an *odd* number of negative factors is negative.

Laws of Exponents

Rule 3–4 Exponential Law of Multiplication
To find the product of two or more exponential terms having the same base, add their exponents.

$$b^p b^q = b^{p+q}$$

Rule 3–5 Law for the Power of a Power

$$(b^p)^q = b^{pq}$$

Rule 3–6 Law for the Power of a Product

$$(bc)^p = b^p c^p$$

Rule for Multiplying Two or More Monomials

Rule 3–7
1. Multiply the numerical coefficients together, paying careful attention to the sign of the result.
2. Multiply the literal factors using the rule for exponents.
3. Write the results of steps 1 and 2 as a single product.

Rule for Multiplying a Multinomial by a Monomial

Rule 3–8 To multiply a multinomial by a monomial, multiply each term of the multinomial by the monomial, according to Rule 3–7, and write each partial product as a term of the new multinomial.

Rule for Multiplying Multinomials

Rule 3–9 To multiply multinomials, multiply every term of one by every term of the other, and write each resulting product as a term of the new multinomial.

Rule 3–10 Law of Exponential Division
To divide exponential terms having the same base, subtract the exponent in the denominator from the exponent in the numerator.

$$\frac{b^p}{b^q} = b^{p-q} \quad (b \neq 0)$$

Rule 3–11 Law of Negative Exponents

$$b^{-p} = \frac{1}{b^p} \quad (b \neq 0)$$

Algebraic Multiplication and Division

Table 3–1 *Powers of ten*

$1{,}000{,}000{,}000{,}000 = 1 \times 10^{12}$	(tera-)	T
$1{,}000{,}000{,}000 = 1 \times 10^{9}$	(giga-)	G
$100{,}000{,}000 = 1 \times 10^{8}$		
$10{,}000{,}000 = 1 \times 10^{7}$		
$1{,}000{,}000 = 1 \times 10^{6}$	(mega-)	M
$100{,}000 = 1 \times 10^{5}$		
$10{,}000 = 1 \times 10^{4}$		
$1{,}000 = 1 \times 10^{3}$	(kilo-)	k
$100 = 1 \times 10^{2}$	(hecto-)	h
$10 = 1 \times 10^{1}$	(deka-)	da
$1 = 1 \times 10^{0}$		
$0.1 = 1 \times 10^{-1}$	(deci-)	d
$0.01 = 1 \times 10^{-2}$	(centi-)	c
$0.001 = 1 \times 10^{-3}$	(milli-)	m
$0.0001 = 1 \times 10^{-4}$		
$0.00001 = 1 \times 10^{-5}$		
$0.000001 = 1 \times 10^{-6}$	(micro-)	μ
$0.0000001 = 1 \times 10^{-7}$		
$0.00000001 = 1 \times 10^{-8}$		
$0.000000001 = 1 \times 10^{-9}$	(nano-)	n
$0.000000000001 = 1 \times 10^{-12}$	(pico-)	p

Rule 3–12 The Law of Zero Exponents

$$b^0 = 1, \quad (b \neq 0)$$

The Sign of a Quotient

Rule 3–13 The quotient of two terms with the same sign is positive.
Rule 3–14 The quotient of two terms with different signs is negative.

Rule for the Division of Monomials

Rule 3–15 To divide one monomial by another:
1. Divide the numerical coefficients, paying careful attention to the sign of the result.
2. Divide the literal factors using the rule for the division of exponents.
3. Form a single quotient using the results of steps 1 and 2.

Rule for Dividing a Multinomial by a Monomial

Rule 3–16 To divide a multinomial by a monomial, divide each term of the multinomial by the monomial according to Rule 3–15, and write each resulting partial quotient as a term of the new multinomial.

Rule for the Division of Multinomials

Rule 3–17
1. Arrange the terms of both multinomials in either an increasing or decreasing pattern of some common literal factor.
2. Divide the first term of the dividend by the first term of the divisor. This is the first term of the quotient.
3. Multiply the first term of the quotient by the entire divisor, and subtract this new expression from the appropriate terms of the dividend.
4. The result of step 3 creates a new dividend, which is then divided by the divisor, and steps 2 and 3 are repeated until either (a) there is no longer a remainder or (b) the remainder can no longer be divided by the divisor.

4 Fundamentals of Linear Equations

Upon completion of this chapter, you will have:
1. Learned how to solve simple linear equations by applying basic axioms.
2. Learned how to solve inequalities.
3. Learned how to approach word problems and solve them effectively.

4-1 INTRODUCTION

An **equation** is a mathematical statement that two quantities are equal. The **equal sign** (=) is used to show this equality between the two quantities. For example, $P = VI + W$ is an equation consisting of a **left member** P and a **right member** $VI + W$. The equal sign states that the two members are equal—that is, P is the same as $VI + W$.

To **solve** an equation means to find the value or values of one of the unknown quantities that make both sides of the equation identical. For example, in the equation $P = VI + W$, if P has the value of 14 and VI has the value of 9, then $14 = 9 + W$, and the only value of W that makes this a true statement is 5. Therefore, 5 is said to be the **root** of the equation $14 = 9 + W$, and substituting this value for W is said to **satisfy** the equation because $14 = 9 + 5$, and both members of the equation are identical.

In this chapter, we will be dealing with only one type of equation called a **linear equation.** These equations are called linear because the picture (graph) of all their points results in a straight line. We discuss linear equations and graphing further in Chapter 7.

4-2 AXIOMS

Certain commonsense rules allow us to manipulate or operate on an equation repeatedly until we find its root. These rules are called **axioms;** axioms are simply statements whose truth is so obvious that no formal proof is required. For example, to state that the whole of something is greater than any of its parts is certainly obvious as well as true. In algebra, there are a few such axioms that allow us to make quick work of solving most simple linear equations. Therefore, you are urged to study these axioms and become familiar with them, even though their relevance may not be immediately apparent. The tremendous value of these simple axioms will become evident as we apply them in solving many linear equations in the next section.

> **Axiom 4-1**
>
> If the same number is added to equal numbers, the sums are still equal.
>
> For example, if $n = n$ and 5 is added to each member, we obtain $n + 5 = n + 5$. Obviously, both members are still equal.

This result is very much like a beam balance with equal weights in both pans (see Figure 4-1). If a 3 lb weight is placed in both pans, the scale is balanced. It should be apparent that if we add 5 lb more to each pan, we have increased the weight to 8 lb, but we have not destroyed the balanced condition.

The beam balance is analogous to an equation, with the pans forming the left and right members. In all cases, nothing should be done that destroys the balance (equality). For example, it makes no sense whatever to write $n = n + 5$. Here, we have no balance, no equality.

In applying this and all subsequent axioms, you are urged to keep this balance idea firmly in mind. An equation *cannot be solved* if anything is done to destroy the balance, or the equality.

Fundamentals of Linear Equations

Figure 4-1

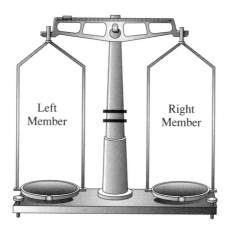

Axiom 4-2

If the same number is subtracted from equal numbers, the remainders are still equal.

For example, if $n = n$, then $n - 5 = n - 5$. Suppose n has a value of 12. Then $12 - 5 = 12 - 5$. In other words, $7 = 7$. The balance of the equation is still maintained.

Axiom 4-3

If equal numbers are multiplied by the same number, the products are equal.

For example, if $n = n$, then $5n = 5n$. If n has a value of 2, then $5 \times 2 = 5 \times 2$, or $10 = 10$. We have not altered the balance of the original equality.

Axiom 4-4

If equal numbers are divided by the same nonzero number, the quotients are still equal.

For example, if $n = n$, then $n/5 = n/5$. If n is 3, then $3/5 = 3/5$, and the equation is still balanced.

Axiom 4-5

If two numbers are equal to the same (or equal) numbers, then they are equal to each other.

For example, if $X = n$ and $n = Y$, then $X = Y$. There is no loss of equality here, and this axiom allows us to substitute any equivalent expression for any term in an equation without disturbing the balance of the equation.

> **Axiom 4–6**
>
> If two numbers are equal, they both may be raised to the same power without altering the equality.
>
> For example, if $n = n$, then raising each to the fifth power does not diminish the equality. If n has the value of 2, then $2^5 = 2^5$, or $32 = 32$.

> **Axiom 4–7**
>
> If two numbers are equal, like roots may be extracted from both members without destroying the equality.
>
> For example, if $n = n$, then $\sqrt{n} = \sqrt{n}$. If $n = 16$, then $\sqrt{16} = \sqrt{16}$, or $4 = 4$. The balance remains intact.

4–3 SOLVING EQUATIONS

In this section, we will apply various axioms to the solution of simple linear equations. Every step will be shown for each sample problem, and the appropriate axiom will be indicated. After you have become proficient in solving equations, you will be able to apply most axioms without giving them very much conscious thought. In the meantime, as you work through the problems in this section, it is strongly recommended that you follow a procedure similar to that shown in the examples and write down every step along with the axiom used. Although this may sound like a good deal of unnecessary work, in the long run it will pay off in terms of greater speed and fewer mistakes.

In the examples that follow, note that more than one approach is possible. Indeed, several other methods might suggest themselves. This versatility is one of the great advantages of algebra; where another approach is evident, you are definitely encouraged to pursue that method, too.

Example 4–1

Given $3 = V - 9$, solve for V.

Step 1 By Axiom 4–1, add 9 to each side of the equation.

$$3 = V - 9$$
$$3 + 9 = V - 9 + 9 \qquad \text{(Axiom 4–1)}$$
$$12 = V$$

Step 2 Check
$$3 = V - 9$$
$$3 \stackrel{?}{=} 12 - 9$$
$$3 = 3$$

Note that the use of Axiom 4–1 has the effect of moving a -9 from the right member and relocating it in the left member as a $+9$. This leads us to the rule:

Fundamentals of Linear Equations

> **Rule 4–1**
> Any term may be **transposed** (moved) from one member to the other simply by changing its sign.

Example 4–2

Given $18 = 4 + P$, solve for P.

Step 1 By Axiom 4–2, subtract 4 from each side of the equation.

$$18 = 4 + P$$
$$18 - 4 = 4 + P - 4 \quad \text{(Axiom 4–2)}$$
$$14 = P$$

Step 2 Check
$$18 = 4 + P$$
$$18 \stackrel{?}{=} 4 + 14$$
$$18 = 18$$

Example 4–3

Given $\dfrac{R}{3} = 4$, solve for R.

Step 1 By Axiom 4–3, multiply both sides of the equation by 3.

$$\frac{R}{3} = 4$$
$$3 \cdot \frac{R}{3} = 3 \cdot 4 \quad \text{(Axiom 4–3)}$$
$$R = 12$$

Step 2 Check
$$\frac{R}{3} = 4$$
$$\frac{12}{3} \stackrel{?}{=} 4$$
$$4 = 4$$

Example 4–4

Given $2x - 5 = 11$, solve for x.

Step 1
$$2x - 5 = 11$$
$$2x - 5 + 5 = 11 + 5 \quad \text{(Axiom 4–1)}$$
$$2x = 16$$

Step 2
$$\frac{2x}{2} = \frac{16}{2} \quad \text{(Axiom 4–4)}$$
$$x = 8$$

Step 3 Check
$$2x - 5 = 11$$
$$2(8) - 5 \stackrel{?}{=} 11$$
$$16 - 5 \stackrel{?}{=} 11$$
$$11 = 11$$

Example 4–5

Given $v - 3 = E - 5 + v$, solve for E.

Step 1 Note that the term $+v$ appears on both sides of the equation. By Axiom 4–2, subtract v from each side, obtaining $-3 = E - 5$. We may now treat this as a new equation and solve for E.

Step 2 By Axiom 4–1, add 5 to each side of the equation, obtaining $2 = E$.

Step 3 Check $-3 = E - 5$
$-3 \stackrel{?}{=} 2 - 5$
$-3 = -3$

Note that the use of either Axiom 4–1 or 4–2, as appropriate, gives us the expedient of simply eliminating any term appearing on both sides of an equation, *provided it is preceded by the same sign*. For example, in the equation $3 - x = 4K - x$, we can eliminate the x terms on both sides and write $3 = 4K$. This fact allows us to state the following rule:

> **Rule 4–2**
> Any term appearing on *both* sides of an equation and preceded by the *same* sign may simply be dropped from the equation.

However, in the equation $3 + x = 4K - x$, we cannot simply remove the x's, since they are preceded by different signs. In this case we might, for example, use Axiom 4–1 to add x to each side of the equation, obtaining $3 + 2x = 4K$. Alternately, Axiom 4–2 can be used to subtract x from each side, obtaining $3 = 4K - 2x$.

Example 4–6

Given $\dfrac{x}{5} - 16 = 3$, solve for x.

Step 1 By Axiom 4–1, add 16 to both sides, obtaining $x/5 = 19$.
Step 2 By Axiom 4–3, multiply both sides by 5, obtaining $x = 95$.
Step 3 Check $\dfrac{x}{5} - 16 = 3$
$\dfrac{95}{5} - 16 \stackrel{?}{=} 3$
$19 - 16 \stackrel{?}{=} 3$
$3 = 3$

Example 4–7

Given $6 = -k + 10$, solve for k.

Step 1 By Axiom 4–2, subtract 10 from both sides, obtaining $-4 = -k$.
Step 2 By Axiom 4–3, multiply both sides of the equation by -1, giving $4 = k$.
Step 3 Check $6 \stackrel{?}{=} -k + 10$
$6 \stackrel{?}{=} -(4) + 10$
$6 = 6$

Fundamentals of Linear Equations

Multiplying both members of an equation by -1 has the effect of changing all the signs of all the terms without destroying the equality. This fact leads to the following rule.

> **Rule 4-3**
>
> The signs of all the terms of an equation may be changed without destroying the equality.

Exercise Set 4-1

Questions

1. In your own words, define the following terms.
   a. Equation
   b. Equal sign
   c. Left and right members
   d. Solve
   e. Root
   f. Satisfy
   g. Linear equation
   h. Axiom
   i. Transpose
2. In your own words, state the seven common axioms used in the solution of linear equations.
3. What is the important idea behind the beam balance shown in Figure 4-1?

Problems

For each of the following linear equations, solve for the literal number indicated. Write out all steps, indicating which axiom or rule has been used for each step, and check your answer.

1. Solve for V: $2V + 12 = 16$
2. Solve for k: $10 = 5 - 2k$
3. Solve for P: $7 = -P + 14$
4. Solve for m: $10m - 4 = 8m + 4$
5. Solve for t: $-9t + 1 = -6t - 2$
6. Solve for Z: $-5Z + 4 = -Z - 100$
7. Solve for x: $x = 18 - 5x + x - 4x$
8. Solve for Y: $Y + 4 = Y(2 - 5)$
9. Solve for C: $C(1.2 + 5) = 2C - C(0.8 - 1)$
10. Solve for f: $f(2 - 0.2) = 5 + f(0.2 - 88)$
11. Solve for x: $3x - 3a - 10x - 3a = 5$
12. Solve for y: $22k - y = 16k + 4y$
13. Solve for x: $(1 \times 10^{-3})x + 0.2x = 0.804$
14. Solve for w: $25w + 4g - 6w + w = -22g + w$
15. Solve for x: $\dfrac{x}{16} + \dfrac{3}{8} = \dfrac{7x}{16}$
16. Solve for z: $\dfrac{3z}{5} - \dfrac{8z}{10} = 5$
17. Solve for V: $15q + a - 2V = 5q + 32a - V$
18. Solve for x: $0.005x - (2 \times 10^{-3})x = 0.03$
19. Solve for G: $\dfrac{G}{8} + \dfrac{3G}{32} - k = \dfrac{5G}{8} + 1$
20. Solve for x: $6ab - 3abx = 3ab$

4–4 INEQUALITIES

Not all expressions are created equal. For example, although it is certainly true that $5 = 10/2$, it is also obvious that 5 is less than 10. Therefore, if we wanted to express this **inequality** using symbols, we would write $5 < 10$, where the symbol $<$ means "is less than." Therefore, we would read $5 < 10$ as "5 is less than 10." Similarly, if we wanted to show that 5 was greater than 2, we would reverse the inequality symbol and write $5 > 2$, read "5 is greater than 2." We could be very economical with our symbolism and write both of these facts in a single expression as $2 < 5 < 10$, which is read in *both* directions from the center term as "5 is greater than 2 and less than 10." In general, we can express the fact that some literal number—for example, x—is sandwiched between two other values A and B by writing $A < x < B$. This expression states that X is simultaneously greater than A and less than B.

Note that in the expression $A < x < B$, x may *not* have either the values A or B. For example, if we write $2 < x < 10$, the x may have any values between 2 and 10 but not 2 and 10 themselves. The value of x might be 2.00001 or 9.999999, for example, but 2 and 10 are excluded. If we want to include the end values, we write $2 \leq x \leq 10$, which is read "x is greater than or equal to 2 and less than or equal to 10." We can write this with two separate expressions as $x \geq 2$ and $x \leq 10$.

Using this new symbolism, we can write statements that involve an *inequality* between members rather than an equality. For example, consider the statement $3x - 1 < 11$. What value of x will make this inequality a true statement? In answering this question, we apply the following simple rules.

Rule 4–4
All axioms that apply to equalities also apply to inequalities, except Axiom 4–4.

Rule 4–5
Whenever both members of an inequality are multiplied or divided by a negative number, *the sense* of the inequality must be reversed.

In applying Rule 4–5, we note, for example, that $5 > 1$, but multiplying by -1 gives us $-5 < -1$. The truth of this statement is obvious, since 5 is to the right of 1 on the number line, but -5 is to the left of -1. Similarly, $10 > 8$, but (dividing by -2) $-5 < -4$.

Returning to our example, we see that adding $+1$ to each side of the inequality gives us $3x < 12$. Dividing by 3 gives $x < 4$. We check the correctness of our answer by substituting *any* value of x less than 4 into the original expression. For example, since 3 is less than 4, we may write $3(3) - 1 < 11$ and see that $8 < 11$ is a true statement.

As another example, consider the inequality $8(V - 3) > 15V - 10$.

$$8V - 24 > 15V - 10 \quad \text{Use the distributive property.}$$
$$8V - 15V > -10 + 24 \quad \text{Transpose similar terms.}$$
$$-7V > 14 \quad \text{Add like terms.}$$
$$V < -2 \quad \text{Divide by } -7.$$

Fundamentals of Linear Equations

Check Since $-3 < -2$, we write

$$8(-3) - 24 > 15(-3) - 10$$
$$-24 - 24 > -45 - 10$$
$$-48 > -55, \text{ or } 48 < 55$$

As another check, we substitute $V = -2$:

$$8(-2) - 24 \stackrel{?}{=} 15(-2) - 10$$
$$-40 = -40$$

By checking $V = -2$ and $V = -3$, we see that both the direction of the inequality and the arithmetic are correct.

Exercise Set 4–2

Questions

1. What is meant by the sense of an inequality?
2. Use the number line to explain Rule 4–5 for inequalities.

Problems

Solve the following inequalities and check each solution.

1. $3V - 1 < 11$
2. $17 \geq 2P - 9$
3. $I + 4 \geq -3$
4. $2f - 16 > 17 + 5f$
5. $7R - 12 < 30$
6. $33 \geq 5 - 4V$
7. $18 - 7Z > -3$
8. $6t + 7 > 4t - 3$
9. $8(L - 3) > 15L - 10$
10. $4E - 22 < 6(E - 7)$
11. $3(2 + 3Q) \geq 5Q - 6$
12. $3i - 2(2i - 7) \leq 2(3 + i) - 4$
13. $18H - 3(8 - 4H) \leq 7(2 - 5H) + 27$
14. $9(2 - 5s) - 4 \geq 13s + 8(3 - 7s)$
15. $3\{2 + 4(B + 5)\} < 30 + 6B$
16. $10 - 5e > 2[3 - 5(e - 4)]$
17. $6d < 2 - 4\{2 - 3(d - 5)\}$
18. $6(10 - 3w) + 25 \geq 4w - 5(3 - 2k)$
19. $\dfrac{X}{3} > 7 - \dfrac{X}{4}$
20. $\dfrac{R - 2}{3} - \dfrac{R + 2}{4} \geq \dfrac{-2}{3}$

4–5 WORD PROBLEMS

A **word problem** is one in which verbal statements are used to express the conditions or circumstances of a situation involving numerical quantities. These problems must be translated into mathematical statements involving the signs and symbols of algebra before they can be solved. Many beginning students find these types of problems difficult because the wording often appears cryptic and confusing:

> If John's sister has seven red marbles and three-fifths of an anchovy pizza, how old is her half-cousin's cat, Marvin, who lives in Detroit?

To many beginners, all word problems seem as absurd as this one. There are several reasons why this confusion exists. First, most beginning students of algebra are accustomed to problems presented in equation form that require only simple—almost mechanical—manipulation in order to obtain an answer. We encountered such problems in the last section. Unfortunately, few situations exist in the real world that require so mindless an activity. Second, it is often difficult for the novice to tell which words or strings of words stand for particular mathematical operations. In order to reduce this confusion, certain common verbal statements are given next, along with their mathematical meanings. These examples are by no means exhaustive.

Fundamentals of Algebra

The following verbal statements are typical of those that mean quantities are to be added:

1. A quantity *x increased* by 5 means: $x + 5$
2. Twelve *more than* V means: $V + 12$
3. The *sum* of P and 15 means: $P + 15$
4. Your age 6 y from now if you are now Y years old means: $Y + 6$
5. The *total* of P_1 and P_2 means: $P_1 + P_2$

The following statements indicate subtraction:

1. A number *N reduced* by 4 means: $N - 4$
2. 16 *decreased* by x means: $16 - x$
3. 7 *less than* k means: $k - 7$
4. Your age 3 y ago if you are now Y years old means: $Y - 3$
5. The amount by which P_2 *exceeds* P_1 means: $P_2 - P_1$

These statements indicate multiplication:

1. V *times* I means: VI
2. The *product* of k and R means: kR
3. The *number of feet in* M miles means: $5280M$
4. The *number of pennies in* D dollars means: $100D$
5. Twenty-eight *percent of* V volts means: $0.28V$

The following statements indicate division:

1. *One-fifth of* R means: $\dfrac{R}{5}$
2. A voltage V *divided* into 7 equal parts means: $\dfrac{V}{7}$

The following words and phrases are often used to mean the same thing as *equals* (=): *is, are, will be, becomes, results in, yields, gives, is the same as,* and *represents*.

The following examples should be studied carefully until you understand the logic underlying the development of the required equation that fulfills all the conditions stated in the problem. You should be able to supply the axiom required in each step toward the solution of the final equation.

Example 4-8

The sum of two resistances is 180 Ω. If one of the resistors is 55 Ω, find the value of the other.

Solution   Let R = the value of the unknown resistor. Then $R + 55 = 180$.

$$R + 55 = 180$$
$$R + 55 - 55 = 180 - 55$$
$$R = 125 \text{ Ω}$$

Fundamentals of Linear Equations

Check $R + 55 \stackrel{?}{=} 180$
$125 + 55 \stackrel{?}{=} 180$
$180 = 180$

Example 4-9

One voltage exceeds another by 200 V. The larger voltage is three times greater than the smaller voltage. Find the value of the two voltages.

Solution If we let the lower voltage equal V, then the first sentence tells us that the higher voltage must be $V + 200$. Moreover, the second sentence tells us that $V + 200 = 3V$. Subtracting V from each side of this equation gives $200 = 2V$. If we now divide each side by 2, we obtain $100 = V$. We now know that the smaller voltage is 100 V. Substituting this fact into the first expression, $V + 200$, results in the value of the higher voltage: $V + 200 = 100 + 200 = 300$ V. Therefore, the lower voltage is 100; the higher voltage is 300.

Check $V + 200 = 3V$
$100 + 200 \stackrel{?}{=} 3(100)$
$300 = 300$

Example 4-10

Twenty-five percent of the capacitors in stock were used to complete a certain project. Only 153 capacitors remain. How many capacitors were there in stock originally?

Solution Let C equal the original number of capacitors. If 25% were used, then $0.25C$ indicates the number used. Therefore, the difference between C (the original number of capacitors) and $0.25C$ (the number used) must equal the number remaining. Therefore, we write $C - 0.25C = 153$. Subtracting like terms in the left member gives $0.75C = 153$. Finally, dividing both sides of the equation by 0.75 results in $C = 204$. Therefore, there were 204 capacitors originally in stock.

Check $C - 0.25C = 153$
$204 - 0.25(204) \stackrel{?}{=} 153$
$204 - 51 \stackrel{?}{=} 153$
$153 = 153$

Example 4-11

The length of a certain printed circuit board is three times its width. The perimeter measures 54 inches (in.). How many square inches of board area are there?

Solution Probably the biggest helper you can have in most problems is a picture of the situation. The simple sketch in Figure 4-2 shows all the information we know about the problem. Since we know that the perimeter is simply the sum of all the sides, we write $54 = 3w + w + 3w + w$. If we add all the like terms in the right member, we obtain $54 = 8w$. Dividing both sides by 8 gives $6.75 = w$. Therefore, the length of the board is $3w$, or $3(6.75) = 20.25$ in. The surface area of the board, then, is the product of the length and width:

$$6.75 \times 20.25 \approx 136.7 \text{ in.}^2$$

Figure 4-2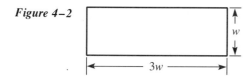

Check $54 = 3w + w + 3w + w$
$54 \stackrel{?}{=} 3(6.75) + 6.75 + 3(6.75) + 6.75$
$54 \stackrel{?}{=} 20.25 + 6.75 + 20.25 + 6.75$
$54 = 54$

Example 4–12 The sum of three consecutive even integers is 66. Find the three numbers.

Solution If we let n equal the smallest integer, then $n + 2$ must be the second *even* integer. (Note that if the word *even* were omitted from the statement of the problem, then the second integer would be $n + 1$.) Since n is the first even integer and $n + 2$ is the second even integer, then $n + 4$ must be the third even integer. Therefore, we write $n + (n + 2) + (n + 4) = 66$. Combining like terms in the left member gives $3n + 6 = 66$. Subtracting 6 from both sides results in $3n = 30$. Dividing by 3 gives $n = 20$. Therefore, the first even integer is 20; the second is $n + 2 = 22$; and the third integer is $n + 4 = 24$.

Check $20 + 22 + 24 \stackrel{?}{=} 66$
$66 = 66$

It would be possible, though not very practical, to continue with examples for a long time and still not cover all the types of simple word problems you might encounter. However, the previous examples should provide you with some insight into the logic of how such problems are to be approached. Therefore, it is time for you to jump into the pool and get wet. Before taking the plunge, however, the following solution hints for word problems are provided for your reference. *Read them often!*

1. Whenever possible, *draw a picture* of the details of the problem, labeling what is known and what is to be found.
2. *Translate* each significant word or phrase into a sign, symbol, or literal number.
3. *Solve* for the required quantity.
4. *Check* your answer against the conditions given in the problem.

Exercise Set 4–3

Questions

1. What is a word problem?
2. What words or phrases indicate each of the following?
   a. Addition
   b. Subtraction
   c. Multiplication
   d. Division
3. Why should a sketch of a word problem be drawn whenever possible?
4. What is the most common difficulty encountered in solving a word problem?
5. Why should your answer to a word problem be checked against the given conditions in the problem?

Problems

1. The sum of two consecutive integers is 31. Find the integers.
2. The sum of two consecutive odd integers is 200. What are the two integers?
3. The total cost of one inductor and three capacitors is $2.83. If the inductor costs $1.05 more than a single capacitor, find the cost of each component.

Fundamentals of Linear Equations

4. A zener diode costs $.22 more than a certain rectifier diode. The cost of one zener and two rectifier diodes is $1.87. What is the cost of each type of diode?
5. The perimeter of a square room is 108 feet (ft). What are the length and width of the room?
6. The perimeter of a certain triangle is 85 in. The longest side is twice the shortest side. The second longest side is 5 in. greater than the shortest side. Find the length of each side.
7. A 20 ft section of antenna mast is cut into two pieces. One of the pieces is 1.5 ft shorter than four times the length of the other. How long is each piece?
8. A 300 meter (m) length of RG-58/U coaxial cable is cut into three lengths. One piece is two-thirds as long as the longest section and twice as long as the shortest piece. How long is each piece of cable?
9. The sum of three consecutive odd integers is 39. What are the three integers?
10. The area of a rectangular room is 450 ft^2. If the room is twice as long as it is wide, find the dimensions of the room.
11. As a technician, John makes $10.85 per hour. If Roberta gets her raise, she will be making $.26 per hour more than John. What percent increase in her hourly rate does this raise represent?
12. The sum of all the angles in any plane triangle is 180°. In a certain triangle, the smallest angle is 33° less than the second largest angle, which is 69° less than the largest angle. Find the angles.
13. Thirty percent of voltage V_1 is the same as 1½ plus voltage V_2. Voltage V_2 is one-fifth of V_1. What are the two voltages?
14. The failure rate of one type of resistor is 10% higher than another. The average combined failure rate is 30%. What is the percent failure rate of each resistor?
15. A certain power transistor operates at 24° centigrade (C) higher than another. The average operating temperature is 120°. What is the operating temperature of each device?

Key Terms

equation	solve	axiom
equal sign	root	transpose
left member	satisfy	inequality
right member	linear equation	word problem

Important Algebraic Axioms, Rules, and Guidelines

Axioms

Axiom 4–1 If the same number is added to equal numbers, the sums are still equal.
Axiom 4–2 If the same number is subtracted from equal numbers, the remainders are still equal.
Axiom 4–3 If equal numbers are multiplied by the same number, the products are equal.
Axiom 4–4 If equal numbers are divided by the same nonzero number, the quotients are still equal.
Axiom 4–5 If two numbers are equal to the same (or equal) numbers, then they are equal to each other.
Axiom 4–6 If two numbers are equal, they both may be raised to the same power without altering the equality.
Axiom 4–7 If two numbers are equal, like roots may be extracted from both members without destroying the equality.

Rules for Solving Equations

Rule 4–1 Any term may be *transposed* (moved) from one member to the other simply by changing its sign.

Rule 4–2 Any term appearing on *both* sides of an equation and preceded by the *same* sign may simply be dropped from the equation.

Rule 4–3 The signs of *all the terms* of an equation may be changed without destroying the equality.

Rules for Solving Inequalities

Rule 4–4 All axioms that apply to equalities also apply to inequalities except;

Rule 4–5 Whenever we multiply or divide both members by a negative number, we must *reverse the sense* of the inequality.

Guidelines for Solving Word Problems

1. Whenever possible, *draw a picture* of the details of the problem, labeling what is known and what is to be found.
2. *Translate* each significant word or phrase into a sign, symbol, or literal number.
3. *Solve* for the required quantity.
4. *Check* your answer against the conditions given in the problem.

5 Factoring

Upon completion of this chapter, you will have:
1. Learned how to deal with fractional exponents.
2. Learned how to use special products.
3. Learned how to factor fundamental algebraic expressions.

5-1 INTRODUCTION

Factoring is the process of resolving an expression into two or more component parts (factors) whose product results in the given expression. For example, the expression $169R^2X^3 - 26RX$ can be resolved into the two factors $13RX$ and $(13RX^2 - 2)$, whose product results in the original expression. That is, $13RX(13RX^2 - 2) = 169R^2X^3 - 26RX$.

In electronics, it is often necessary to simplify a given algebraic expression so that it becomes easier to apply in other parts of a given problem. Factoring is one process whereby expressions are simplified.

One of the more fundamental operations of the factoring process is writing an integer as a product of its prime numbers. You may recall from arithmetic that a **prime number** is one evenly divisible only by itself and 1. For example, 5 is a prime number, since it cannot be evenly divided by any whole number other than itself and 1. The first eight primes are 2, 3, 5, 7, 11, 13, 17, and 19. Note that the only even prime number is 2; all other prime numbers are odd.

Another type of number is a **composite number,** which is not a prime number but rather is the product of prime numbers. For example, 105 is a composite number, since it is the product of $3 \times 5 \times 7$, which are all prime numbers. Another way of defining a composite number is to say that it is divisible by numbers other than itself and 1. Note that all even numbers (except 2) are composite numbers.

By convention the number 1 is regarded as neither prime nor composite. Moreover, we do not consider zero as either prime or composite, since division by zero is not permissible, as we learned in Chapter 1.

Prime factorization is the process of writing a given number as a product of its prime factors. This process for integers is relatively simple and begins by attempting to divide the given number by some convenient prime number. We say *attempting* because it is sometimes difficult to tell by inspection whether a number is, in fact, composite or prime. For example, it may not be readily apparent that 3187 is a prime number. And obviously, as stated before, a prime number cannot be divided evenly by any number other than itself and 1.

Example 5-1 Find the prime factors of 288.

Solution We begin by dividing 288 by the prime number 2, obtaining the quotient 144. We continue dividing by some prime number until a quotient is obtained that is itself a prime number.

$$\begin{array}{r|r} 2 & 288 \\ \hline 2 & 144 \\ \hline 2 & 72 \\ \hline 2 & 36 \\ \hline 2 & 18 \\ \hline 3 & 9 \\ \hline & 3 \end{array}$$

It is apparent, then, that the prime factorization of 288 is $2 \times 2 \times 2 \times 2 \times 2 \times 3 \times 3$, which can be written as $2^5 3^2$. Note that it was not necessary to

Factoring

begin the factorization process with division by 2. We could have started with division by 3 and obtained the same result, as shown next.

$$
\begin{array}{r|l}
3 & 288 \\
\hline
3 & 96 \\
\hline
2 & 32 \\
\hline
2 & 16 \\
\hline
2 & 8 \\
\hline
2 & 4 \\
\hline
& 2
\end{array}
$$

We see that the factors are the same.

Example 5-2 Find the prime factors of 2,156,000.

Solution

$$
\begin{array}{r|l}
2 & 2{,}156{,}000 \\
\hline
2 & 1{,}078{,}000 \\
\hline
2 & 539{,}000 \\
\hline
2 & 269{,}500 \\
\hline
2 & 134{,}750 \\
\hline
5 & 67{,}375 \\
\hline
5 & 13{,}475 \\
\hline
5 & 2{,}695 \\
\hline
7 & 539 \\
\hline
7 & 77 \\
\hline
& 11
\end{array}
$$

The prime factorization is $2^5 5^3 7^2 11$.

Exercise Set 5-1

Questions

1. In your own words, define each of the following terms.
   a. Factoring
   b. Prime number
   c. Composite number
   d. Prime factorization

Problems

1. Which of the following numbers are prime and which are composite?

   3, 13, 15, 17, 27, 48, 49, 51, 53, 81, 83, 99, 101, 105

2. Determine the prime factorization of the following numbers.
   a. 136   b. 208   c. 315
   d. 1002   e. 2688   f. 4505
   g. 10,688   h. 15,266   i. 24,608

5-2 FACTORS OF A RADICAND

It was mentioned in the previous section that one of the purposes of factoring is in the simplification of algebraic expressions. The simplification process itself often relies on the removal of common and higher-order factors from the given expression. For example, the removal of the common factors 5 and 13 from both terms of the fraction 130/325 reduces the fraction to the simpler form 2/5. In other words, since the fraction 130/325 can be written as

$$\frac{130}{325} = \frac{2 \times 5 \times 13}{5 \times 5 \times 13}$$

cancellation of the factors common to both numerator and denominator greatly simplifies the fraction without changing its value—that is,

$$\frac{2 \times \cancel{5} \times \cancel{13}}{5 \times \cancel{5} \times \cancel{13}} = \frac{2}{5}$$

In a similar way, radical expressions can be simplified by removing factors from the radicand. For example, the radical expression $\sqrt{3825}$ can be simplified by removing factors that are themselves squared integers. If we were to perform the prime factorization of this radicand, we would obtain $3^2 5^2 17$. The radicand can now be written as $\sqrt{3^2 5^2 17}$, or $\sqrt{3^2} \times \sqrt{5^2} \times \sqrt{17}$. Since the first two radicands are squared integers, we can write this last expression as $3 \times 5 \times \sqrt{17} = 15\sqrt{17}$, which is a much simpler version of the original radical. We can treat radicands composed of literal factors in the same way. For example, $\sqrt{a^2 b^2 c}$ can be written as $\sqrt{a^2} \times \sqrt{b^2} \times \sqrt{c}$, which, upon extracting the square roots, results in $ab\sqrt{c}$.

You will recall from Chapter 1 that the process of extracting the square root of a number amounted to finding two identical factors—*exactly* alike in absolute value and sign—whose product results in the given radicand. For example, $\sqrt{25} = \pm 5$, since $(+5)(+5) = 25$ and $(-5)(-5) = 25$. We shall now extend this idea to include higher-order roots so that we can simplify radicands involving other than squared literal factors.

The **index of the root** is the number written in the valley of the radical sign, such as the 3 in $\sqrt[3]{64}$, indicating the root to be extracted. For example, $\sqrt[4]{625}$ indicates that the fourth root of 625 is to be extracted. Hence, $\sqrt[4]{625} = 5$, since $5^4 = 625$. Note that the index 2 is omitted from the radical sign in the case of the square root, since it is obviously impossible to have a product of fewer than two terms.

With the foregoing in mind, you have probably figured out by now that the extraction of such roots as $\sqrt{a^2}$ and $\sqrt{b^2}$ in the preceding examples simply amounts to dividing the exponent by the index of the root. In other words, $\sqrt{a^2} = a^{2/2} = a^1 = a$ and $\sqrt{b^2} = b^{2/2} = b$. Similarly, $\sqrt[3]{x^3} = x^{3/3} = x$ and $\sqrt[4]{z^6} = z^{6/4}$, or $z^{3/2}$. In order to provide a strict mathematical basis for this simple operation, however, we must discuss the idea of a **fractional exponent**.

In Chapter 3, the law of multiplicative exponents was stated as $b^p b^q = b^{p+q}$. Therefore, $b^{1/2} b^{1/2} = b^{1/2 + 1/2} = b^1$, or simply b. Also, $b^{1/3} b^{1/3} b^{1/3} = b^{1/3 + 1/3 + 1/3} = b$. In

Factoring

other words, $b^{1/2}$ is one of *two* equal factors of b. Similarly, $b^{1/3}$ is one of *three* equal factors of b. We are lead to conclude that the denominator of a fractional exponent equals the index of the root and the numerator is the power of the base. This allows us to state the law of fractional exponents as follows.

Rule 5–1

$$b^{p/q} = \sqrt[q]{b^p} \qquad q \neq 0$$

Example 5–3 Simplify the radical expression $\sqrt{50R^5X^6}$.

Solution Observing that $50 = 5^2 2$ and $R^5 = R^4 R$, we rewrite the radical as $\sqrt{5^2 2 R^4 R X}$. Then, by applying the law of fractional exponents to each factor of the radicand, we write

$$\sqrt{5^2 2 R^4 R X} = 5R^2 \sqrt{2RX}$$

Example 5–4 Simplify the radical expression $\sqrt{289 v^2 i^3}$.

Solution Rewriting the radical as $\sqrt{17^2 v^2 i^2 i}$, we may apply the law of fractional exponents to obtain

$$\sqrt{17^2 v^2 i^2 i} = 17vi\sqrt{i}$$

Example 5–5 Simplify the radical expression $\sqrt[4]{256 H^2}$.

Solution The radical may be written in exponential form as $(2^8 H^2)^{1/4} = 4\sqrt{H}$.

Exercise Set 5–2

Questions

1. In your own words, define what is meant by the index of the root.
2. State the law of fractional exponents.
3. Why is the index 2 omitted from the square root?

Problems

Simplify the following expressions.

1. $\sqrt{289 c^3 v^2 x^4}$
2. $\sqrt{225 X^3}$
3. $\sqrt{729 I^2 R}$
4. $\sqrt[3]{125 P^6}$
5. $\sqrt[4]{1024 V^2}$

6. $\sqrt[5]{243Y^6Z^3}$
7. $\sqrt[6]{64Z^m}$
8. $\sqrt[4]{32k^2m^7}$
9. $\sqrt[4]{243Q^8P^2}$
10. $\sqrt[7]{256V^8}$
11. $(3125x^5y^{12})^{1/4}$
12. $(\sqrt{8y^6})^{1/3}$
13. $(216x^3)^{1/6}$
14. $(64^{1/6}I^{1/3}R^3)^6$
15. $(169x^{1/2}y^4)^{1/2}$

5–3 THE COMMON FACTOR

Many algebraic multinomials can be simplified by removing factors that are common to all the terms in the expression. Consider the multinomial $256V^3I^4 + 12V^5I^2 - 4V^2I^2$. If we first obtain the prime factors of all the numerical coefficients, we can rewrite the multinomial as $2^8V^3I^4 + 2^23V^5I^2 - 2^2V^2I^2$. It is now apparent that each of these three terms contains the common factors 2, V, and I. Moreover, it should be apparent that each of these multinomial terms is evenly divisible by some *greatest common factor* (abbreviated GCF). Furthermore, if such a division is actually carried out, the quotient obtained, along with the GCF itself, will form the simplest factors of the given expression. In the present example, a brief inspection reveals that the greatest numeral that will evenly divide the numerical coefficients is 2^2. Similarly, the highest common V factor is V^2, and the highest I factor is I^2. By forming the product of these factors, we obtain the GCF of $2^2V^2I^2$. Finally, by dividing the original multinomial by this GCF, we obtain

$$\frac{2^8V^3I^4 + 2^23V^5I^2 - 2^2V^2I^2}{2^2V^2I^2} = 2^6VI^2 + 3V^3 - 1$$

Therefore, the product of the GCF and the quotient form the factors of the required product. We write

$$256V^3I^4 + 12V^5I^2 - 4V^2I^2 = 4V^2I^2(64VI^2 + 3V^3 - 1)$$

We now formally define the **greatest common factor (GCF)** as the greatest factor that will evenly divide two or more terms. Many textbooks refer to the GCF as the highest common factor (HCF) or the greatest common divisor (GCD). Each term means the same thing.

To remove the common monomial factor from a multinomial, we use the following rule:

Rule 5–2

1. Determine the GCF of all terms by inspection.
2. Divide the given multinomial by the GCF.
3. Indicate the product of the quotient and the GCF as the required factors of the given expression.

Factoring

Example 5-6

Factor $32IR + 24IZ$.

Solution We rewrite this expression as $2^5IR + 2^33IZ$. By inspection, the GCF is 2^3I. Dividing the multinomial by the GCF gives $2^2R + 3Z$. Therefore, the given multinomial is factored as $8I(4R + 3Z)$.

Example 5-7

Determine the factors of $f^2L^2/72 - f^2L^3/18 - f^3L^2/36$.

Solution It is seen by inspection that 18 is the greatest common numerical factor of all three denominators. Moreover, f^2L^2 is the greatest literal factor. Combining these facts gives us the GCF of $f^2L^2/18$. Dividing each term by the GCF results in the quotient $1/4 - L - f/2$. Finally, we write the factors as $(f^2L^2/18)(1/4 - L - f/2)$.

Example 5-8

Determine the factors of $0.07x^3y^5 - 0.003x^2y^2$.

$$\frac{7x^3y^5}{100} - \frac{3x^2y^2}{1000} = \frac{7x^3y^5}{2^25^2} - \frac{3x^2y^2}{2^35^3} = \frac{x^2y^2}{2^25^2}\left(7xy^3 - \frac{3}{10}\right)$$

$$= \frac{x^2y^2}{100}(7xy^3 - 0.3) = 0.01x^2y^2(7xy^3 - 0.3)$$

Exercise Set 5-3

Questions

1. What is meant by the greatest common factor?
2. What other terms are used to mean GCF?

Problems

Determine the GCF and factor the following expressions.

1. $51Z^2 - 34Z$
2. $64v^3 + 256v^2$
3. $10^2k^2g + 1000k^2g^2$
4. $0.01VI + 0.002V^2I$
5. $\frac{1}{4}P^2Q^2 - \frac{1}{8}P^2Q + \frac{1}{32}PQ^2$
6. $\frac{V^3e^3}{15} + \frac{V^2e^2}{3}$
7. $\frac{ax^3}{5} - \frac{a^2x^2}{20} + \frac{ax^3}{80}$
8. $\frac{Q^5}{18} + \frac{Q^3}{6} - \frac{Q^6}{108}$
9. $0.005I^2R^2 - 0.0005I^3R^3$
10. $0.03ax^3 + 0.06ax^5$
11. $625k^3M^2 - 15625k^2M^3 + 50000k^4M^2$
12. $2048C^3 + 2^9C^2 + 256C^2$
13. $\frac{1}{6a^5} - \frac{1}{10a^7}$
14. $54V^3e^2 + 81V^2e - 108Ve^3$
15. $0.5X^5 - \frac{X^3}{4} + 0.0625X^2$

5-4 THE PERFECT TRINOMIAL SQUARE

Certain products occur so frequently in mathematics that they have become known as *special products*. Two such products are the squares of the binomials $(a + b)$ and $(a - b)$. We apply the rule for multinomial products from Chapter 3 in order to find the terms of each of these squares. For example,

$$(a + b)^2 = (a + b)(a + b) = a^2 + ab + ab + b^2$$

On combining like terms we obtain

$$(a + b)^2 = a^2 + 2ab + b^2$$

Similarly,

$$(a - b)^2 = (a - b)(a - b) = a^2 - ab - ab + b^2$$

By combining like terms we get

$$(a - b)^2 = a^2 - 2ab + b^2$$

Note that the only difference between the two resulting expressions is the sign of the term $2ab$. Note, too, that the first and last terms are perfect squares and that the middle term is twice the product of the end terms of the original binomial.

Since these two forms are so common in the mathematics of electronics, it will be well worth your while to memorize these expressions. They will be very useful in factoring trinomials of this type.

Example 5-9

Factor the expression $16 + 24V + 9V^2$.

Solution We observe that the two outermost terms are perfect squares; 16 is the square of 4 and $9V^2$ is the square of $3V$. Moreover, we note that the middle term is exactly twice the product of 4 and $3V$. And since the sign of the middle term is positive, we conclude that this expression must be the square of $(4 + 3V)$. In other words,

$$16 + 24V + 9V^2 = (4 + 3V)^2$$

Example 5-10

Find the factors of $-2x^4 + x^8 + 1$.

Solution At first glance, this trinomial may not seem to fit either of the conventional forms given earlier. It sometimes happens, though, that we must attempt to rearrange terms in order to see the pattern emerge. First, we observe that 1 is, in fact, a perfect square. We do not generally think of 1 as a perfect square because it is the only nonzero number having the property that its square equals itself. There is no other nonzero number having the property $a^2 = a$.

We also observe that x^8 is really the square of x^4. And finally, by rearranging the expression as $1 - 2x^4 + x^8$, we see that it fits the form $a^2 - 2ab + b^2$. We can now write

$$-2x^4 + x^8 + 1 = 1 - 2x^4 + x^8 = (1 - x^4)^2$$

Factoring

Example 5-11

Factor the trinomial $3 + 2\sqrt{3}R + R^2$.

Solution Although 3 is not the square of an integer, it is the square of $\sqrt{3}$, which is part of the middle term. Recognizing this fact is the only difficult part of the problem. We now write

$$3 + 2\sqrt{3}R + R^2 = (\sqrt{3} + R)^2$$

Exercise Set 5-4

Questions

1. What makes the special products special?
2. Using words only, state the results of squaring the two binomials $(a + b)$ and $(a - b)$.

Problems

Factor the following expressions.

1. $V^2 - 2Ve + e^2$
2. $4I^2 + 4Ii + i^2$
3. $1 + 2k + k^2$
4. $r^2 - 2r + 1$
5. $\dfrac{V^4}{4} - V^2P + P^2$
6. $a^2 - ab + \dfrac{b^2}{4}$
7. $(1 \times 10^{-4}) + (4 \times 10^{-3})b + \dfrac{b^2}{25}$
8. $a^2 - 2ab + b^2 + 2c(a - b) + c^2$
9. $\dfrac{1}{9} + \dfrac{4x^2}{3} + 4x^4$
10. $4r^4 + r^8 + 4$
11. $4 - 2\sqrt{4v} + v^2$
12. $\dfrac{1}{2} + \dfrac{2m}{\sqrt{2}} + m^2$
13. $\dfrac{a^2}{32} - \dfrac{ab}{32} + \dfrac{b^2}{128}$
14. $5z^2 - 10zx + 5x^2$
15. $24x^2y + 24xyz + 6yz^2$

5-5 THE DIFFERENCE OF SQUARES

Another special product that occurs frequently in electronics is the product of the sum and difference of the same two terms. Expressed mathematically, $(a + b)(a - b) = a^2 - b^2$. This can be verified by applying the rules for binomial multiplication learned in Chapter 3. We write

$$(a + b)(a - b) = a^2 - ab + ab - b^2 = a^2 - b^2$$

Example 5-12 Factor the expression $0.01X^2 - R^2$.

Solution If we rewrite this binomial as $X^2/100 - R^2$, it becomes apparent that we are dealing with the difference of squares. By applying the standard form, we obtain

$$\frac{X^2}{100} - R^2 = \left(\frac{X}{10} - R\right)\left(\frac{X}{10} + R\right)$$

Example 5-13 Find the factors of $-1 + x^4$.

Solution If we rearrange the given factors as $x^4 - 1$ and recognize that 1 is a squared term and x^4 is the square of x^2, it becomes evident that we are dealing with the difference of squares. Therefore, we obtain

$$x^4 - 1 = (x^2 - 1)(x^2 + 1)$$

On looking more closely at this result, we see that $x^2 - 1$ is also the difference of two squares. It is obvious that this expression can also be factored as $(x + 1)(x - 1)$. Therefore, our final answer is written as

$$(x - 1)(x + 1)(x^2 + 1)$$

Exercise Set 5-5

Problems

Factor the following expressions.
1. $1 - w^2$
2. $-3 + a^2$
3. $0.01Z^2 - 1$
4. $Q^4 - 1$
5. $-1 + X^4$
6. $-1 + \dfrac{c^4}{16}$
7. $\dfrac{1}{R^2} - \dfrac{2}{r^2}$
8. $-\left(\dfrac{1}{X^2} - \dfrac{1}{Y^2}\right)$
9. $-1 \times 10^{-2} + V^2$
10. $P^8 - 16$
11. $-0.25 + k^4$
12. $\dfrac{-1}{8} + \dfrac{x^2}{2}$
13. $\dfrac{t^2}{3} - \dfrac{1}{27}$
14. $\dfrac{1}{5} - 5e^2$
15. $0.0625X^4 - 256.0$

Factoring

16. A student worked the following problem using the axioms given in Chapter 4 and the factoring method of *difference of squares* presented in this section. What error did the student make?

1. Given.	$V = k$
2. Square each side.	$V^2 = k^2$
3. Substitute.	$V^2 = Vk$
4. Subtract k^2 from each side.	$V^2 - k^2 = Vk - k^2$
5. Factor.	$(V - k)(V + k) = k(V - k)$
6. Divide each side by $V - k$.	$(V + k) = k$
7. Substitute.	$V + V = V$
8. Combine like terms.	$2V = V$
9. Divide each side by V.	$2 = 1$

5-6 FACTORING TRINOMIALS OF THE TYPE $x^2 + (a + b)x + ab$

In the previous section, we saw that the product of the two binomials $(a + b)(a - b)$ resulted in another binomial, $a^2 - b^2$. In general, however, the product of any two binomials *with a common term* results in a trinomial, not a binomial. In this section, we discuss strategies for factoring trinomial expressions formed by binomial factors sharing a common term.

Although general guidelines for the factorization process exist, none can be classed as true algorithms. Most methods involve a considerable amount of trial-and-error testing with tentative factors. And where attempts have been made to set forth strict rules, their degree of complexity often exceeds the difficulty level of the problem itself. The memorization of such convoluted rules does not appear to offer much in the way of real understanding. Therefore, our approach will be entirely empirical, based only on certain fundamental facts of the trinomial structure itself.

Consider the product $(x + a)(x + b) = x^2 + ax + bx + ab$. Note that the first term of the product (x^2) is simply the square of the common term. Note, too, that the last term (ab) is the product of the other two binomial terms. Finally, observe that since the two middle terms are like terms, they may be combined into the single term $(ax + bx)$, thereby forming the middle term of the trinomial. Therefore, the product of any two binomials having a common term will, in general, produce a trinomial of the form $x^2 + (ax + bx) + ab$, which may also be written as $x^2 + (a + b)x + ab$.

A diagram of the structure of the various terms is shown next. There are several important things to observe:

1. The *first* term of the trinomial is the product of the *first* terms of the binomials.

$$(x + a)(x + b) = x^2 + (ax + bx) + ab$$

2. The *last* term of the trinomial is the product of the *last* terms of the binomials.

$$(x + a)(x + b) = x^2 + (ax + bx) + ab$$

3. The *middle* term of the trinomial is the *sum* of the products of the *innermost* and *outermost* terms of the binomials.

$$(x + a)(x + b) = x^2 + (ax + bx) + ab$$

The physical symmetry of the product is easy to see and should provide a simple, visual mnemonic for the factorization process. Let us now apply what we know about the trinomial structure to a practical example. Consider the trinomial $x^2 + 5x + 6$. In order to factor this trinomial, we need to find two binomials whose terms meet the criteria just outlined. We begin by providing an empty set of parentheses for the binomial terms:

$$x^2 + 5x + 6 = (\quad)(\quad)$$

According to the structural diagram, we know that the first term of the trinomial is the square of the common term. Therefore, we can start filling in the parentheses:

$$x^2 + 5x + 6 = (x\quad)(x\quad)$$

The last term of the trinomial is the product of the last terms of the binomials. Here is where the real-world factorization process can get a little tricky. For there are obviously several combinations of numbers whose product is 6. However, only one of those combinations is useful in forming the middle term of the trinomial. Here are your choices:

$$6 = 1 \times 6$$
$$6 = -1 \times (-6)$$
$$6 = 2 \times 3$$
$$6 = -2 \times (-3)$$

At this point most factorization algorithms break down, leaving trial-and-error methodology. With the structure of the trinomial in mind, however, the amount of guesswork can be kept to a minimum. For example, it probably wouldn't be a good idea to consider the first two choices, since their sums are either $+7$ or -7, and the middle term of the trinomial clearly demands a $+5$. This also rules out the fourth set of factors, since their sum is -5, not $+5$. Therefore, let us try the third pair of factors.

$$(x + 2)(x + 3) = x^2 + 2x + 3x + 6 = x^2 + 5x + 6$$

Since this gives us *both* the required middle *and* end terms of the trinomial, we conclude that this is the correct factorization.

We now consider a slightly more difficult problem. Consider the trinomial $x^2 - x - 12$. We begin by providing an empty set of parentheses for the binomial terms.

$$x^2 - x - 12 = (\quad)(\quad)$$

Next, we fill in the common term.

$$x^2 - x - 12 = (x\quad)(x\quad)$$

Now comes the more challenging part. There are six combinations of factors whose product is -12, but only one will provide the required middle term of the trinomial. The choices are

$$-12 = -1 \times 12$$
$$-12 = 1 \times (-12)$$
$$-12 = -2 \times 6$$
$$-12 = 2 \times (-6)$$
$$-12 = -3 \times 4$$
$$-12 = 3 \times (-4)$$

Since the middle term indicates that the sum of the end terms equals -1, it is obvious that the first five choices can be eliminated, since their sums, in order, are 11, -11, 4, -4, and $+1$. Only the last choice meets the middle-term requirement of summing to -1.

Factoring

Therefore, let us test this combination.
$$(x + 3)(x - 4) = x^2 + 3x - 4x - 12 = x^2 - x - 12$$

We see that this combination of terms has resulted in the correct trinomial.

It is evident from the foregoing that the real difficulty in factoring these types of trinomials arises from the need of having to find number combinations whose products as well as sums result in particular values. Naturally, there will be some guesswork involved. If you keep the trinomial structure firmly in mind, however, the number of guesses can be made fairly small. We now take up some additional examples.

Example 5-14

Find the binomial factors of $Z^2 + 24Z - 81$.

Solution First, we provide parentheses to contain the terms of the binomial factors.

$$Z^2 + 24Z - 81 = (\quad)(\quad)$$

Secondly, we write in the common factor.

$$Z^2 + 24Z - 81 = (Z\quad)(Z\quad)$$

Next, we find two numbers whose *product* equals the end term of the trinomial and whose *sum* gives the numerical coefficient of the middle term. We list the various combinations here:

$$-81 = 9 \times (-9)$$
$$-81 = -9 \times 9$$
$$-81 = 3 \times (-27)$$
$$-81 = -3 \times 27$$

We can immediately eliminate the first two combinations, since they have a sum of zero. The third combination leads to a sum of -24 and can also be eliminated. Finally, we see that the last combination yields a $+24$ as the required coefficient of x. Therefore, we write

$$Z^2 + 24Z - 81 = (Z - 3)(Z + 27)$$

Example 5-15

Find the factors of $P^4 - P^2/4 - 1/8$.

Solution As a start, we write the partial binomial factors of $(P^2\quad)(P^2\quad)$. The next step requires us to look for two numbers whose product equals $-1/8$ and whose sum is $-1/4$. We list the various possible combinations as:

$$-\frac{1}{8} = -1 \times \frac{1}{8}$$
$$-\frac{1}{8} = 1 \times \left(-\frac{1}{8}\right)$$
$$-\frac{1}{8} = \frac{1}{2} \times \left(-\frac{1}{4}\right)$$
$$-\frac{1}{8} = -\frac{1}{2} \times \frac{1}{4}$$

The sums of the first three combinations are $-7/8$, $+7/8$, and $+1/4$. Since the numerical coefficient of the middle term is $-1/4$, we see that this requirement is fulfilled only by the last combination. Therefore, we write

$$P^4 - \frac{P^2}{4} - \frac{1}{8} = \left(P^2 - \frac{1}{2}\right)\left(P^2 + \frac{1}{4}\right)$$

Exercise Set 5–6

Problems

Factor the following expressions.
1. $V^2 - 9V + 14$
2. $i^2 - i - 6$
3. $I^2 - Ie - 6e^2$
4. $-9a^2 - x^2 + ax$
5. $w^2 + 9w - 22$
6. $v^2 + 9v - 10$
7. $i^2 - 17i - 60$
8. $-24 + W^2 - 10W$
9. $k^4 - 4k^2t - 12t^2$
10. $-2P^2 + Q^4 - PQ^2$
11. $m^4 - \dfrac{m^2}{4} - \dfrac{1}{8}$
12. $Z^2 + 16 - \dfrac{5Z}{8}$
13. $r^2 + \dfrac{2rR}{x} - \dfrac{24R^2}{x^2}$
14. $i^2 - I - \dfrac{6I^2}{i^2}$
15. $\dfrac{1}{p^2} + \dfrac{3}{P} - 10$
16. $\dfrac{1}{x^2} + \dfrac{1}{x} + 12$
17. $Z^4 - \dfrac{3t^2}{8} - 0.03125$
18. $t^4 + t^2 - 6$
19. $a^2b^2 - 4abc + 3c^2$
20. $V^2I^2 + VIW - 6W^2$

5–7 FACTORING TRINOMIALS OF THE TYPE $acx^2 + (ad + bc)x + bd$

In the previous section, we learned how to factor trinomials whose binomial factors were of the type $(x + a)(x + b)$. It is often the case, however, that the coefficients of the x terms in the factors are not equal to 1. In fact, each x term may have an entirely different numerical coefficient. Consider the product

$$(ax + b)(cx + d) = acx^2 + adx + bcx + bd$$

We see that this expression is also a trinomial, since adx and bcx are like terms and can, therefore, be combined into the single term $(ad + bc)x$.

Factoring

You have probably noticed that the structure of this trinomial is not too unlike the structures discussed previously. That is, the first term of the trinomial is the product of the first terms of the binomials. The last term of the trinomial is the product of the last terms of the binomials. And finally, the middle trinomial term has a numerical coefficient that depends on all four numerical values of the binomials. The numerical coefficient of the middle term has a structure of its own that is worth examining more closely. A diagram of the structure of this term is shown next. Note that the numerical coefficient of the middle term is the sum of the products of the innermost and outermost numerical coefficients of the binomials.

$$(ax + b)(cx + d) = acx^2 + (ad + bc)x + bd$$

The following steps describe this process.

1. The *first* term of the trinomial is the product of the *first* terms of the binomials.

$$(ax + b)(cx + d) = acx^2 + (ad + bc)x + bd$$

2. The *last* term of the trinomial is the product of the *last* terms of the binomials.

$$(ax + b)(cx + d) = acx^2 + (ad + bc)x + bd$$

3. The numerical coefficient of the *middle* term is the *sum* of the products of the *innermost* and *outermost* numerical coefficients of the binomials.

$$(ax + b)(cx + d) = acx^2 + (ad + bc)x + bd$$

The best way to demonstrate the factoring process involving these types of trinomials is with examples.

Example 5–16

Factor the trinomial $5R^2 + 13R + 6$.

Solution We begin by providing an empty set of parentheses for the binomial terms.

$$5R^2 + 13R + 6 = (\quad)(\quad)$$

From this point on, intuition more than mathematical prowess becomes an important part in selecting the proper combination of terms. The first binomial terms are not too difficult to see, and so we write

$$5R^2 + 13R + 6 = (5R\quad)(R\quad)$$

In trying to decide on the second terms, however, we are faced with the problem of simultaneously identifying values whose product results in the last term of the trinomial and whose sum of products with terms already selected forms the middle term.

Since this process is essentially a trial-and-error undertaking, we can dispense with the mathematical niceties and get right down to trying those combinations that "look good." We are often wrong. For example, suppose the factors $(5R - 2)(R + 3)$ look promising because we recognize that the product of the innermost terms, $-2R$, and the outermost terms, $15R$, have the required sum, $13R$, of the trinomial. Moreover, we see that the product of 2 and 3 forms the last trinomial term. Unfortunately, despite all our cleverness, we have neglected the small matter of the sign of the 6, which is to be positive in the trinomial, not negative. We have obviously guessed incorrectly. We need to try another combination.

In our second attempt, we see that dropping the minus sign in the first binomial solves the problem of the -6 but produces $17R$ rather than $13R$ for the middle term of the trinomial. However, we note that by merely exchanging the 2 and 3 in the binomials, we obtain $(5R + 3)(R + 2)$, which gives us the needed $+13R$ for the middle trinomial term as well as the $+6$ for the last term. These, then, are the correct binomial factors, and so $5R^2 + 13R + 6 = (5R + 3)(R + 2)$.

Most students find it helpful in factoring these types of trinomials to make a list of trial factors so that various combinations can be tried quickly and easily. In the foregoing problem, our list might look something like the following:

Trial Factors	Second Term	
(5 6)(1 1)	$5 + 6 = 11$	(wrong)
(5 2)(1 3)	$15 + 2 = 17$	(wrong)
(5 3)(1 2)	$10 + 3 = 13$	(correct)

Example 5-17   Determine the factors of $16v^2 - 22v - 3$.

Solution   We note that since the coefficient of the v^2 term is not prime, several possible combinations exist as choices for the first binomial term. We may be uncertain whether to form the product as 1×16, $-2 \times (-8)$, 4×4, or $-16 \times (-1)$. However, our problem is simplified somewhat if we observe that the last term of the trinomial is prime, and this helps us narrow the possibilities for the first terms. We form a list of some of the more promising trial factors as follows:

Trial Factors	Second Term	
(16 1)(1 -3)	$-48 + 1 = -47$	(wrong)
(4 1)(4 3)	$12 + 4 = 16$	(wrong)
(2 -3)(8 1)	$2 - 24 = -22$	(correct)

We conclude that the factors are $(2v - 3)(8v + 1)$.

Note that the first set of trial factors probably would never have been chosen in the first place by an observant student, since this combination is obviously too extreme.

Factoring

Example 5-18

Factor $15Z^2 + 24Z - 12$.

Solution We begin by setting up a list of the more inviting trial factors:

Trial Factors	Second Term	
(3 2)(5 6)	$18 + 10 = 28$	(wrong)
(3 6)(5 −2)	$-6 + 30 = 24$	(correct)
(15 −6)(1 2)	$30 - 6 = 24$	(correct)

We see that the last two combinations, perhaps somewhat to our dismay, are both factors of the original expression. We may begin to wonder if the factorization process is, after all, unique. Observe, however, that the trinomial as given contains the common factor 3. On removing this factor and rewriting the expression as $3(5Z^2 + 8Z - 4)$, we see that this may be factored as $3(5Z - 2)(Z + 2)$.

The lesson to be learned here is to factor out common factors before attempting to factor the trinomial.

Exercise Set 5-7

Problems

Factor each expression.
1. $4t^2 - 8tu - 21u^2$
2. $3V^2 - 2Ve - 8e^2$
3. $8e^2 + 4e - 24$
4. $12Z^2 - 23Z + 10$
5. $18Z^2 + 31Z + 6$
6. $15P^2 + 24P - 12$
7. $24x^4 - 30x^2y + 9y^2$
8. $6C^4 - 2C^2Q - 8Q^2$
9. $18V^2e^2 + 57Vek + 35k^2$
10. $8m^2n^2 - 8mnR - 6R^2$
11. $27Q^2 + 15PQ - 2P^2$
12. $10X^2 + 17XZ - 20Z^2$
13. $10m^2n^2 - 17mnt + 3t^2$
14. $6k^2t^2 + 16kts - 22s^2$
15. $42e^2 + 11et - 20t^2$
16. $36E^2 + 66EV + 28v^2$
17. $48a^2b^2 - 5ab + \frac{1}{8}$
18. $250C^2 - 4C - 0.02$
19. $X^2Z^2 - \frac{5XZ}{6} + \frac{1}{6}$
20. $S^2 + \frac{S}{8} - 0.03125$

5-8 FURTHER TOPICS IN FACTORING

Often it is possible to factor expressions that, on the surface, do not appear to fit any of the types we have been discussing. It is sometimes possible, however, to rearrange these expressions so that they fall into one of the common categories that are readily factorable. For example, find the factors of $3t^3 - 27t + 5t^2 - 45$.

Our first reaction might be to conclude that this expression will not yield to any of the factoring methods we have developed. On second glance, we discover that the factor $3t$ is common to the first two terms and 5 is common to the second two terms. If we then rewrite the expression as $3t(t^2 - 9) + 5(t^2 - 9)$, it becomes apparent that the binomial $(t^2 - 9)$ is itself common to the two remaining terms. Factoring out this term leaves us with $(t^2 - 9)(3t + 5)$. We also recognize that $(t^2 - 9)$ is the difference of squares and can be factored as $(t - 3)(t + 3)$. Finally, we combine these various results into the completely factored expression $(3t + 5)(t - 3)(t + 3)$.

The following examples will serve to clarify further several of the methods used in factoring expressions that are not in a conventional form.

Example 5-19

Factor $35 - 18gr - 42g + 15r$.

Solution We see that 7 is common to the first and third terms and $3r$ is common to the second and fourth terms. It is possible to rearrange the given expression to make removal of common terms more convenient, and we write $35 - 42g - 18gr + 15r$. Removing the common terms from each set gives

$$7(5 - 6g) - 3r(6g - 5)$$

The two parenthetical binomials are alike except the signs of the terms are reversed. From the laws of grouping discussed in Chapter 2, we know we can rewrite this expression as

$$7(5 - 6g) + 3r(5 - 6g)$$

At this point we recognize that $5 - 6g$ is a common binomial factor. Therefore, upon factoring, we obtain

$$(7 + 3r)(5 - 6g)$$

Example 5-20

Factor the expression $a(b + c)^2 - a(b - c)^2$.

Solution On removing the common term a, we recognize this as the difference of squares, and we can now write

$$a[(b + c)^2 - (b - c)^2]$$

Factoring the bracketed terms, we obtain

$$a[(b + c) - (b - c)][(b + c) + (b - c)]$$

Removing parentheses and combining like terms, we have

$$a[b + c - b + c][b + c + b - c] = a[2c][2b]$$
$$= 4abc$$

In factoring these somewhat complex expressions, it is often a good idea to check our answers. This can be done by substituting simple numerical values for the literal numbers in both the original and factored expressions and comparing the results. In Example 5-20,

Factoring

we let $a = 2$, $b = 3$, and $c = 4$. On substituting in the given expression $a(b + c)^2 - a(b - c)^2$ we obtain $2(3 + 4)^2 - 2(3 - 4)^2 = 2(49) - 2(1) = 96$.

Upon substitution of these same values into the factored expression, we obtain $4(2)(3)(4) = 96$. Therefore, we conclude that our factorization is correct.

Example 5–21

Completely factor the expression $2(V^2 - I^2) + (V + I)^2$.

Solution Since the first parenthetical expression is the difference of squares, we write

$$2(V - I)(V + I) + (V + I)^2$$

We now factor out $(V + I)$, obtaining

$$(V + I)[2(V - I) + (V + I)] = (V + I)[2V - 2I + V + I]$$

On combining similar terms, we have

$$(V + I)(3V - I)$$

We can check this solution by numerical substitution. For example, let $V = 2$, and $I = 3$. In the original expression, this gives us $2(4 - 9) + (2 + 3)^2 = 2(-5) + 25 = -10 + 25 = 15$.

On substituting in the factored expression, we obtain $(2 + 3)(6 - 3) = 5(3) = 15$, which is the same as the value obtained from the given expression.

Exercise Set 5–8

Problems

Completely factor the following expressions, and check your answers using numerical substitution.
1. $Z^4 - X^4$
2. $\dfrac{1}{v^4} - \dfrac{81}{E^4}$
3. $2(P - 3)^2 + 3(P^2 - 9)$
4. $5(V^2 - 4) - 3(V + 2)^2$
5. $3r^3 - 3r^2g - 36rg^2$
6. $6s^3 + 20s^2t - 16st^2$
7. $3(V + 2)^2 - 12(V + 2) - 36$
8. $29(t - ¼) + 5(t - ¼)^2 - 42$
9. $10ir + 5xe - 6ir - xe$
10. $11Vv - 12Ee + Vv + 48Ee$
11. $6ab - 4cd + 15ad - 5dc$
12. $5Z(X + 4Y) - 10XY$
13. $3z^2 - 27$
14. $\dfrac{P^2}{2} - \dfrac{1}{18}$
15. $6x^2 + 6x - 72$
16. $11k^2 - \dfrac{33k}{2} - 11$
17. $C^2 - 2CV + V^2 - Q^2$
18. $m^2 + n^2 - 169 - 2nm$
19. $6Z^4 - 4aZ^2 - 2a^2$
20. $8D^4 + 12D^2 - 8$

Key Terms

factoring
prime number
composite number

prime factorization
index of the root

greatest common factor (GCF)
fractional exponent

Important Algebraic Laws and Rules

Law of Fractional Exponents
Rule 5–1

$$b^{p/q} = \sqrt[q]{b^p} \qquad q \neq 0$$

Removal of Common Monomial Factor
Rule 5–2

1. Determine the GCF of all terms by inspection.
2. Divide the given multinomial by the GCF.
3. Indicate the product of the quotient and the GCF as the required factors of the given expression.

Special Products and Their Factors

Square of the sum: $(a + b)^2 = a^2 + 2ab + b^2$
Square of the difference: $(a - b)^2 = a^2 - 2ab + b^2$
Product of the sum and difference: $(a + b)(a - b) = a^2 - b^2$

Factoring Trinomials of the Form $x^2 + (a + b)x + ab$

1. The *first* term of the trinomial is the product of the *first* terms of the binomials.

$$(x + a)(x + b) = x^2 + (ax + bx) + ab$$

2. The *last* term of the trinomial is the product of the *last* terms of the binomials.

$$(x + a)(x + b) = x^2 + (ax + bx) + ab$$

3. The *middle* term of the trinomial is the *sum* of the products of the *innermost* and *outermost* terms of the binomials.

$$(x + a)(x + b) = x^2 + (ax + bx) + ab$$

Factoring Trinomials of the Form $acx^2 + (ad + bc)x + bd$

1. The *first* term of the trinomial is the product of the *first* terms of the binomials.

$$(ax + b)(cx + d) = acx^2 + (ad + bc)x + bd$$

Factoring

2. The *last* term of the trinomial is the product of the *last* terms of the binomials.

$$(ax + b)(cx + d) = acx^2 + (ad + bc)x + bd$$

3. The numerical coefficient of the middle term is the sum of the products of the innermost and outermost numerical coefficients of the binomials.

$$(ax + b)(cx + d) = acx^2 + (ad + bc)x + bd$$

6 Algebraic Fractions

Upon completion of this chapter, you will have:
1. Learned how to add algebraic fractions.
2. Learned how to subtract algebraic fractions.
3. Learned how to multiply algebraic fractions.
4. Learned how to divide algebraic fractions.

6-1 INTRODUCTION

As mentioned in Chapter 1, a fraction is an *indicated* division. That is, the division of x by y is indicated by the fraction x/y, but we are not expected to carry out the actual division process. For example, the fraction $5/8$ indicates the division of 5 by 8, but we are not necessarily interested in the fact that the actual division results in the decimal 0.625.

In the fraction x/y, x is called the **numerator,** and y is referred to as the **denominator.** Usually, these quantities are called the **terms** of the fraction. The diagonal line between the two terms is a **vinculum,** which, as you may recall from Chapter 2, is a symbol of grouping. In a fraction, the vinculum has the function of grouping together all the individual parts in the numerator as well as the denominator. For example, in the algebraic fraction

$$\frac{a - b}{c + d}$$

the vinculum indicates that the binomial $(a - b)$ is to be divided by the binomial $(c + d)$, where the parentheses here have the same grouping function as the vinculum.

The rules that apply to arithmetic fractions apply to algebraic fractions as well. In addition, certain new rules, ideas, and procedures will be introduced. Therefore, it is important that you understand the various operations associated with simple arithmetic fractions before proceeding. It is recommended, then, that you take the time at this point to review the material in Chapter 1 regarding arithmetic fractions.

6-2 PROPERTIES OF FRACTIONS

There are three signs associated with any fraction: the sign of the numerator, the sign of the denominator, and the sign of the entire fraction.

> **Rule 6-1**
> Any *two* of the three signs of a fraction can be changed without changing the value of the fraction.

Each of the fractions shown next would have the same value, $+0.6$, if we were actually to carry out the indicated division. You should verify this assertion for yourself.

$$+\frac{+3}{+5} = \frac{-3}{-5} = -\frac{-3}{+5} = -\frac{+3}{-5}$$

Example 6-1

Write three fractions equivalent to $\dfrac{V - v}{I + i}$, each having different signs.

$$\frac{-V + v}{-I - i} = -\frac{-V + v}{I + i} = -\frac{V - v}{-I - i}$$

Note that changing the sign of either numerator or denominator is equivalent to multiplying either binomial expression by negative one. That is, $(-1)(V - v) = -V + v$, and $(-1)(I + i) = -I - i$.

Algebraic Fractions

> **Rule 6–2**
>
> The same *nonzero* number can be used either to multiply or divide *both terms* of a fraction without changing its value.

Consider the fraction $6/32$. If we were to multiply *both* terms by 3, for example, we would obtain the equivalent fraction $18/96$. This can be verified by noting that $6/32 = 0.1875 = 18/96$. Moreover, if we divided *both* terms of $6/32$ by 2, we would obtain the fraction $3/16$, which is also equal to 0.1875. It is clear, then, that all three fractions are equal. That is,

$$\frac{6}{32} = \frac{18}{96} = \frac{3}{16} = 0.1875$$

You should be aware that operating on both terms of a fraction produces equivalent fractions only in the case of multiplication or division by a nonzero number. Equivalent fractions do not result if the terms are operated on by addition, subtraction, exponentiation, or root extraction. For example, consider the fraction $4/9$. Adding 2 to each term produces the fraction $6/11$. Note, however that whereas $4/9 \approx 0.444$, the fraction $6/11 \approx 0.545$. Obviously, the two fractions are not equivalent.

As a further example, if we extracted the square root of each term, we would obtain the new, nonequivalent fraction $2/3$, the approximate value of which is 0.667. It is apparent that this fraction is considerably larger than the original fraction, $4/9$.

> **Rule 6–3**
>
> Factors common to both numerator and denominator can be canceled.

For example, in the fraction $3r/5r$, the factor r may be canceled from both terms, leaving $3/5$. Cancellation is often indicated by drawing a diagonal line through the common factors, as shown below.

$$\frac{4V^2E^5}{12V^3E^2} = \frac{\cancel{2} \cdot \cancel{2} \cdot \cancel{V} \cdot \cancel{V} \cdot \cancel{E} \cdot \cancel{E} \cdot E \cdot E \cdot E}{\cancel{2} \cdot \cancel{2} \cdot 3 \cdot \cancel{V} \cdot \cancel{V} \cdot V \cdot \cancel{E} \cdot \cancel{E}} = \frac{E^3}{3V}$$

You should be aware that the word *cancellation* actually refers to the process of dividing both terms of the fraction by the same number. Thus, in the preceding example, we canceled the 2s by dividing both terms by 2, twice. Moreover, such cancellation results in the value of 1, and not 0, as many students erroneously believe. That is, $a/a = 1$, not 0.

Students frequently make the mistake of canceling *terms* in the numerator or denominator rather than *factors*. For example, it would be *incorrect* to cancel the R term in the fraction shown next, since R is a term of the numerator rather than a factor.

$$\frac{5\cancel{Z} + \cancel{R}}{\cancel{R}} = 5Z \qquad \text{Incorrect}$$

However, in the following fraction, R is a factor of the numerator and can be canceled as shown.

$$\frac{5Z\cancel{R}}{\cancel{R}} = 5Z$$

> **Rule 6-4**
> Fractions can be *reduced to lowest terms* by factoring out and canceling all factors common to both numerator and denominator.

For example, in the fraction

$$\frac{15X^2 - 3X}{5X - 1}$$

we note that the terms in the numerator have a common factor $3X$, which may be factored out, giving

$$\frac{3X(5X - 1)}{(5X - 1)}$$

We now observe that both numerator and denominator have the binomial expression $(5X - 1)$ as a common factor, which can be canceled, leaving $3X$ as the fraction reduced to lowest terms. Strictly speaking, a fraction is said to be reduced to lowest terms when the numerator and denominator have no factors in common other than 1.

Example 6-2

Reduce $\dfrac{V^2 - 4}{2V^2 - V - 6}$ to lowest terms.

Solution Since the numerator is the difference of squares, we can factor this expression as $(V - 2)(V + 2)$. Moreover, we note that the denominator can be factored as $(2V + 3)(V - 2)$. We now rewrite the fraction in factored form as

$$\frac{\cancel{(V-2)}(V + 2)}{(2V + 3)\cancel{(V-2)}} = \frac{V + 2}{2V + 3}$$

Therefore, since there are no factors common to both terms, the fraction is said to be in lowest terms. You should remember that a reduced fraction has a value exactly equivalent to the original fraction. This can be verified by letting $V = 3$, for example, in both the original and reduced fractions:

$$\frac{V^2 - 4}{2V^2 - V - 6} = \frac{(3)^2 - 4}{2(3)^2 - 3 - 6} = \frac{5}{9}$$

and

$$\frac{(V + 2)}{(2V + 3)} = \frac{(3 + 2)}{(6 + 3)} = \frac{5}{9}$$

Exercise Set 6-1

Questions

1. Using your own words, define the following terms.
   a. Numerator
   b. Denominator

Algebraic Fractions

    c. Terms
    d. Factors
    e. Vinculum
2. Why is a fraction referred to as an indicated division?
3. Look up the definition of the word *vinculum* in an unabridged dictionary. What is the etymology (derivation) of the word, and why is the term appropriate to fractions?

Problems

Write three fractions having different signs that are equivalent to the given fraction.

1. $\dfrac{1}{8}$

2. $\dfrac{-3}{8}$

3. $\dfrac{-R_2 R_t}{-R_t + R_2}$

4. $\dfrac{-I^2 R}{P_1 - P_2}$

5. $\dfrac{-V - e}{I - i}$

6. $-\dfrac{t - k}{tk}$

7. $-\dfrac{k}{2 - p}$

8. $-\dfrac{L - 3}{L + 3}$

Reduce the following fractions to lowest terms.

9. $\dfrac{600}{180}$

10. $\dfrac{1875}{10{,}000}$

11. $\dfrac{144 V^2 R^3}{6 V^3 R}$

12. $\dfrac{512 s^2 t^3}{1024 s^3 t^5}$

13. $\dfrac{V^2 + 2Ve + e^2}{V^2 - e^2}$

14. $\dfrac{1 + 2x + x^2}{1 - x^2}$

15. $\dfrac{x^2 - 4}{5x^2 + 7x - 6}$

16. $\dfrac{4 + 4a + a^2}{24 + 10a - a^2}$

17. $\dfrac{18 I^2 - 3Ii - 3i^2}{6I - 3i}$

18. $\dfrac{10a^2 - 90}{5a^2 - 10a - 75}$

19. $\dfrac{(a + k)^2 - (a - k)^2}{-4ak}$

20. $\dfrac{(1 - V)^2 - (1 + V)^2}{4V^2}$

6-3 ADDITION AND SUBTRACTION

Algebraic fractions having a common denominator can be added or subtracted in the same way as arithmetic fractions. That is, the sum or difference is written over the common denominator. For example, to add the two fractions $3x/b$ and $7x/b$, we write

$$\frac{3x}{b} + \frac{7x}{b} = \frac{3x + 7x}{b} = \frac{10x}{b}$$

Similarly, to subtract $8V/k$ from $2V/k$, we write

$$\frac{2V}{k} - \frac{8V}{k} = \frac{2V - 8V}{k} = \frac{-6V}{k}$$

Example 6-3 Find the difference $\dfrac{P}{t} - \dfrac{Q - R}{t}$.

Solution Since we already have a common denominator, we write

$$\frac{P}{t} - \frac{Q - R}{t} = \frac{P - (Q - R)}{t} = \frac{P - Q + R}{t}$$

Many students forget that the vinculum in the second fraction in Example 6-3 is a symbol of grouping and mistakenly write the difference as

$$\frac{P - Q - R}{t}$$

You will recall from Chapter 2 that any symbol of grouping preceded by a negative sign can be removed *only* if the sign of each term enclosed is changed. Remembering that the vinculum of a fraction is a grouping symbol will prevent serious errors.

In those cases of addition and subtraction where a common denominator does not already exist, one must be supplied. For example, consider the sum of the two fractions

$$\frac{R}{4} + \frac{G}{5}$$

We can supply the needed common denominator simply by multiplying the terms of each fraction by the denominator of the other fraction. That is

$$\frac{R}{4} + \frac{G}{5} = \left(\frac{R}{4} \times \frac{5}{5}\right) + \left(\frac{G}{5} \times \frac{4}{4}\right) = \frac{5R + 4G}{20}$$

Algebraic Fractions

Example 6–4

Find the sum $\dfrac{3Z}{16Y^3} + \dfrac{4X}{64Y^2}$.

Solution Here we see that the fractions do not have the same denominator. However, we can proceed in either of two ways. In the first method, we can form a common denominator by multiplying the terms of each fraction by the denominator of the other, as we did in Example 6–3. Although this method will work, we end up having to handle some rather large, cumbersome numbers and must finally remove all common factors in order to get a simplified version of the correct answer. Proceeding in this way, we obtain an answer of

$$\frac{192Y^2Z + 64XY^3}{1024Y^5}$$

which can be factored and written as

$$\frac{16Y^2(12Z + 4XY)}{16Y^2(64Y^3)}$$

When common factors are canceled, we obtain the simplified and equally correct answer

$$\frac{12Z + 4XY}{64Y^3}$$

In the second method, we eliminate the need of working with redundant factors and large, cumbersome numbers by selecting a common denominator that is the smallest possible quantity evenly divisible by both denominators of the given fractions. In our example, it can be seen by inspection that this number is $64Y^3$. We call the denominator formed in this way the **lowest common denominator (LCD)**. Its use, while not required, saves considerable time and reduces the possibility of computational errors. Therefore, we will adopt this method in general for all algebraic fractions encountered in this book.

Using the LCD of $64Y^3$ allows us to write

$$\frac{3Z}{16Y^3} + \frac{4X}{64Y^2} = \frac{4(3Z) + Y(4X)}{64Y^3} = \frac{12Z + 4XY}{64Y^3}$$

The same mechanics are used to handle algebraic fractions as arithmetic fractions, as shown by Rule 6–5.

Rule 6–5

1. Find the LCD.
2. Divide the LCD by the denominator of the first fraction and then multiply the quotient by the numerator of that same fraction.
3. Write the product formed in step 2 as the first term of the new numerator.
4. Repeat steps 2 and 3 for the other fractions involved.
5. Find the algebraic sum or difference of the products formed in steps 2 through 4 and write the result over the LCD.

Example 6-5

Find the sum $\dfrac{2P}{P^2 - Q^2} + \dfrac{3Q}{2P^2 + 3PQ + Q^2}$.

Solution We begin by factoring both denominators, thus making it easier to determine the LCD.

$$\dfrac{2P}{(P - Q)(P + Q)} + \dfrac{3Q}{(P + Q)(2P + Q)}$$

The smallest quantity that is evenly divisible by both denominators (i.e., the LCD) is seen to be $(P - Q)(P + Q)(2P + Q)$. Therefore, we write

$$\dfrac{2P(2P + Q) + 3Q(P - Q)}{(P - Q)(P + Q)(2P + Q)} = \dfrac{4P^2 + 5PQ - 3Q^2}{2P^3 + P^2Q - 2PQ^2 - Q^3}$$

The solution to fractional addition or subtraction can always be checked by substituting simple numerical quantities for the literal numbers in both the original and final fractions and comparing the results. For example, if $P = 2$, and $Q = 3$, upon substituting in the original fraction, we obtain

$$\dfrac{2P}{P^2 - Q^2} + \dfrac{3Q}{2P^2 + 3PQ + Q^2} = \dfrac{4}{4 - 9} + \dfrac{9}{8 + 18 + 9}$$

$$= \dfrac{4}{-5} + \dfrac{9}{35}$$

$$= \dfrac{-28 + 9}{35} = \dfrac{-19}{35}, \text{ or } -\dfrac{19}{35}$$

On substituting in the final sum, we obtain

$$\dfrac{4P^2 + 5PQ - 3Q^2}{2P^3 + P^2Q - 2PQ^2 - Q^3} = \dfrac{16 + 30 - 27}{16 + 12 - 36 - 27}$$

$$= \dfrac{19}{-35}, \text{ or } -\dfrac{19}{35}$$

Since the two answers are equivalent, we conclude that our sum is correct.

Exercise Set 6-2

Questions

1. In your own words, define the term LCD.
2. Is it necessary to use the LCD to add or subtract fractions? Explain your answer.

Problems

Perform the indicated operations on the following fractions and check your results using numerical substitution.

1. $\dfrac{V}{2e} - \dfrac{-e + V}{2e}$

2. $\dfrac{V^2}{R} - \dfrac{2V^2 - 25}{R}$

Algebraic Fractions

3. $\dfrac{P^2}{R} - \dfrac{9P^2 - 30}{R}$

4. $\dfrac{Q^2 - 5}{P} + \dfrac{12 - 2Q^2}{P}$

5. $\dfrac{2V^2}{15I} + \dfrac{3V}{12I^2}$

6. $\dfrac{2K}{4H} + \dfrac{2K^3}{5H^2}$

7. $\dfrac{1}{8m^3n} - \dfrac{3}{100m^2n^2} + \dfrac{5}{40mn^3}$

8. $\dfrac{4}{10x^2y} + \dfrac{3}{25xy^3} - \dfrac{-2}{150x^3y^2}$

9. $\dfrac{2}{3Z^5Y} + \dfrac{4}{2Z^2Y^3} - \dfrac{3}{14ZY^3}$

10. $\dfrac{-1}{5a^2b^3} - \dfrac{2}{12a^3b^3} + \dfrac{-4}{3ab^2}$

11. $\dfrac{1}{V^2 - Ve} + \dfrac{1}{e^2 - Ve}$

12. $\dfrac{3}{t^2 - t} - \dfrac{-1}{1 - t^2}$

13. $\dfrac{2}{Z^3} + \dfrac{1}{Z^2 - 9Z} + \dfrac{1}{9Z^2 - Z^3}$

14. $\dfrac{4}{a - b} - \dfrac{1}{b - a} + \dfrac{5}{a^2 - b^2}$

15. $\dfrac{R - r}{R + r} + \dfrac{r - R}{R - r} + \dfrac{R + r}{r - R}$

16. $\dfrac{u + v}{u - v} - \dfrac{v - u}{-v + u} + \dfrac{u - v}{v - u}$

17. $\dfrac{x^2 - 2x + 1}{x^2 + 2x + 1} + \dfrac{x + 1}{x - 1}$

18. $\dfrac{4 - 4p + p^2}{4 + 4p + p^2} + \dfrac{2 - p}{p - 2}$

19. $\dfrac{5t + 1}{2 - 2t} + \dfrac{11t - 2}{3t - 3}$

20. $\dfrac{x - 1}{5x + 5} - \dfrac{x + 1}{-3 + 3x}$

6-4 MULTIPLICATION AND DIVISION

In Chapter 3, we stated that multiplication could be viewed as a time-saving tool for addition. For example, 2 × 6 can be regarded as 2 added 6 times. In other words, 2 + 2 + 2 + 2 + 2 + 2 = 12 has the same meaning as 2 × 6 = 12. This idea can also be extended to include multiplication of fractions. For example, consider the multiplication given by $\tfrac{2}{3} \times \tfrac{5}{8}$. Although it makes no sense to regard this as two-thirds added five-eighths of a time, it does make sense to say that this is equivalent to $\tfrac{1}{24}$ added 10 times, so $\tfrac{2}{3} \times \tfrac{5}{8} = \tfrac{10}{24}$.

The rules for multiplication of arithmetic fractions apply to multiplication of algebraic fractions. That is, to multiply two (or more) fractions, multiply all their numerators together and then multiply all their denominators together. The resultant fraction is composed of the products of the numerators written over the products of the denominators. For example, $\tfrac{a}{b} \times \tfrac{c}{d} = \tfrac{ac}{bd}$.

> **Rule 6–6**
> 1. Find the product of all the numerators.
> 2. Find the product of all the denominators.
> 3. Form the final answer by writing the result of step 1 over the result of step 2.

Example 6–6 Find the product $\dfrac{18YZ^5}{6Y^2Z} \times \dfrac{5Y^2}{30Z}$.

Solution We multiply the numerators together, forming $90Y^3Z^5$. Next, we obtain the product of the denominators, $180Y^2Z^2$. Finally, the answer is written as

$$\frac{90Y^3Z^5}{180Y^2Z^2} = \frac{YZ^3}{2}$$

Note that the laws of exponents presented in Chapter 3 were used in forming the products of the new terms as well as simplifying the resulting fraction.

It was mentioned in an earlier section that both terms of a fraction could be multiplied by the same nonzero quantity without changing the value of the fraction. We use this fact to derive a simple method for dividing one fraction by another. For example, consider this division:

$$\frac{\dfrac{a}{b}}{\dfrac{c}{d}}$$

If we multiply both terms of this complex fraction by the product of the individual denominators bd, we obtain the following interesting result.

$$\frac{\dfrac{a}{\cancel{b}}(\cancel{b}d)}{\dfrac{c}{\cancel{d}}(b\cancel{d})}$$

Algebraic Fractions

Since the factors of b cancel out in the numerator and the factors of d cancel in the denominator, we are left with the simple fraction

$$\frac{ad}{bc}$$

which can be written as the product of the two fractions

$$\frac{a}{b} \times \frac{d}{c}$$

Careful observation of this result shows that this product is formed from the original terms of the complex fraction with the denominator inverted. Because of this simple relationship, we can state the already-familiar rule: To divide fractions, invert the denominator and multiply.

> **Rule 6–7**
> 1. Invert the fraction used as the divisor.
> 2. Multiply according to the rules for fractions.

Example 6–7 Find the quotient of these two fractions:

$$\frac{35Q^2C}{18V} \div \frac{7QC}{3V^2}$$

Solution From Rule 6–7, we write

$$\frac{35Q^2C}{18V} \times \frac{3V^2}{7QC} = \frac{5QV}{6}$$

Example 6–8 Divide: $\dfrac{1}{r-g} \div \dfrac{r^2 - g^2}{r^2 + 2rg + g^2}$.

Solution We can factor the terms of the second fraction and then invert the divisor to obtain

$$\frac{1}{r-g} \times \frac{(r+g)(r\cancel{+g})}{(r-g)(r\cancel{+g})}$$

Canceling common factors and expanding the denominator results in

$$\frac{r+g}{r^2 - 2rg + g^2}$$

Example 6–9 Simplify: $\dfrac{1 - \dfrac{Z-X}{Z+X}}{1 + \dfrac{Z-X}{Z+X}}$.

Solution Since this is a **complex fraction** (i.e., both terms are fractions), we begin by simplifying the numerator and denominator.

$$\frac{\frac{Z+X-Z+X}{Z+X}}{\frac{Z+X+Z-X}{Z+X}} = \frac{\frac{2X}{Z+X}}{\frac{2Z}{Z+X}}$$

$$= \frac{2X}{Z+X} \times \frac{Z+X}{2Z} = \frac{X}{Z}$$

Exercise Set 6–3

Problems

Perform the indicated operation, and simplify your answer where necessary.

1. $\dfrac{2V^2 e}{5I} \times \dfrac{3Ve^3}{2i^2}$

2. $\dfrac{25PQ}{39C} \times \dfrac{3PQ}{5C}$

3. $\dfrac{1}{2}(R + r) \times \dfrac{1}{4}(R + r)$

4. $\dfrac{3}{5}(b + y) \times \dfrac{1}{3}(b - y)$

5. $\dfrac{68P^2}{t^3} \times \dfrac{t^2}{17P}$

6. $\dfrac{105Z}{V^3} \times \dfrac{V}{35Z^3}$

7. $\dfrac{m^2 + 2mn + n^2}{m^2 - n^2} \times \dfrac{n - m}{n + m}$

8. $\dfrac{a^2 - b^2}{a^2 - 2ab + b^2} \times \dfrac{a - b}{a + b}$

9. $\dfrac{5C^2 + 11C - 12}{2C - 8} \div \dfrac{3 + C}{-4 + C}$

10. $\dfrac{3a^2 + 5ax - 2x^2}{a^2 - x^2} \div \dfrac{a + 2x}{a + x}$

11. $\dfrac{3y^2 + 14y - 5}{1} \div \dfrac{y + 5}{-1 + y}$

12. $\dfrac{-v^2 - v + 2}{4} \div \dfrac{12 - 8v - 4v^2}{1}$

13. $\dfrac{3h^2 - 3k^2}{9h^2 + 18hk + 9k^2} \div \dfrac{h - k}{h + k}$

14. $\dfrac{I^2 - i^2}{I^2 + 2Ii + i^2} \div \dfrac{i + I}{(i + I)^3}$

Algebraic Fractions

15. $\dfrac{\dfrac{49a^2bc^2}{24xy^2}}{\dfrac{14abc}{9y}}$

16. $\dfrac{\dfrac{63V^3R^2}{36r}}{\dfrac{9VR^2}{6r^2}}$

17. $\dfrac{\dfrac{1}{V^2-1} - \dfrac{1}{V+1}}{\dfrac{1}{V-1} + \dfrac{1}{V^2-1}}$

18. $\dfrac{\dfrac{1}{t^2-1} + \dfrac{1}{t+1}}{\dfrac{1}{(t-1)^2} - \dfrac{1}{t-1}}$

19. $\dfrac{1 - \dfrac{m^2}{n^2}}{1 - \dfrac{2m}{-n} + \dfrac{m^2}{n^2}}$

20. $\dfrac{1 - \dfrac{P^2}{Q^2}}{\left(\dfrac{P}{Q} + 1\right)^2}$

6-5 OPERATIONS WITH FRACTIONS CONTAINING RADICALS

Frequently, radicals are contained in the terms of fractions. These radicals tend to complicate the ordinary operations with fractions that were previously discussed. Therefore, various strategies have been developed to lessen their effect and make computation easier. The following represent a few such strategies.

Rationalizing the Denominator

For a variety of reasons that will become evident later, it is easier to operate on fractions that do not have radicals in their denominators. A common method of clearing such radicals is called **rationalizing the denominator.** It works as follows. Consider the fraction

$$\frac{3}{\sqrt{5}}$$

Fundamentals of Algebra

By multiplying both numerator and denominator by $\sqrt{5}$, we obtain a fraction whose denominator does not contain a radical. That is,

$$\frac{3}{\sqrt{5}} \times \frac{\sqrt{5}}{\sqrt{5}} = \frac{3\sqrt{5}}{5}$$

Example 6-10 Simplify the fraction $\dfrac{V^2}{\sqrt{12V^2}}$.

Solution From the methods of Chapter 5, we can remove factors from the radicand and write

$$\frac{V^2}{\sqrt{12V^2}} = \frac{V^2}{2V\sqrt{3}} = \frac{V}{2\sqrt{3}}$$

We now rationalize the denominator, obtaining

$$\frac{V}{2\sqrt{3}} \times \frac{\sqrt{3}}{\sqrt{3}} = \frac{V\sqrt{3}}{6}$$

Example 6-11 Simplify $\dfrac{2R}{k}\sqrt{\dfrac{k^3X}{R}}$.

Solution We can write the fraction within the radical symbol as the quotient of the two radicals

$$\frac{\sqrt{k^3X}}{\sqrt{R}}$$

This is made possible by using the laws of exponents developed in Chapters 3 and 5 and may be shown as follows:

$$\frac{\sqrt{a}}{\sqrt{b}} = \frac{a^{1/2}}{b^{1/2}} = a^{1/2}b^{-1/2} = a^{1/2}(b^{-1})^{1/2} = (ab^{-1})^{1/2} = \sqrt{ab^{-1}} = \sqrt{\frac{a}{b}}$$

Removing factors from the radicand of the original fraction, we write

$$\frac{2R}{k}\sqrt{\frac{k^3X}{R}} = \frac{2R}{\not{k}} \times \frac{\not{k}\sqrt{kX}}{\sqrt{R}} = \frac{2R}{1} \times \frac{\sqrt{kX}}{\sqrt{R}}$$

Upon rationalizing the denominator, we get

$$\frac{2R}{1} \times \frac{\sqrt{kX}}{\sqrt{R}} \times \frac{\sqrt{R}}{\sqrt{R}} = \frac{2\not{R}\sqrt{kRX}}{\not{R}} = 2\sqrt{kRX}$$

Addition and Subtraction

Radicals having the same index and radicand can be treated as like terms and added or subtracted in the same way as literal numbers. For example, $8\sqrt{3} + 2\sqrt{3} - 4\sqrt{3} = 6\sqrt{3}$.

Algebraic Fractions

Example 6-12

Simplify $2\sqrt{5} + \sqrt{20} + \dfrac{\sqrt{80}}{2}$.

Solution We can remove appropriate factors from the radicands of the last two terms and write

$$2\sqrt{5} + \sqrt{20} + \dfrac{\sqrt{80}}{2} = 2\sqrt{5} + 2\sqrt{5} + \dfrac{4\sqrt{5}}{2}$$
$$= 2\sqrt{5} + 2\sqrt{5} + 2\sqrt{5} = 6\sqrt{5}$$

Example 6-13

Simplify $4\sqrt{\dfrac{1}{8}} + 5\sqrt{\dfrac{1}{2}} + 3\sqrt{2}$.

$$4\sqrt{\dfrac{1}{8}} + 5\sqrt{\dfrac{1}{2}} + 3\sqrt{2} = \dfrac{4}{2}\sqrt{\dfrac{1}{2}} + 5\sqrt{\dfrac{1}{2}} + \dfrac{3\sqrt{2}}{1} \times \dfrac{\sqrt{2}}{\sqrt{2}}$$
$$= 2\sqrt{\dfrac{1}{2}} + 5\sqrt{\dfrac{1}{2}} + \dfrac{6}{\sqrt{2}}$$
$$= 2\sqrt{\dfrac{1}{2}} + 5\sqrt{\dfrac{1}{2}} + 6\sqrt{\dfrac{1}{2}}$$
$$= 13\sqrt{\dfrac{1}{2}} = \dfrac{13}{\sqrt{2}} \times \dfrac{\sqrt{2}}{\sqrt{2}} = \dfrac{13\sqrt{2}}{2}$$

Multiplication and Division

The multiplication of radicals presents no special problems. You may recall from Chapter 5 the assertion that factors contained within a radicand can be written as the product of individual radicals. That is, $\sqrt{abc} = \sqrt{a} \cdot \sqrt{b} \cdot \sqrt{c}$. This rather simple conclusion follows from the exponential law for the power of a product developed in Chapter 3.

Example 6-14

Find the product $(5\sqrt{5} - 3\sqrt{2})(2\sqrt{5} + 4\sqrt{2})$.

Solution We proceed as in the ordinary multiplication of binomials.

$$(5\sqrt{5} - 3\sqrt{2})(2\sqrt{5} + 4\sqrt{2}) = 10(5) + 20\sqrt{10} - 6\sqrt{10} - 12(2)$$
$$= 50 + 14\sqrt{10} - 24$$
$$= 26 + 14\sqrt{10}$$

In Chapter 5, it was stated that one of the special products was of the form $(a - b)(a + b) = a^2 - b^2$. We can use this fact to simplify fractions with radical terms.

Example 6-15

Simplify the fraction $\dfrac{1}{3 - \sqrt{2}}$.

Solution If we multiply both numerator and denominator by $3 + \sqrt{2}$, we get

$$\dfrac{1}{3 - \sqrt{2}} \times \dfrac{3 + \sqrt{2}}{3 + \sqrt{2}} = \dfrac{3 + \sqrt{2}}{9 - 2} = \dfrac{3 + \sqrt{2}}{7}$$

Exercise Set 6-4

Problems

Simplify the following expressions.

1. $\sqrt{\dfrac{5}{4e}}$
2. $\dfrac{9k^3}{\sqrt{27k^2}}$
3. $\sqrt{\dfrac{3}{2a^2}}$
4. $\dfrac{cx^2 - 2c^2x + c^3}{\sqrt{(x - c)^2 c}}$
5. $2Z\sqrt{5} - Z\sqrt{5} + 5Z\sqrt{5}$
6. $\sqrt{8Q} - 3\sqrt{2Q} + \sqrt{18Q}$
7. $A\sqrt{2A^3} - 3\sqrt{8A^5} + A\sqrt{72A^3}$
8. $P\sqrt{9P^3Q^2} - 5P^2\sqrt{PQ^2} + 6Q\sqrt{P^5}$
9. $(6 + \sqrt{2e})(4 + \sqrt{2e})$
10. $(\sqrt{5} + 2\sqrt{2})^2$
11. $(3\sqrt{I} + \sqrt{i})^2$
12. $\dfrac{3}{\sqrt{3} - 5}$
13. $\dfrac{2\sqrt{3} + \sqrt{2}}{\sqrt{3} - \sqrt{2}}$
14. $\dfrac{2\sqrt{m} - n}{\sqrt{m} - n}$
15. $\dfrac{x + 2\sqrt{y}}{x - 2\sqrt{y}}$

Key Terms

numerator
denominator
terms

vinculum
lowest common denominator (LCD)

complex fraction
rationalizing the denominator

Algebraic Fractions

Important Properties of Fractions

Rule 6-1 Any two of the three signs of a fraction can be changed without changing the value of the fraction.

Rule 6-2 The same *nonzero* number can be used either to multiply or divide *both terms* of a fraction without changing its value.

Rule 6-3 Factors common to both numerator and denominator can be canceled.

Rule 6-4 Fractions can be reduced to lowest terms by factoring out and canceling all factors common to both numerator and denominator.

Important Rules for Operating on Fractions

Rule for Adding or Subtracting Fractions

Rule 6-5

1. Find the LCD.
2. Divide the LCD by the denominator of the first fraction and then multiply the quotient by the numerator of that same fraction.
3. Write the product formed in step 2 as the first term of the new numerator.
4. Repeat steps 2 and 3 for the other fractions involved.
5. Find the algebraic sum or difference of the products formed in steps 3 and 4 and write the result over the LCD.

Rule for Multiplying Fractions

Rule 6-6

1. Find the product of all the numerators.
2. Find the product of all the denominators.
3. Form the final answer by writing the result of step 1 over the result of step 2.

Rule for Dividing Fractions

Rule 6-7

1. Invert the fraction used as the divisor.
2. Multiply according to the rules for fractions.

PART THREE
Elementary Applications of Algebra to Electronics

7
Series and Parallel Circuits

7 Series and Parallel Circuits

Upon completion of this chapter, you will have:

1. Learned how to apply elementary algebraic techniques to the analysis of a variety of electronic circuits.
2. Learned how to deal with concepts of direct and inverse variation.
3. Learned the method of analysis of units.

7-1 INTRODUCTION

Our primary purpose here is to utilize fundamental electrical-circuit theory to enhance our understanding of algebraic principles. *It is not our intention to teach electronic circuit theory.* Any such learning is incidental to our main objective. Nonetheless, you will discover that as you approach the study of electronics from a mathematical point of view, your understanding of both subjects will increase in breadth as well as depth. It is assumed that either you are taking or have taken a first course in basic electronics at the college level.

7-2 THE SERIES CIRCUIT

Perhaps the most fundamental relationship in elementary electronics was that discovered in 1827 by the German physicist Georg Simon Ohm. This relationship is referred to as *Ohm's law* and is often expressed mathematically by the equation $I = V/R$, where I represents the current; V (formerly E), the voltage across the resistance; and R, the ohmic value of the circuit resistance. Expressed in words, the relationship states that the current in a circuit changes in direct proportion with the voltage, *provided the resistance remains constant.* This last proviso is important, since if V and R were *both* increased or decreased together in the same ratio, the value of I would remain constant. If, for example, both V and R were increased fivefold, we would have the fraction $5V/5R$, and the 5s would cancel, leaving the same value of I as we had originally. Moreover, if R changed at all as V changed, the current would not change in direct proportion but would obey some other law. In stating Ohm's law, then, many beginning students fail to mention this important restriction, thus showing that they do not completely understand Ohm's discovery.

Whenever a scientific relationship like Ohm's law is expressed by an equation, we call the equation a **formula.** Formulas can be regarded as a shorthand method of expressing the relationship between various physical quantities. From our discussion of equations in Chapter 4, it is apparent that Ohm's law can be expressed in either of two additional ways as $V = IR$ or $R = V/I$. We will now see how these formulas are applied to a simple series circuit.

As you are probably aware, a series circuit can be defined as one in which there is only one path for the current (electrons) to flow. This is another way of saying that the current is the same in every part of the circuit. Figure 7–1 shows a simple series circuit and illustrates the fact that there is but one single path through which the current can flow. The voltage source (shown here as a battery) provides the electrical force (V) that pushes the current (I) out of the negative battery terminal, through the resistor (R), and finally back to the positive terminal of the battery.

Figure 7–1

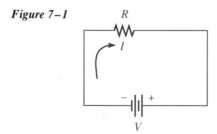

Series and Parallel Circuits

Let us assume for the moment that the value of the resistor R is 25 Ω, and the magnitude of the voltage source V is 4 V. We can calculate the current through the circuit using Ohm's law as follows:

$$I = \frac{V}{R} = \frac{4}{25} = 0.16 \text{ A}$$

In electronics, we use the uppercase Greek letter omega (Ω) to represent the unit ohms. The unit for direct current (DC) volts is abbreviated using the uppercase letter V. Finally, the unit of direct current intensity I is frequently abbreviated with the capital letter A, for amperes. These ideas are summarized in Table 7-1.

Table 7-1
Basic electrical quantities and their units

Electrical Quantity	Symbol	Units	Unit Abbreviation
Voltage	V	Volts	V
Resistance	R	Ohms	Ω
Current	I	Amperes	A

We can now rewrite the preceding expression using the appropriate units for each quantity.

$$I = \frac{4 \text{ V}}{25 \text{ }\Omega} = 0.16 \text{ A}$$

Example 7-1

Find the value of the resistance in Figure 7-1 if the voltage is 10 V and the current is 2.5 A.

Solution As stated earlier, Ohm's law can also be expressed as $R = V/I$. Therefore, we write

$$R = \frac{V}{I} = \frac{10 \text{ V}}{2.5 \text{ A}} = 4 \text{ }\Omega$$

Example 7-2

What voltage is required to send a current of 150 mA through a resistance of 20 kΩ?

Solution The unit mA stands for milliamperes, and the unit kΩ means kilohms. (See Table 3-1 for metric prefixes and their powers of ten.) Using yet another expression for Ohm's law, we write $V = IR$, remembering that juxtaposition of literal terms means to multiply. That is, the right member of the equation (IR) indicates that the literal factors I and R are to be multiplied together. Therefore, using the appropriate power of ten along with scientific notation, we write

$$V = IR = 150 \text{ mA} \times 20 \text{ k}\Omega = 1.5 \times 10^{-1} \text{ A} \times 2 \times 10^4 \text{ }\Omega$$
$$= 3 \times 10^{-1+4} = 3000 \text{ V}$$

We can also write the values out completely without powers of ten as 0.15 A × 20,000 Ω = 3000 V.

Figure 7–2 illustrates the fact that the elements of a series circuit are additive. In other words, the total circuit resistance (R_t) is the sum of the individual resistances, and the total voltage (V_t) is the sum of the separate sources.

Figure 7–2

$$R_t = R_1 + R_2$$

$$V_t = V_1 + V_2$$

If you want to find the total current I using the formula $I = V/R$, you have to take into consideration the total voltage ($V_1 + V_2$) as well as the total resistance ($R_1 + R_2$). That is, you would write

$$I = \frac{V_t}{R_t} = \frac{V_1 + V_2}{R_1 + R_2}$$

Note that since the vinculum is a symbol of grouping, it is not necessary to use parentheses to group the individual resistances or voltages together. If, however, you wanted to write this expression in a single line, as would be the case in a BASIC program, you would use a **solidus fraction** (fraction written with a slant line /) and make appropriate use of the parentheses. Therefore, the preceding fraction would be written in solidus form as

$$I = (V_1 + V_2)/(R_1 + R_2)$$

Note in Figure 7–2 that the two voltage sources (V_1 and V_2) are connected such that the negative terminal (−) of V_1 is connected to the positive terminal (+) of V_2. The two sources are in series and are said to be connected **series-aiding.** That is, the voltages aid one another in producing the total circuit voltage. Suppose, for example, that $V_1 = 3$ V and $V_2 = 5$ V. Since they are connected series-aiding, the total circuit voltage is $V_t = V_1 + V_2 = 3 + 5 = 8$ V. Consider now the circuit shown in Figure 7–3a.

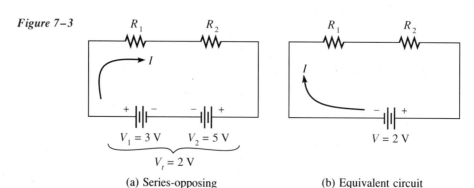

Figure 7–3

(a) Series-opposing   (b) Equivalent circuit

Series and Parallel Circuits

We note that the only difference between this circuit and the one shown in Figure 7–2 is that voltage source V_2 has been reversed. That is, the negative terminals of both sources are connected together. Since the electrons attempting to leave the negative terminal of V_1 are repelled by the negative charge of V_2 and vice versa, it is apparent that the two voltages oppose each other. This type of series connection is said to be **series-opposing,** and the total voltage is the *arithmetic difference* of the two voltage sources. Note, too, that the direction of current flow is opposite that shown in Figure 7–2. This is to be expected, since V_2 is stronger than V_1 by 2 V. The equivalent circuit is shown in Figure 7–3b.

Example 7–3 Find the total current in the circuit shown in Figure 7–4.

Figure 7–4

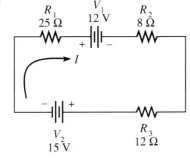

Solution The sum of the resistances is 45 Ω. Since the two voltage sources are connected series-aiding, their voltages add to 27 V. From Ohm's law,

$$I = \frac{V_t}{R_t} = \frac{V_1 + V_2}{R_1 + R_2 + R_3}$$

$$= \frac{27 \text{ V}}{45 \text{ }\Omega} = 0.6 \text{ A, or } 600 \text{ mA}$$

Note that since this is a series circuit, it does not matter where the resistors are placed relative to one another in the series string. Moreover, as long as the relative polarity of the individual batteries is maintained, it is immaterial where they are located in the circuit. Figure 7–4 might just as easily have been drawn as Figure 7–5 is.

Figure 7–5

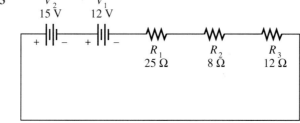

Example 7–4

In Example 7–3, if V_1 and R_2 are reversed, find (a) the total circuit resistance; (b) the total circuit voltage; and (c) the total current.

Solution a. Since resistors are not polarized and are passive circuit elements, reversing any of the resistors does not affect the circuit in any way. The total resistance is still 45 Ω.

  b. V_1 and V_2 are now connected series-opposing. Therefore, the total circuit voltage is 3 V, and the equivalent circuit is shown in Figure 7–6. Note that we have combined ("lumped") all the individual resistors into a single equivalent resistance of 45 Ω.

  c. From Figure 7–6, we see that the circuit current can be obtained from Ohm's law as $I = V_t/R_t = (3\text{ V})/(45\text{ Ω}) \approx 0.067$ A, or 67 mA.

Figure 7–6

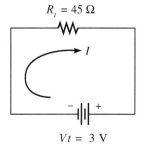

$R_t = 45\ \Omega$

$Vt = 3\text{ V}$

7–3 POWER IN SERIES CIRCUITS

Resistors dissipate power in the form of heat. The power, measured in watts, is given by $P = VI$. For example, if 2 A flow through a resistance under an electromotive force (voltage) of 5 V, the power is $P = VI = 5\text{ V} \times 2\text{ A} = 10\text{ W}$. The power dissipated in a series circuit is the sum of the individual powers.

From Axiom 4–5, we know that we can substitute an equivalent expression into an equation without changing its equality. Therefore, in the equation $P = VI$, we may substitute the expression $I = V/R$ for I, giving

$$P = VI = V\left(\frac{V}{R}\right) = \frac{V^2}{R}$$

Similarly, if we substitute the relationship $V = IR$ for V in $P = VI$, we obtain

$$P = VI = (IR)I = I^2R$$

We see, then, that by making use of some of our simple algebraic ideas learned earlier, we can derive several useful expressions for the power dissipated in a circuit.

Example 7–5

In Figure 7–7, the 8 Ω resistor dissipates 4.5 W. Find the voltage of the battery.

Solution From the equation $P = V^2/R$, we may solve for V, obtaining

$$V = \sqrt{PR} = \sqrt{(4.5\text{ W}) \times (8\text{ Ω})} = 6\text{ V}$$

Series and Parallel Circuits

Note that only the positive (principal) square root of the result is used, since a negative voltage has no meaning in this context.

Figure 7-7

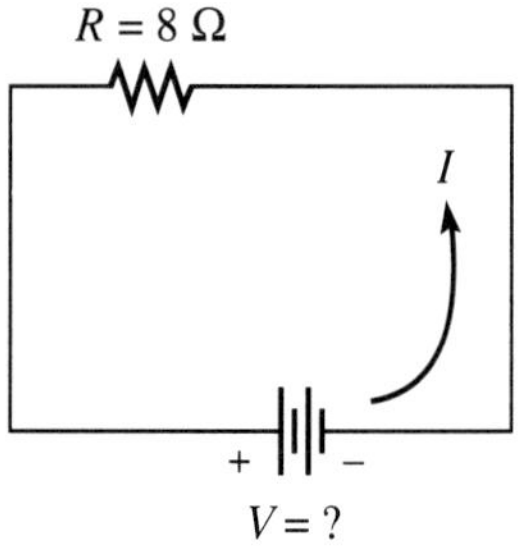

$R = 8\,\Omega$

$V = ?$

Example 7-6

In the circuit of Figure 7-7, find the value of the current.

Solution From $P = I^2R$, we solve for I, obtaining

$$I = \sqrt{\frac{P}{R}} = \sqrt{\frac{4.5\text{ W}}{8\,\Omega}} = 0.75\text{ A}$$

Again, we see that the negative square root has no meaning in the context of this problem. Therefore, we use only the principal square root of 0.75 as our answer.

As a check, note that the product of the voltage (6 V) obtained in Example 7-5 and the current (0.75 A) results in a power of 4.5 W, as originally specified.

7-4 SUMMATION NOTATION

As we have seen, the total resistance of a series circuit is the sum of the individual resistances. Expressed mathematically, $R_t = R_1 + R_2 + \cdots + R_n$, where n is the number of the last resistor in the string. If we have a circuit with a very large number of resistors in series, this equation can be somewhat cumbersome. For example, suppose there were 25 resistors in series. The preceding expression would require us to write their sum as $R_t = R_1 + R_2 + R_3 + R_4 + R_5 + R_6 + R_7 + R_8 + R_9 + R_{10} + R_{11} + R_{12} + R_{13} + R_{14} + R_{15} + R_{16} + R_{17} + R_{18} + R_{19} + R_{20} + R_{21} + R_{22} + R_{23} + R_{24} + R_{25}$. Fortunately, another method, called **summation notation**, provides us with a shorthand way to specify the sum of a large number of elements of any kind. Using this notation, the uppercase Greek letter sigma (Σ) stands for the summing operation. In particular, we write Rule 7-1 as shown.

Rule 7-1

$$R_t = \sum_{k=1}^{n} R_k = R_1 + R_2 + \cdots + R_n$$

which is read, "The sum of resistances from k equals 1 to n." In the example of 25 resistors, we write

$$R_t = \sum_{k=1}^{25} R_k$$

which has the same meaning as $R_t = R_1 + R_2 + \cdots + R_{25}$.

It is common in many situations to abbreviate this notation even further and omit the limits of summation (k and n) by writing

$$R_t = \Sigma R$$

In the previous section, it was shown that the total power dissipated in a series circuit was the sum of the individual powers. Moreover, in connection with Figure 7–2, it was shown that the voltages in a series circuit are additive. Using the summation notation, we can state these facts very compactly as

$$P_t = \sum_{k=1}^{n} P_k = P_1 + P_2 + \cdots + P_n \quad \text{or} \quad P_t = \Sigma P$$

and

$$V_t = \sum_{k=1}^{n} V_k = V_1 + V_2 + \cdots + V_n \quad \text{or} \quad V_t = \Sigma V$$

This last expression applies equally to the individual voltage drops across the resistors in a series string. For example, in Example 7–3, we had a total applied voltage of 27 V ($V_1 + V_2$) causing a current of 600 mA to flow through the three circuit resistors. We can use the expression $V = IR$ to find the individual voltage drops across the resistors.

$$V_1 = IR_1 = 0.6 \text{ A} \times 25 \text{ }\Omega = 15 \text{ V}$$
$$V_2 = IR_2 = 0.6 \text{ A} \times 8 \text{ }\Omega = 4.8 \text{ V}$$
$$V_3 = IR_3 = 0.6 \text{ A} \times 12 \text{ }\Omega = 7.2 \text{ V}$$

Therefore, we write $V_R = \sum_{k=1}^{n} V_k = V_1 + V_2 + \cdots + V_n$, where V_R represents the total voltage drop across all resistors. In the present case, $k = 1$, $n = 3$, and $V_R = \sum_{k=1}^{3} V_k = V_1 + V_2 + V_3 = 15 \text{ V} + 4.8 \text{ V} + 7.2 \text{ V} = 27 \text{ V}$. You will note that $V_R = V_t$.

That is, the sum of the individual voltage drops equals the total applied circuit voltage. This is always the case in a series circuit, and we will have occasion to refer to this fact in more detail when we study Kirchhoff's voltage law.

7–5 DIRECT AND INVERSE VARIATION

Ohm's law states, in part, that the current flowing in a circuit varies directly with the applied voltage. In other words, as the voltage increases, the current increases. This fact

Series and Parallel Circuits

is evident from the expression $I = V/R$, where, as V gets bigger, so does I. Conversely, as V gets smaller, so does I. We will assume that R remains constant.

Another way of expressing the fact that I **varies directly** with V (i.e., I is directly proportional to V) is with the use of the **proportionality symbol** ($\propto$). We can write $I \propto V$, which is read "I varies directly with V," or "I is directly proportional to V." The only way we can replace the proportionality symbol ($\propto$) with an equality symbol ($=$) is if we supply the **constant of proportionality** (k). In the statement of Ohm's law, $k = 1/R$. Indeed, the very definition of this law requires that R remain constant (see the statement of Ohm's law given earlier in this chapter). Therefore, we can now write

$$I \propto V \quad \text{or} \quad I = kV \quad \text{or} \quad I = \left(\frac{1}{R}\right)V \quad \text{or} \quad I = \frac{V}{R}$$

Example 7–7

For a given diameter of copper wire, the resistance R varies directly with (is directly proportional to) the length L. Hence, we write $R \propto L$. If we wish to place this general rule into the form of an equation, we need to know the value of the constant of proportionality k. We can then write $R = kL$. For example, suppose that for a given diameter of copper wire, the value of k is 0.0007 Ω/ft. We write

$$R = 0.0007L \ \Omega$$

The problem with Example 7–7 is that the resistance of a given type (material) of conductor not only depends on its length but also **varies inversely** with (is inversely proportional to) its cross-sectional area (A). We indicate these two facts, respectively, as

$$R \propto L \quad \text{and} \quad R \propto \frac{1}{A}$$

Note that in the case of direct variation, the determining factor (often called the *independent variable*) appears as a term in the numerator—that is, the L in the implied fraction $L/1$. On the other hand, in the case of inverse variation, the independent variable appears in the denominator of the fraction $1/A$. These fractional characteristics are always true with direct and inverse variation, and you should remember these properties.

We can combine the fact that R varies with both L and A in a single expression by writing

$$R \propto \frac{L}{A}$$

Moreover, if we supply the constant of proportionality (k), we can replace the proportionality symbol with an equal sign:

$$R = k\frac{L}{A}$$

For wires of all materials, k is called the *resistivity* (or specific resistance) and is generally symbolized by the lowercase Greek letter rho (ρ). We can now write the actual value of resistance for any type of wire of any length whose area and specific resistance are known:

$$R = \rho \frac{L}{A}$$

Note that since ρ is sensitive to temperature, tables of ρ are commonly specified at a particular ambient temperature, often 20°C, as shown in Table 7–2.

Table 7–2
Resistivity of various materials

Material	ρ at 20°C (CMIL·Ω/ft)
Aluminum	17
Constantan	295
Copper	10.4
Gold	14
Iron	58
Nichrome	676
Nickel	52
Silver	9.8
Tungsten	33.8

From Table 7–2, we notice that the units of ρ (CMIL·Ω/ft) appear somewhat unfamiliar. We can easily clarify this problem by observing that (1) wire sizes are commonly expressed in circular mils (CMIL), where one CMIL is the area of a wire 0.001 in. in diameter and (2) a particular diameter wire has a certain resistance in terms of ohms per foot. We conclude that any material 1/1000 in. in diameter and 1 ft long must have a resistance of so many ohms (Ω). Combining these facts, we are lead to the units of resistivity, CMIL·Ω/ft. Notice in the following example that all these units cancel out in the multiplication process, leaving only the unit ohms in the final answer.

Example 7–8

What is the resistance of a 300 ft length of no. 18 gage copper wire at 20°C?

Solution From Table 7–3, we see that no. 18 wire has a diameter of 40.30 mils. The circular-mill area (CMIL) is simply the square of the diameter in mils, or about 1624 CMIL.

From Table 7–2, we see that the resistivity of copper is 10.4 CMIL·Ω/ft. We may now write

$$R = \rho \frac{L}{A} = \frac{10.4\ \cancel{\text{CMIL}}\cdot\Omega}{\cancel{\text{ft}}} \times \frac{300\ \cancel{\text{ft}}}{1} \times \frac{1}{1624\ \cancel{\text{CMIL}}}$$

$$\approx 1.921\ \Omega$$

Note that all the units cancel out except for the ohms.

Series and Parallel Circuits

Table 7–3
Wire sizes

Gage No.	Diameter, Mils	Circular-Mil Area
1	289.3	83,690
2	257.6	66,370
3	229.4	52,640
4	204.3	41,740
5	181.9	33,100
6	162.0	26,250
7	144.3	20,820
8	128.5	16,510
9	114.4	13,090
10	101.9	10,380
11	90.74	8234
12	80.81	6530
13	71.96	5178
14	64.08	4107
15	57.07	3257
16	50.82	2583
17	45.26	2048
18	40.30	1624
19	35.89	1288
20	31.96	1022
21	28.46	810.1
22	25.35	642.4
23	22.57	509.5
24	20.10	404.0
25	17.90	320.4
26	15.94	254.1
27	14.20	201.5
28	12.64	159.8
29	11.26	126.7
30	10.03	100.5
31	8.928	79.70
32	7.950	63.21
33	7.080	50.13
34	6.305	39.75
35	5.615	31.52
36	5.000	25.00
37	4.453	19.83
38	3.965	15.72
39	3.531	12.47
40	3.145	9.88

7-6 THE VOLTAGE DIVIDER

We can apply much of what we have learned thus far to a simple and common circuit called a *voltage divider*. Figure 7–8 shows two series resistors connected across a voltage source (V_s). If we want to find the voltage drop across R_1, for example, we multiply R_1

Elementary Applications of Algebra to Electronics

by the total current I_t. That is, the voltage across R_1 is given as $V_1 = R_1 \times I_t$. Similarly, the voltage across R_2 is given as $V_2 = R_2 \times I_t$. Of course, it is first necessary to find the total current by applying Ohm's law. We write

$$I_t = \frac{V_s}{R_t} = \frac{V_s}{\Sigma R} = \frac{3\ \text{V}}{3\ \Omega} = 1\ \text{A}$$

Figure 7-8

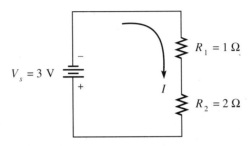

Therefore, $V_1 = R_1 \times I_t = 1\ \Omega \times 1\ \text{A} = 1\ \text{V}$. Likewise, $V_2 = R_2 \times I_t = 2\ \Omega \times 1\ \text{A} = 2\ \text{V}$. We verify the correctness of this answer by noting that the sum of the individual voltage drops must add up to the supply voltage:

$$V_s = \Sigma V_R = V_1 + V_2$$
$$= 1\ \text{V} + 2\ \text{V} = 3\ \text{V}$$

Throughout the foregoing discussion, you may have noticed that we did considerably more work than necessary in arriving at our answer. It may now be apparent that the supply voltage was simply divided between the two resistors in direct proportion to their individual values: 1 Ω, 1 V; 2 Ω, 2 V. By viewing the problem as a proportion, we eliminate the need to first calculate the total current, as we did earlier. These results are entirely consistent with Ohm's law, which states that the voltage drops are proportional to the resistances for a given current flow—that is, $V \propto R$, or $V = IR$.

Using the language and symbols of mathematics, let us see exactly how this idea of proportion works in the case of the voltage divider circuit shown in Figure 7-4. Suppose we are interested in knowing the voltage drop across R_2. We know from our earlier studies that a proportion is simply a statement of equality between two ratios. For example, $a/b = c/d$ is the statement of a proportion. In the case of our voltage divider, we can write a proportion stating that *part* of the voltage is to the *total* voltage as *part* of the resistance is to the *total* resistance. In symbols, we can write this statement as

$$\frac{V_2}{V_s} = \frac{R_2}{R_1 + R_2}$$

Plugging in some numbers from our previous results, we get

$$\frac{2\ \text{V}}{3\ \text{V}} = \frac{2\ \Omega}{3\ \Omega}$$

which is certainly no big surprise. And upon solving our original proportion for V_2, we get

$$V_2 = V_s \times \frac{R_2}{R_1 + R_2} = 3\ \text{V} \times \frac{2\ \Omega}{3\ \Omega} = 2\ \text{V}$$

Series and Parallel Circuits

We also could have set up a statement of proportionality for V_1 as

$$\frac{V_1}{V_s} = \frac{R_1}{R_1 + R_2}$$

Solving this equation for V_1, we have

$$V_1 = V_s \times \frac{R_1}{R_1 + R_2} = 3 \text{ V} \times \frac{1 \, \Omega}{3 \, \Omega} = 1 \text{ V}$$

which is exactly the value obtained earlier for V_1.

We are now in a position to present a formal statement of the voltage divider principle based entirely on the laws of mathematical proportion. We write such a statement as

$$\frac{V_x}{V_s} = \frac{R_x}{\Sigma R}$$

where x is any integer, so that

$$V_x = V_s \times \frac{R_x}{\Sigma R}$$

This expression is commonly referred to as the **voltage divider formula.**

Rule 7-2

$$V_x = V_s \frac{R_x}{\Sigma R}$$

Let us now move to a more involved example and see how much more easily the individual voltages are determined using the laws of proportion. Consider the circuit shown in Figure 7-9. Suppose we want to find the voltage drop across R_2. From

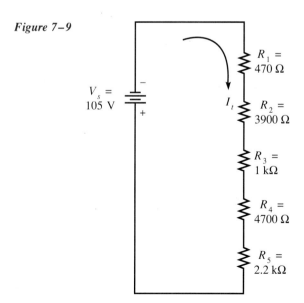

Figure 7-9

$V_s = 105 \text{ V}$

$R_1 = 470 \, \Omega$

$R_2 = 3900 \, \Omega$

$R_3 = 1 \text{ k}\Omega$

$R_4 = 4700 \, \Omega$

$R_5 = 2.2 \text{ k}\Omega$

140 Elementary Applications of Algebra to Electronics

$$V_x = V_s \times \frac{R_x}{\Sigma R}$$

we write

$$V_2 = 105 \text{ V} \times \frac{3900 \text{ }\Omega}{\Sigma R} = 105 \text{ V} \times \frac{3900 \text{ }\Omega}{12{,}270 \text{ }\Omega} \approx 33.37 \text{ V}$$

Note that this result is the same we would have obtained by first finding the current using Ohm's law:

$$I = \frac{V_s}{R_t} = \frac{105}{12{,}270} \approx 8.557 \text{ mA}$$

$$V_2 = I_t \times R_2 = 8.557 \times 10^{-3} \times 3900 \approx 33.37 \text{ V}$$

Example 7–9 In the circuit of Figure 7–10, what voltage will be read at point P?

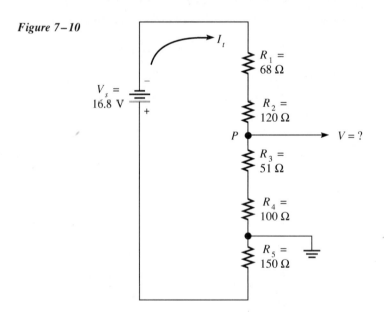

Figure 7–10

Solution The first question that arises is, "Where are the voltmeter leads connected?" The ground (symbol ⏚) was introduced in Chapter 2 in connection with signed numbers. At that time, the ground was defined as (1) an *arbitrary point of reference* from which all other voltage measurements are taken and (2) the *arbitrary point of zero voltage*. With these definitions in mind, we connect the voltmeter as shown in Figure 7–11 and realize that (1) the voltage at P will be negative with respect to ground and (2) the problem reduces to finding the

Series and Parallel Circuits

voltage drop across the $R_3 + R_4$ series combination. Applying the voltage divider formula, we get

$$V_x = V_s \times \frac{R_x}{\Sigma R} = 16.8 \text{ V} \times \frac{R_3 + R_4}{\Sigma R}$$

$$= 16.8 \text{ V} \times \frac{151 \text{ }\Omega}{489 \text{ }\Omega} \approx -5.188 \text{ V}$$

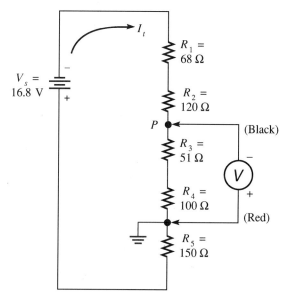

Figure 7–11

Example 7–10 What is the voltage measured at P in the circuit shown in Figure 7–12?

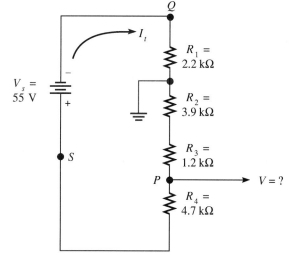

Figure 7–12

142   Elementary Applications of Algebra to Electronics

Solution   The voltmeter is connected as shown in Figure 7–13. In this case, the voltage at P is positive with respect to ground and is the drop across the $R_2 + R_3$ series combination. Applying the voltage divider formula, we get

$$V_x = V_s \times \frac{R_x}{\Sigma R} = 55 \text{ V} \times \frac{R_2 + R_3}{\Sigma R}$$

$$= 55 \text{ V} \times \frac{5.1 \text{ k}\Omega}{12 \text{ k}\Omega} = +23.375 \text{ V}$$

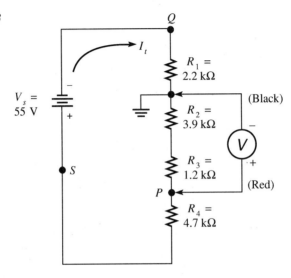

Figure 7–13

Example 7–11

In Example 7–10, what voltages are read at points Q and S?

Solution   The voltage at Q is negative relative to ground and is simply the voltage drop across R_1, as given by

$$V_x = V_s \times \frac{R_x}{\Sigma R} = 55 \text{ V} \times \frac{2.2 \text{ k}\Omega}{12 \text{ k}\Omega} \approx -10.08 \text{ V}$$

The voltage at S is positive with respect to ground and is the voltage drop across the $R_2 + R_3 + R_4$ series combination. The voltage drop is given by

$$V_x = V_s \times \frac{R_x}{\Sigma R} = 55 \text{ V} \times \frac{R_2 + R_3 + R_4}{\Sigma R}$$

$$= 55 \text{ V} \times \frac{9.8 \text{ k}\Omega}{12 \text{ k}\Omega} \approx +44.92 \text{ V}$$

Exercise Set 7–1

Questions

1. Using your own words, define the following terms.
   a. Formula

Series and Parallel Circuits

b. Ohm's law
c. Series circuit
d. Solidus fraction
e. Series-aiding
f. Series-opposing
g. Summation notation
h. Direct variation
i. Inverse variation

Problems

1. What current is read on the ammeter in Figure 7–14?

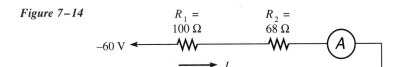

Figure 7–14

2. In Problem 1, determine the voltage drops across R_1 and R_2. What is the sum of these voltages, and what simple fact of series circuits can you infer from this result?
3. A 350 Ω resistive load consumes 1200 W of power. What current flows in the load?
4. In the circuit of Figure 7–15, the sum of the readings on A_1 and A_2 is 124 mA. What is the value of the supply voltage V_s?

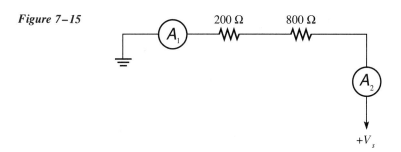

Figure 7–15

5. A voltmeter is intentionally connected into the circuit as shown in Figure 7–16. The reading on the voltmeter is 100 V. Shorting out any of the individual resistors does not affect the meter reading. However, shorting out the voltmeter causes the fuse to blow. Explain.

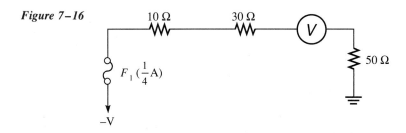

Figure 7–16

6. With the ground (reference point) shown in Figure 7–17 connected at A, the voltage at V_1 is +14.8 V. When the ground is relocated to B, what voltage will be read at V_1?

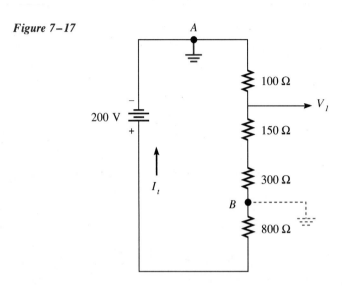

Figure 7–17

7. In Figure 7–18, (a) what voltage would be read on a voltmeter connected between the negative terminal of the 8 V battery and ground? (b) What current is read on the ammeter?

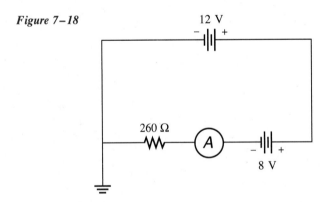

Figure 7–18

8. Refer to Figure 7–19. When S_1 is open, the ammeter reads 160 mA. With S_1 closed, the current is 200 mA. What is the value of resistor R_2?

Hint: Let I_o be the value of the current when S_1 is open and I_c be the current with S_1 closed. Then, set up an appropriate proportion using the ratios I_c/I_o and $(R_1 + R_2)/R_1$.

Figure 7-19

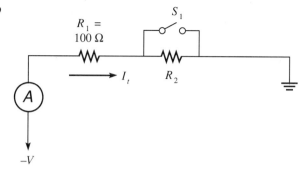

9. The power-supply voltage in the circuit shown in Figure 7-20 was increased from 80 V to a point where the 100 Ω resistor was consuming 23% more power. By what percent was the voltage increased?

Figure 7-20

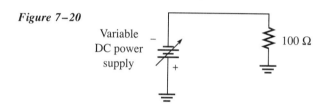

10. Write a BASIC program that computes the total resistance of any number of resistors in series. Extend the program to calculate the line current and power dissipated for a given value of applied voltage.
11. The resistance of most metals increases as they are heated. What type of variation exists in the relationship between resistance R and temperature T? Write an equation relating these two variables.
12. The rate you pay for electrical energy actually decreases as your energy usage becomes greater. Write an expression of proportion relating cost and energy.
13. Materials containing mostly carbon are used for brushes in DC motors and generators because of the negative temperature coefficient of resistance. Write an equation (not a statement of proportionality) that relates the resistance of the brush to its temperature.
14. Conductance (G) is the reciprocal of resistance (R). What type of variation does this represent? Write an equation for G in terms of R. What is the numerical value of the constant of proportionality in the equation?
15. What is the resistance of 1562 yd of no. 26 gage iron wire at 68°F?
16. As the current through a fixed value of resistance R increases, the power increases. This fact is represented by the familiar formula $P = I^2R$. In this case, we say that the power varies *directly* as the *square* of the current. For a fixed value of resistance, how is the power related to the voltage? Write the relationship between P and V as (a) a proportion and (b) an equation.
17. The expression for the total series-circuit resistance was given earlier by the formula

$$R_t = \sum_{k=1}^{n} R_k = R_1 + R_2 + \cdots + R_n$$

Basically, this formula states that the greater the number (n) of resistors (R) in the circuit, the higher the total resistance. Write a simple statement of proportion that expresses this idea.

18. For a given material and length, the resistance of a wire conductor varies inversely with its cross-sectional area. Write an equation showing that the resistance also varies inversely with the square of the diameter. What is the numerical value of the constant of proportionality in this equation? What happens to the resistance if the diameter is made three times as large?

7-7 THE PARALLEL CIRCUIT

As we said earlier, a series circuit has the same current flowing in all parts of the system, whereas the voltage drops across the various circuit elements may differ from one another. In the parallel circuit, however, these conditions are reversed. That is, the voltage is common to all branches, whereas the current may be different in each individual branch.

Consider the parallel circuit shown in Figure 7–21. Note that although the total voltage (20 V) is across each branch, the currents flowing in the individual branches are different. The current flowing through R_1 is given by Ohm's law as $I_1 = V/R_1 = (20\ \text{V})/(8\ \Omega) = 2.5$ A. The current through R_2 is $I_2 = V/R_2 = (20\ \text{V})/(14\ \Omega) \approx 1.43$ A. The total current I_t (often called the line current) is the sum of the individual branch currents and is given by $I_t = \sum_{k=1}^{2} I_k$. In Figure 7–21, the line current is

Figure 7–21

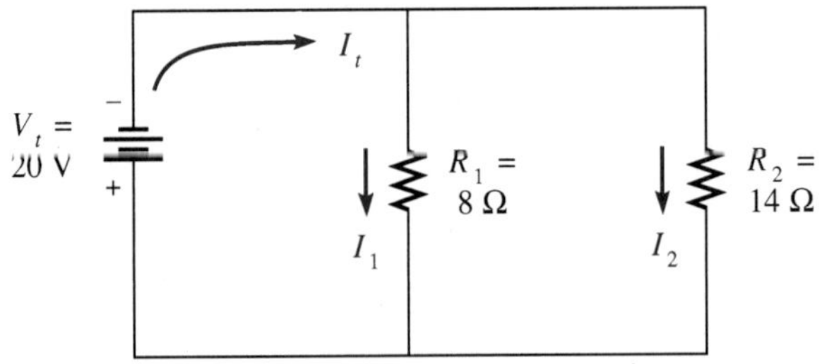

$$I_t = I_1 + I_2 = 2.5\ \text{A} + 1.43\ \text{A} = 3.93\ \text{A}$$

As in series circuits, the total power in a parallel circuit is the sum of individual powers.

Rule 7-3

$$P_t = \sum_{k=1}^{n} P_k = P_1 + P_2 + \cdots + P_n$$

7-8 RESISTANCES IN PARALLEL

We can use the fact that the total line current is the sum of the branch currents in order to derive a formula for resistances in parallel. Refer to Figure 7–22. Note that since the individual branch currents are given by $I = V_t/R$, we can write the total current as

$$I_t = \frac{V_t}{R_1} + \frac{V_t}{R_2} + \cdots + \frac{V_t}{R_n}$$

Series and Parallel Circuits

Figure 7-22

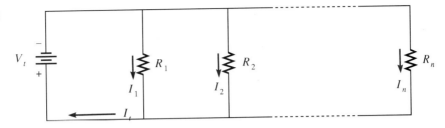

On factoring out V_t, we obtain

$$I_t = V_t \left(\frac{1}{R_1} + \frac{1}{R_2} + \cdots + \frac{1}{R_n} \right)$$

Dividing both sides of the equation by V_t gives us I_t/V_t, which we recognize as the reciprocal of resistance—that is, the total circuit conductance G. This allows us to write

$$G_t = \frac{1}{R_1} + \frac{1}{R_2} + \cdots + \frac{1}{R_n}$$

Realizing that resistance may also be expressed as the reciprocal of conductance ($R = 1/G$), we write

$$R_t = \frac{1}{G_t} = \frac{1}{\frac{1}{R_1} + \frac{1}{R_2} + \cdots + \frac{1}{R_n}}$$

The total resistance of any number of resistors in parallel is thus given by Rule 7-4.

Rule 7-4

$$R_t = \frac{1}{\frac{1}{R_1} + \frac{1}{R_2} + \cdots + \frac{1}{R_n}}$$

Note that if only two resistors are connected in parallel, we can derive a simple formula for this situation, as follows. Using Rule 7-4, we write

$$R_t = \frac{1}{\frac{1}{R_1} + \frac{1}{R_2}}$$

Considering only the denominator of this complex fraction for the moment, we write

$$\frac{1}{R_1} + \frac{1}{R_2} = \frac{R_2 + R_1}{R_1 R_2}$$

Therefore,

$$\frac{1}{\frac{R_2 + R_1}{R_1 R_2}} = \frac{R_1 R_2}{R_1 + R_2} = R_t$$

Example 7-12

In Figure 7-23, find (a) the total current and (b) the total power dissipated in the circuit.

Figure 7-23

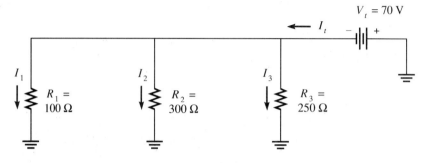

Solution a. Since $I_t = \sum_{k=1}^{3} I_k$, we must first determine the individual branch currents as follows:

$$I_1 = \frac{V_t}{R_1} = \frac{70 \text{ V}}{100 \text{ }\Omega} = 0.7 \text{ A}$$

$$I_2 = \frac{V_t}{R_2} = \frac{70 \text{ V}}{300 \text{ }\Omega} \approx 0.233 \text{ A}$$

$$I_3 = \frac{V_t}{R_3} = \frac{70 \text{ V}}{250 \text{ }\Omega} = 0.28 \text{ A}$$

Therefore, $I_t = \sum_{k=1}^{3} I_k = 0.7 \text{ A} + 0.233 \text{ A} + 0.28 \text{ A} = 1.213 \text{ A}$.

b. We may find the total power dissipated in the three resistor circuit in any of three ways:

1. $P = VI = 70 \text{ V} \times 1.213 \text{ A} \approx 84.9 \text{ W}$
2. $P = I^2 R_t$, where R_t is given by

$$R_t = \frac{1}{\dfrac{1}{R_1} + \dfrac{1}{R_2} + \cdots + \dfrac{1}{R_n}}$$

$$= \frac{1}{\dfrac{1}{100} + \dfrac{1}{300} + \dfrac{1}{250}} \approx 57.69 \text{ }\Omega$$

Therefore, $P = I^2 R_t = (1.213)^2(57.69) \approx 84.9 \text{ W}$

3. $P = \dfrac{V^2}{R} = \dfrac{70^2}{57.69} \approx 84.9 \text{ W}$

As you might have guessed, all three methods for determining the power yield the same result.

Series and Parallel Circuits

Example 7-13 Find the resistance between points A and B in Figure 7-24.

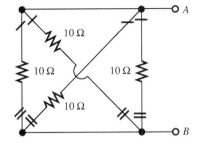

Figure 7-24

Solution Note that one end of each resistor has been marked with a single hash mark (/) and the other ends have been marked with a double hash mark (//). Observe, further, that if all the (/) ends are connected together and all the (//) ends are similarly tied together, we may redraw the circuit as shown in Figure 7-25. It is clear, now, that this is a parallel circuit whose total resistance may be determined by the usual formula.

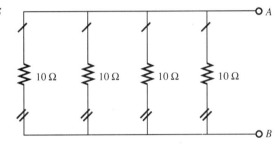

Figure 7-25

$$R_t = \frac{1}{\frac{1}{10} + \frac{1}{10} + \frac{1}{10} + \frac{1}{10}}$$

$$= 2.5 \, \Omega$$

Note that when all the resistors in parallel have the *same* value, it is not necessary to apply this formula. The total resistance can be found simply by dividing an individual resistance by the number of resistors in parallel. In the previous example, $R_t = 10/4 = 2.5 \, \Omega$.

This strategy works for the following reason. For n resistors in parallel, each one carries the same fraction of the total current $1/n$. This obvious fact implies that the total resistance must be $1/n$ that of a single resistor, namely, R/n. For example, if four 20 Ω resistors are connected in parallel, the total resistance is $20/4 = 5 \, \Omega$. You should verify this for yourself using the formula given earlier.

7-9 THE CURRENT DIVIDER

In the previous section, we saw that the voltage drop across an individual resistor in a series string was directly proportional to its resistance. That is, the greater the resistance, the greater the voltage drop. Of course, the current in such a circuit remains constant for a given applied voltage. Consider now the case of resistances in parallel as shown in Figure 7–26. Since the voltage is the same across each branch, it makes sense to inquire about how the currents divide in proportion to the individual branch resistances. It is apparent, after a little thought, that the greater the resistance, the smaller the value of current flowing through that particular branch. This conclusion is evident from Ohm's law ($I = V/R$), since as R increases in the denominator, the value of the fraction (I) decreases. However, the greater the conductance, the greater the current. And since conductance (G) is the reciprocal of resistance ($G = 1/R$), the smaller the conductance, the smaller the current and vice versa. This fact makes it simple for us to write a statement of direct proportionality similar to that used in deriving the voltage divider formula. We write

$$\frac{I_x}{I_t} = \frac{G_x}{G_t}$$

where I_x is the current through the particular branch under consideration, and G_x is the conductance ($1/R_x$) of that branch.

Figure 7–26

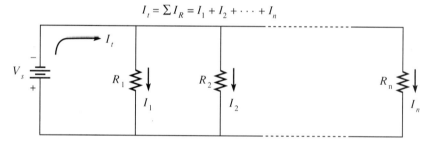

$$I_t = \Sigma I_R = I_1 + I_2 + \cdots + I_n$$

We may now solve this simple proportion for I_x to get the **current divider formula.**

Rule 7–5

$$I_x = I_t \frac{G_x}{G_t}$$

For example, consider the circuit shown in Figure 7–27. Suppose the total current (I_t) is 35 mA. What value of current flows through R_2? Our first task is to find the total conductance. We already have this formula, as given in a previous section.

Series and Parallel Circuits

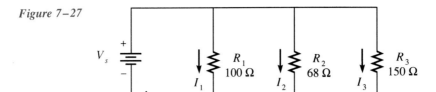

Figure 7–27

$$G_t = \frac{1}{R_1} + \frac{1}{R_2} + \cdots + \frac{1}{R_n}$$

In the present example, $G_t = \frac{1}{100} + \frac{1}{68} + \frac{1}{150} \approx 31.37 \times 10^{-3}$ siemens (S). Observe how easily this computation is done using a scientific calculator:

$$100 \; \boxed{\text{INV}} \; \boxed{1/x} \; \boxed{+}$$
$$68 \; \boxed{\text{INV}} \; \boxed{1/x} \; \boxed{+}$$
$$150 \; \boxed{\text{INV}} \; \boxed{1/x} \; \boxed{=}$$
$$\boxed{\text{ENG}} \; 31.37254902{-}03$$

Since $G_2 = \frac{1}{R_2} = \frac{1}{68} \approx 14.706 \times 10^{-3}$, we may write the value of I_2 as

$$I_2 = I_t \times \frac{G_x}{G_t} = 0.035 \times \frac{14.706 \times 10^{-3}}{31.372 \times 10^{-3}}$$
$$\approx 16.407 \text{ mA}$$

Some textbooks give the current divider formula in terms of individual branch resistances as

$$I_x = I_t \frac{R_t}{R_x}$$

Although this formula gives the same results, it causes some students difficulty for two reasons. First, its use involves the additional step of computing R_t rather than simply G_t. Second, the x term appears in the denominator (rather than in the numerator) due to the inverse relationship between current and resistance. Consequently, getting R_t and R_x in the wrong part of the fraction will lead to a completely erroneous answer. In applying this formula, use caution in order not to confuse it with the similar-appearing formula for direct variation used with the voltage divider circuit.

Exercise Set 7–2

Problems

1. The residential rate for electrical power in the city of Los Angeles is 8.546 cents/kilowatt-hour (¢/kWh). How much would it cost to operate five 120 V lamps connected in parallel for 72 h if each bulb has a hot resistance of 98 Ω?

2. For the circuit shown in Figure 7–28, find (a) the total resistance, (b) the total line current, and (c) the total power using three different methods.

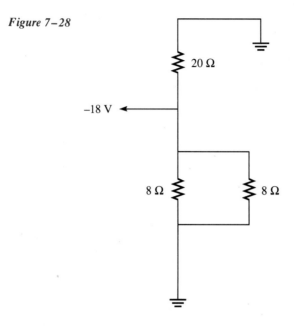

Figure 7–28

3. Each condition indicated refers to one or more malfunctions in the circuit shown in Figure 7–29. For each symptom, describe the possible cause.
   a. Fuse F_1 blows.
   b. Reading on ammeter suddenly drops to zero.
   c. Reading on ammeter drops to 81 mA.
   d. Ammeter reads about 112 mA, then drops to 44.2 mA, and then returns to 112 mA. This pattern repeats several times on a random basis.
   e. Ammeter reads 67.5 mA. All resistors are good.
   f. Ammeter reads zero. Voltmeter across fuse reads about 68 V.

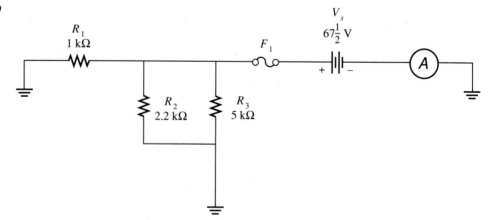

Figure 7–29

Series and Parallel Circuits

4. A technician wishes to determine the amperage of the fuse required to protect the circuit shown in Figure 7–30. She will include a 22% overload factor in the fuse rating in order to allow for transient effects (surges). What amperage rating should the fuse have if the load characteristics are as follows?
   a. Load 1: 28 Ω resistive load
   b. Load 2: Resistive load dissipating 255 W
   c. Load 3: Consists of two number 1 loads in series

Figure 7–30

5. In electronics, we often use the symbol ‖ to indicate that two or more resistors are in parallel. For example, if we wish to indicate that a 5 Ω and a 15 Ω resistor are connected in parallel, we may write 5 ‖ 15. In a similar manner, resistors in series are often symbolized using a plus (+) sign. For example, if a 3 Ω and an 18 Ω resistor are in series, we write 3 + 18.

   How would you symbolize the combinations shown in Figure 7–31a and b? What is the total resistance in each case?

Figure 7–31

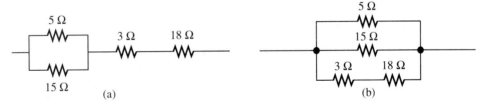

6. In the equation $R_t = \dfrac{R_1 R_2}{R_1 + R_2}$, solve for R_1 in terms of the other variables. Solve for R_2 in terms of the other variables.

7. For the circuit shown in Figure 7–32, use Ohm's law to verify the total current flow as determined by the current divider formula.

Figure 7–32

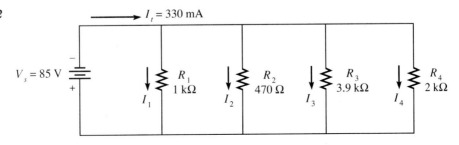

8. Determine the voltage read at point P in Figure 7-33.

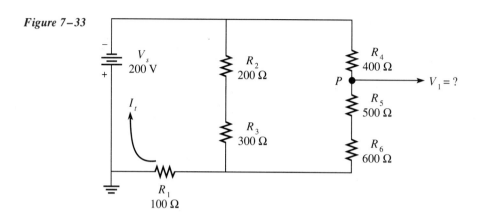

Figure 7-33

9. Write a BASIC program that will compute the total resistance of any number of resistances in parallel. Extend this program to calculate the total line current and power dissipated in the circuit for a given applied voltage.
10. Write a BASIC program that applies the current divider formula to a selected branch in a parallel circuit consisting of any number of resistances.

7-10 ANALYSIS OF UNITS

We showed in connection with Example 7-8 that units common to both numerator and denominator may be canceled as if they were literal factors. Recall from Chapter 6 that common literal factors appearing in a fraction can be canceled. This property leads to a useful topic that allows us to perform a variety of practical mathematical operations ranging from conversion of units to simplification of algebraic expressions.

Consider the following example. A man wants to sell you a racehorse whose track speed is allegedly 10,000 furlongs per fortnight. His asking price is $5000. Is this necessarily a good investment? Unfortunately, most of us have no real idea just how fast 10,000 furlongs per fortnight is, and so we must resort to converting these nebulous units into units more familiar to us. This example will introduce the method of **analysis of units.** In this strategy, we treat the units as if they were literal numbers, and write each pertinent factor as a fraction in such a way that common units cancel.

Conversion factors are to be found in a variety of scientific and engineering handbooks and, in the present example, an ordinary dictionary was used to determine that a furlong is ⅛ mi and a fortnight is 2 wk (14 d). We can now proceed with the analysis of the given problem.

$$\frac{10,000 \text{ furlongs}}{\text{fortnight}} \times \frac{\text{fortnight}}{14 \text{ days}} \times \frac{0.125 \text{ mi}}{\text{furlong}} \times \frac{\text{day}}{24 \text{ h}}$$

Note that every factor is written as a fraction. Furthermore, observe that all the units cancel except for the miles in the numerator of the third fraction and hours in the fourth fraction. Therefore, performing the indicated multiplications and divisions gives us a final answer of 3.72 mi/h, or slightly faster than my grandmother. This is definitely not a thoroughbred animal.

Series and Parallel Circuits

The beautifully simple part of this procedure is that it requires virtually no thought (if that may be viewed as an asset!) in setting up the required fractions. All you need to do is arrange the fractions one way or another such that the unwanted units cancel, as indicated in Rule 7–6.

Rule 7–6

Arrange each pertinent factor as a fraction in such a way that unwanted terms cancel, leaving the desired units intact.

Another example might make the point clearer. Suppose we want to convert 26 km to miles. We begin by writing the metric distance as a fraction:

$$\frac{26 \text{ km}}{1}$$

We may not know how many miles there are in a kilometer, but we may be able to make use of smaller bits of information. For example, we know that there are 1000 m in a kilometer. If we then write the next fraction properly, we obtain

$$\frac{26 \cancel{\text{ km}}}{1} \times \frac{1000 \text{ m}}{\cancel{\text{km}}}$$

At this point we may convert meters to feet and, finally, feet to miles. Suppose we determine that there are 3.28 ft per meter; then we write our next step as

$$\frac{26 \cancel{\text{ km}}}{1} \times \frac{1000 \text{ m}}{\cancel{\text{km}}} \times \frac{\text{m}}{3.28 \text{ ft}}$$

Something must be wrong here, since we now have units of m²/ft rather than simply feet. No problem; we merely invert the fraction and write

$$\frac{26 \cancel{\text{ km}}}{1} \times \frac{1000 \cancel{\text{ m}}}{\cancel{\text{km}}} \times \frac{3.28 \text{ ft}}{\cancel{\text{m}}}$$

Since the meters cancel out, we are left with the simple task of converting feet to miles. We know that there are 5280 ft in a statute mile, and so our final fraction is written as

$$\frac{26 \cancel{\text{ km}}}{1} \times \frac{1000 \cancel{\text{ m}}}{\cancel{\text{km}}} \times \frac{3.28 \cancel{\text{ ft}}}{\cancel{\text{m}}} \times \frac{\text{mi}}{5280 \cancel{\text{ ft}}}$$

Finally, performing the indicated operations yields 16.15 mi.

Exercise Set 7–3

Questions

1. Explain the method of analysis of units using your own words.
2. What makes this method so simple to use?

Problems

1. Make the following conversions. Conversion factors may be found in Appendix A.
   a. 6.8 nautical miles to statute miles.
   b. The speed of light in meters per second (m/s) to feet per hour (ft/h).
   c. 36,000 km to miles.

d. 55 mi/h to kilometers per hour (km/h).
   e. 67 km/h to miles per hour.
   f. 53.2 rev/min to radians per second.
   g. 32 ft/s² to meters per second squared.
2. How long does it take a pulse of SHF energy from a radar antenna to travel to a target 1 nautical mile away and back? (The speed of the pulse is the same as light.)
3. The mean diameter of the earth is approximately 8000 mi. Determine the earth's circumference in meters.
4. A sailboat is moving through the water at 7.5 knots. What is its speed in statute miles per hour?
5. A simple dipole antenna has an overall length of 5 ft 4 in. To what length is this equivalent in centimeters?
6. A cylindrical water tank has a diameter of 12 ft and a depth of 22 ft. What is the weight of the water if the tank is full to the top?
7. A piece of steel plate measures ½ in. thick, 6 in. wide, and 22.4 cm long. If the density of steel is 0.01 lb/in.³, how much does the plate weigh?
8. In Problem 6, what pressure in pounds per square inch does the water in the tank exert on the bottom of the tank?
9. The commercial FM (frequency modulation) band extends from 88 to 108 MHz. What is the width of this band in kHz?
10. The contacts on a certain relay close at the rate of 3050 m/s. What speed does this represent in feet per minute?

Key Terms

formula	summation notation	varies inversely
solidus fraction	varies directly	voltage divider formula
series-aiding	proportionality symbol	current divider formula
series-opposing	constant of proportionality	analysis of units

Important Rules and Formulas

Total Resistance of a Series Circuit
Rule 7–1

$$R_t = \sum_{k=1}^{n} R_k = R_1 + R_2 + \cdots + R_n$$

The Voltage Divider Formula
Rule 7–2

$$V_x = V_s \frac{R_x}{\Sigma R}$$

Total Power in a Series or a Parallel Circuit
Rule 7–3

$$P_t = \sum_{k=1}^{n} P_k = P_1 + P_2 + \cdots + P_n$$

Series and Parallel Circuits

Total Resistance of a Parallel Circuit

Rule 7-4

$$R_t = \frac{1}{\dfrac{1}{R_1} + \dfrac{1}{R_2} + \cdots + \dfrac{1}{R_n}}$$

The Current Divider Formula

Rule 7-5

$$I_x = I_t \frac{G_x}{G_t}$$

Rule for Analysis of Units

Rule 7-6 Arrange each pertinent factor as a fraction in such a way that unwanted terms cancel, leaving the desired units intact.

PART FOUR
Advanced Topics in Algebra

8
Graphs and Equations

9
Simultaneous Linear Equations

10
Quadratic Equations

8 Graphs and Equations

Upon completion of this chapter, you will have:
1. Learned how to plot points on a coordinate system.
2. Learned how to find the distance between points in a plane.
3. Learned the properties of special triangles.
4. Learned additional properties of linear equations and how to graph them.
5. Learned how to extract information from graphs and interpret the meaning of a graph.

8-1 INTRODUCTION

The ability to make inferences based on graphical data is paramount among the functional skills of the modern electronics technician. From interpreting the displays on an oscilloscope graticule to extracting needed information from complex performance curves of electronic devices, the technician must be able to understand quickly and accurately what graphs are trying to communicate. In this chapter, we look at the methods of interpreting such graphical information as well as gain a general understanding of the manner in which equations may be depicted graphically.

8-2 TWO-DIMENSIONAL REALITY

As you are probably aware, the physical world in which we reside is essentially a three-dimensional phenomenon. (There are four dimensions if you include time.) Fortunately for the electronics technician, we are usually concerned with only two of those dimensions in the interpretation of graphical data. On an oscilloscope screen, for example, we may see a trace that occupies a specific space vertically and a specific space horizontally. Any phenomenon that exists in two directions only is called a **plane** entity. For example, the CRT displays shown in Figure 8-1 have width as well as height. We refer to the width (left to right) as the **horizontal** dimension and the height (up and down) as the **vertical** dimension. Note, however, that nothing sticks out of the page. That is, there is no "page-to-you" dimension. Consequently, we say that the displays shown in Figure 8-1 are plane, or two-dimensional, representations of some real phenomena. Likewise, the printing on this page is essentially two-dimensional, since the "thickness" of the printing ink is all but nonexistent. If these pages were in Braille, however, then the printing would, indeed, be three-dimensional in nature.

As you may know already, the horizontal axis of the oscilloscope represents time if we are viewing a phenomenon in the time domain and frequency if we are viewing an event in the frequency domain. The vertical scale ordinarily represents amplitude in volts or decibels. In a later section, we will return to CRO displays and how they are interpreted. In the meantime, we must gain some mathematical insights into the graphical system used to portray the magnitudes and relationships of physical events.

8-3 THE COORDINATE SYSTEM

Referring to Figure 8-2, how would you (1) describe the location of the point A relative to the point B and (2) determine the distance between the two points?

You might try saying something like "Point A is almost in the middle of the figure, and B is to the upper left of A." Although this is a fairly accurate description of the *relative* locations of these points, we have failed to *quantify* (give a measurement of) their *exact* locations.

In Figure 8-3, we have superimposed a reference grid, or network, of small squares onto the figure.

The use of a grid certainly gives us a better feel for the location of the two points, but it leaves much to be desired. Using a suitable measuring scale calibrated in units of

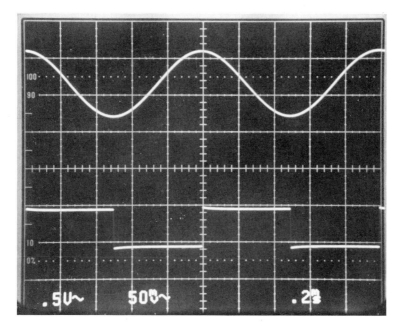

Figure 8–1
(Courtesy of Tektronix, Inc., Beaverton, OR)

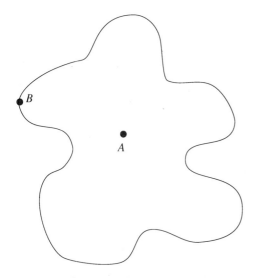

Figure 8–2

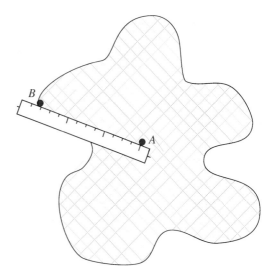

Figure 8–3

163

squares, we may, by direct measurement, see that A and B are separated by about 11¼ squares. It should be apparent to you that through the simple expedient of superimposing a completely *artificial* and *arbitrary* network of little squares, we have taken our first steps toward quantifying the locations of our two points. We say that the grid network is *artificial* because it was not originally a part of the figure. Moreover, we say that the network was *arbitrary* because the distance measurements can be anything we desire: inches, centimeters, feet, nautical miles, or just plain squares.

A more refined and useful grid system was developed in the seventeenth century by the French mathematician and philosopher René Descartes and is called the **Cartesian coordinate system** in his honor.* This reference system is shown in Figure 8–4.

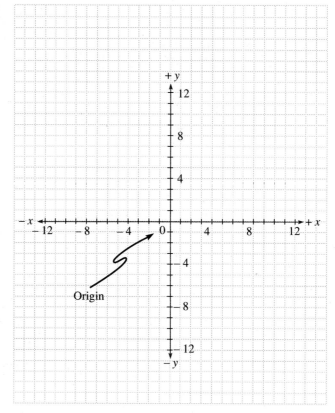

Figure 8–4

Note that the Cartesian system also consists of small squares, but there are two major differences between this system and that shown in Figure 8–3. First, note that there are two reference lines, called *axes*. The *horizontal axis* is called the *x*-axis (or **abscissa**), and the *vertical axis* is called the *y*-axis (or **ordinate**). Secondly, note that these axes are

---

*This system is also commonly referred to as a *rectangular* coordinate system.

Graphs and Equations

actually two number lines* that intersect at their zero points. This point of intersection is called the **origin.**

In Figure 8-5, we have redrawn our original figure with the two points A and B clearly labeled. Note, however, that these points have been identified with the **coordinates** of their x and y magnitudes. That is, point A(3, 3) is located +3 units along the x-axis and +3 units along the y-axis. Similarly, B(−8, 6) is −8 units along the x-axis and +6 units along the y-axis. These coordinates, then, locate A and B precisely relative to some zero position that we have called the origin. It is customary to give the x-coordinate first, followed by the y-coordinate. For example, a point P is identified as P(x, y).

Figure 8-5

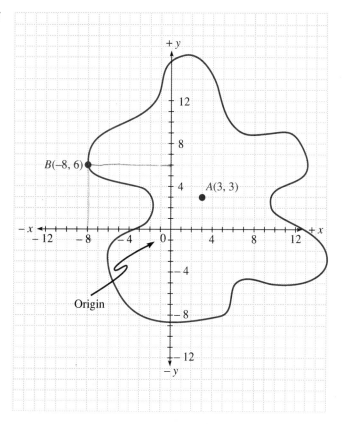

Since these points are now precisely located with their rectangular, or Cartesian, coordinates, it is no longer necessary to either guess or make crude measurements as to the distance between them. For, as we shall see in the next section, their exact distance may now be calculated mathematically with great precision. Incidentally, points A and B are separated by 11.4018 units, which is slightly more than what was measured in Figure 8-3. This same result would have been obtained *no matter where the irregular figure was located or what its relative orientation might have been.*

---

*For a review of the number line and its properties, see Chapter 1.

Advanced Topics in Algebra

Note that in addition to providing the *x*- and *y*-coordinates of any point in a plane (two-dimensional space), the rectangular system also divides that plane into four **quadrants,** as shown in Figure 8–6.

Figure 8–6

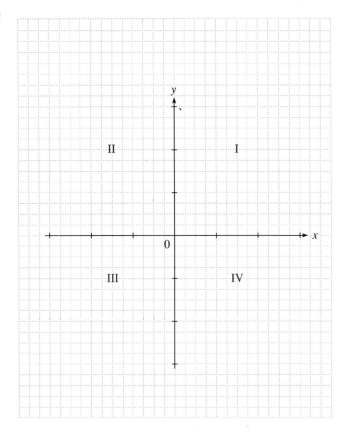

By convention, these quadrants are numbered using Roman numerals, beginning with the quarter in the upper right portion of the plane, quadrant I, and continuing counterclockwise through quadrants II, III, and IV. In our example, point $A(3, 3)$ is said to lie in the first quadrant, and point $B(-8, 6)$ is said to be a second-quadrant point.

Exercise Set 8–1

Questions

1. Define the following terms or phrases in your own words.
   a. Plane
   b. Quantify
   c. Cartesian coordinate system
   d. Abscissa
   e. Ordinate
   f. Coordinates of a point

Graphs and Equations

   g. Origin
   h. Quadrant
   i. Rectangular coordinate system

Problems

1. On standard graph paper (graph paper with a grid of squares), locate the following points.
   a. $A(2, -4)$
   b. $B(12, 1)$
   c. $C(-3, 6)$
   d. $D(4, -8)$
   e. $E(3, 9)$
   f. $F(-6, -5)$
   g. $G(-6, 5)$
   h. $H(-3, -7)$
2. Construct a table showing the quadrant in which each point in Problem 1 is located.
3. In Problem 1, use a piece of graph paper to measure the distance between the following pairs of points:
   a. A to B
   b. C to E
   c. G to B
   d. F to A
   e. H to D
   f. C to B
   Record these measurements in a table for later use.

8-4 THE THEOREM OF PYTHAGORAS

Perhaps one of the most useful ideas in all of fundamental mathematics is that attributed to the Greek mathematician Pythagoras of Samos (ca. 580–496 B.C.). There are more than one hundred proofs of his theorem (one is presented in Appendix C), which may be stated as follows (refer to Figure 8–7).

Figure 8–7

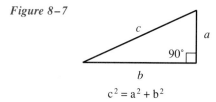

$$c^2 = a^2 + b^2$$

Rule 8-1 The Pythagorean Theorem

In any right triangle (one that contains a 90° angle), the lengths of the three sides are related by the equation

$$c^2 = a^2 + b^2$$

Advanced Topics in Algebra

We can apply Axiom 4–7 to solve for the length of side c, as shown in Rule 8–2.

Rule 8–2

$$c = \sqrt{a^2 + b^2}$$

Remember, the Pythagorean theorem applies *only* to a right triangle. For example, in Figure 8–8, if side a is 7 units long and side b is 4 units in length, we can use Rule 8–2 to find c as follows:

$$c = \sqrt{a^2 + b^2} = \sqrt{7^2 + 4^2} = \sqrt{65} \approx 8.06$$

Figure 8–8

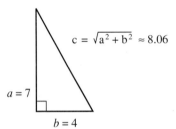

8–5 THE DISTANCE BETWEEN TWO POINTS

Figure 8–8 has been redrawn on a system of Cartesian coordinates and is shown as Figure 8–9. Unlike Figure 8–8, where the lengths of the sides were given, Figure 8–9 gives only the coordinates of the end points of the side opposite the origin. The side of a right triangle opposite the right angle is called the **hypotenuse** and its endpoints in Figure 8–9 are labeled $A(0, 7)$ and $B(4, 0)$. In applying Rule 8–2, we see that the length of the base (side b in Figure 8–8) is simply the x-coordinate (4) of point B. Moreover, the length of the altitude (side a in Figure 8–8) is the y-coordinate of point A. Therefore, we can write the length of the hypotenuse as the distance between the two points $A(0, 7)$ and $B(4, 0)$ as $\sqrt{x^2 + y^2} = \sqrt{4^2 + 7^2} \approx 8.06$.

Figure 8–9

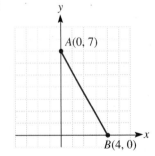

Graphs and Equations

Suppose the triangle shown in Figure 8–9 is relocated to some other region of quadrant I, as shown in Figure 8–10. We note that the length of side b is simply the difference between the x-coordinates of points Q and R—that is, $7 - 3 = 4$. Similarly, the length of side a is the difference between the y-coordinates of points P and R—that is, $9 - 2 = 7$. The distance between P and Q, then, can be written as $\sqrt{(7-3)^2 + (9-2)^2} = \sqrt{4^2 + 7^2} \approx 8.06$. Note that since the x-coordinate (3) of point R is the x-coordinate of P and the y-coordinate (2) of R is the y-coordinate of Q, we do not actually need point R at all in order to obtain the distance between points P and Q (see Figure 8–11).

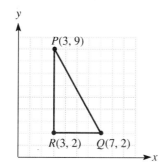

Figure 8–10

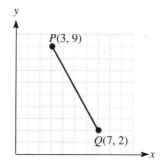

Figure 8–11

Observe that the horizontal distance between P and Q is the difference between their x-coordinates. Note, further, that the vertical distance between P and Q is the difference between their y-coordinates. Therefore, we may write a general expression for the distance (D) between any two points in a Cartesian coordinate system as follows.

Rule 8–3 Distance Between Two Points

$$D = \sqrt{(x_1 - x_2)^2 + (y_1 - y_2)^2}$$

Note that since the x and y differences are squared, the order of subtraction is unimportant. For example, $(x_1 - x_2)^2 = (7 - 3)^2 = (3 - 7)^2 = 16$, and $(y_1 - y_2)^2 = (9 - 2)^2 = (2 - 9)^2 = 49$.

Advanced Topics in Algebra

Example 8–1 Find the distance between the points $M(-6, 2)$ and $N(-1, 6)$, as shown in Figure 8–12.

Figure 8–12

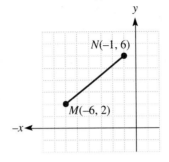

Solution Applying Rule 8–3, we obtain

$$D = \sqrt{(x_1 - x_2)^2 + (y_1 - y_2)^2} = \sqrt{(-6 + 1)^2 + (2 - 6)^2}$$
$$= \sqrt{(-5)^2 + (-4)^2}$$
$$= \sqrt{25 + 16} \approx 6.40$$

or

$$D = \sqrt{(-1 + 6)^2 + (6 - 2)^2} \approx 6.40$$

Example 8–2 Determine the length of the line segment between the points $R(-3, 8)$ and $S(4, -6)$, as shown in Figure 8–13.

Figure 8–13

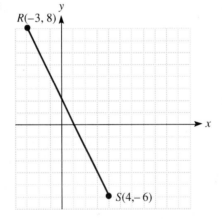

Solution

$$D = \sqrt{(x_1 - x_2)^2 + (y_1 - y_2)^2} = \sqrt{(-3 - 4)^2 + (8 + 6)^2}$$
$$= \sqrt{(-7)^2 + 14^2} \approx 15.65$$

or

$$D = \sqrt{(4 + 3)^2 + (-6 - 8)^2}$$
$$= \sqrt{7^2 + (-14)^2} \approx 15.65$$

Graphs and Equations

Exercise Set 8-2

Questions

1. In your own words, state the theorem of Pythagoras.
2. a. What is meant by the hypotenuse of a right triangle?
   b. Can the hypotenuse ever be shorter than either of the other two sides?
3. Explain why the order of subtraction is not important in Rule 8–3.

Problems

1. Determine the actual distance between points A and B of Figure 8–5.
2. a. In Exercise Set 8–1, Problem 3, use Rule 8–3 to verify your measured distances between the pairs of points listed.
   b. Calculate the percent of error for each set of points as follows:

$$\% \text{ error} = \frac{|C - M|}{C} \times 100$$

   where C = distance calculated and M = distance measured.
3. a. Solve the equation in Rule 8–1 for a.
   b. Solve the equation in Rule 8–1 for b.
4. For each set of points listed, plot the points on graph paper and determine the distance between them.
   a. $P(1, -6), Q(1, -15)$
   b. $M(3, -9), N(5, -14)$
   c. $R(-4, -6), S(5, 7)$
   d. $T(-8, 16), U(3, 9)$
   e. $V(-2.5, 11/3), W(5.7, 14.75)$
   f. $J(-3.66, 16.2), K(1.3, -3/4)$
   g. $E(100, 2014), F(-600, 974.8)$
      (*Hint:* Let each square stand for 200.)
   h. $G(-7/8, 0.5), H(11/16, 0.25)$
      (*Hint:* Let each square stand for 0.1.)
5. In the drawing of the triangle in Figure 8–14, determine the length of side x.

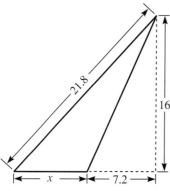

Figure 8–14

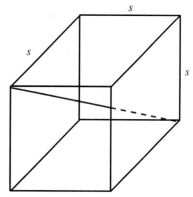

Figure 8–15

6. Write a BASIC computer program for Rule 8–3, and verify your answers to Problems 4 and 5.
7. Figure 8–15 is that of a cube whose sides are s units in length. A thread is stretched between opposite vertices (corners). Show that the length of the thread is $s\sqrt{3}$.

8-6 SPECIAL TRIANGLES

In Chapter 5, we discussed two of the so-called special products. At that time we indicated that these products were special because they occurred so frequently in mathematics. Similarly, there are two right triangles that occur so often in mathematics and electronics that they have come to be known as special triangles. One of these triangles is the **45-45 triangle;** the other is the **30-60 triangle.** We now discuss each of these in turn.

The 45-45 Triangle

The 45-45 right triangle shown in Figure 8–16 has three important properties.

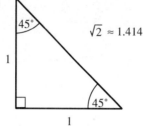

Figure 8–16

Rule 8-4 The 45-45 Triangle
1. Both acute angles (those less than 90°) are the same and equal 45°.
2. Both the legs (sides) are equal in length.
3. The length of the hypotenuse is always $\sqrt{2} \approx 1.414$ times longer than either leg.

From Rule 8–2, we see that the length of the hypotenuse of a 45-45 triangle whose legs are of unit length (i.e., one unit long) is $\sqrt{1^2 + 1^2} = \sqrt{2} \approx 1.414$. It should be apparent, then, that (1) if one leg of a 45-45 triangle is 5, for example, then the other leg is also 5; and (2) the length of the hypotenuse is $5\sqrt{2} \approx 7.07$.

Example 8–3

What is the length of the diagonal of the square shown in Figure 8–17?

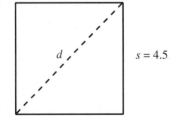

Figure 8–17

Graphs and Equations

Solution Since the figure is a square, all the sides (s) are the same length (i.e., 4.5 units), and the diagonal (d) cuts the 90° corner angles into two equal parts, each equal to 45°. Therefore, the diagonal divides the square into two equal 45-45 triangles.

As stated, the diagonal of a 45-45 triangle is always $\sqrt{2}$ times as long as a side. Therefore, we may write

$$d = s\sqrt{2} \approx 4.5(1.414) \approx 6.36$$

Example 8-4 What is the length of a side in a 45-45 triangle whose hypotenuse is 13?

Solution Since the hypotenuse (h) of a 45-45 triangle is always $\sqrt{2}$ times as long as either side (s), we may write the general expression for the relationship among the sides as

$$h = s\sqrt{2}$$

We can now solve this equation for s as follows.

$$s = \frac{h}{\sqrt{2}} \approx \frac{13}{1.414} \approx 9.194$$

The 30-60 Triangle

The 30-60 triangle is shown in Figure 8–18. The 30-60 triangle has the following important properties.

Figure 8–18

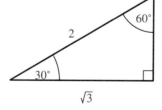

Rule 8–5 The 30-60 Triangle
1. The side opposite the 30° angle is half the hypotenuse.
2. The side opposite the 60° angle is $\sqrt{3} \approx 1.732$ times longer than the side opposite the 30° angle.

Advanced Topics in Algebra

Example 8–5

In Figure 8–19, the sun is shown at an elevation of 60°. If the length of the shadow is measured as 30 ft, what is the height of the flagpole?

Solution From the figure, we see that the shadow (s) is opposite the 30° angle, and the flagpole (f) is opposite the 60° angle. From the relationship among the sides of a 30-60 triangle, we can set up the following proportion:

$$\frac{s}{1} = \frac{f}{\sqrt{3}} = \frac{30 \text{ ft}}{1}$$

On solving this equation for f, we obtain

$$f = 30\sqrt{3} \approx 30(1.732) \approx 52 \text{ ft}$$

It is always easy to establish a proportion for any given problem by making use of the method of similar triangles, as shown in Figure 8–20. Note that we can form at least the following proportions among the sides of the two similar triangles:

$$\frac{a}{1} = \frac{b}{\sqrt{3}} \qquad \frac{a}{1} = \frac{c}{2} \qquad \frac{b}{\sqrt{3}} = \frac{c}{2}$$

Example 8–6

In Figure 8–21, a vertical antenna is held erect by guys. If the antenna is 530 ft high and one of the guys makes an angle of 60° with the ground, how long is the guy?

Solution From the conditions of the problem, we know that situation involves a 30-60 triangle. Therefore, we can set up the following proportion:

$$\frac{h}{\sqrt{3}} = \frac{g}{2}$$

where h = antenna height = 530 ft and g = length of guy.
On substituting the known height of the tower (h = 530) and solving for g, we obtain

$$g = 2\frac{h}{\sqrt{3}} \approx 2\frac{530}{1.732} \approx 612 \text{ ft}$$

Exercise Set 8–3

Questions

1. What are the three important properties of a 45-45 triangle?
2. What are the two important properties of a 30-60 triangle?
3. What makes the two triangles discussed in this section *special*?

Problems

1. A 22 in. diagonal line drawn on a rectangular metal chassis creates two 30° angles, as shown in Figure 8–22. What is the area of the surface shown?
2. In Figure 8–23, angles A, B, C, and D are all equal. The area of the rectangular figure is 169 square centimeters (cm²). What are the lengths of sides M and N?

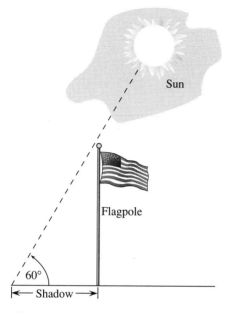

Figure 8–19

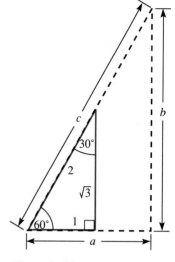

Figure 8–20

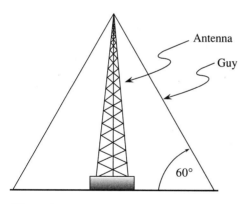

Figure 8–21

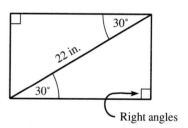

Figure 8–22

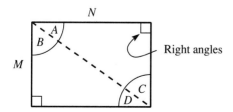

Figure 8–23

175

3. Ropes R_1 and R_2 are both 4 ft in length and make equal angles of 60° with the ceiling. What is the distance from the top of the suspended weight W to the ceiling? (See Figure 8–24.)
4. Figure 8–25 shows a cube with a volume of 512 in.^{3} What is the total length of the three lines drawn between the corners labeled A, B, and C?
5. The largest of the ancient Egyptian pyramids in the southern outskirts of Cairo near Giza has a square base. If the distance from center to corner is 160.5 m, as shown in Figure 8–26, what is the area of the base of the pyramid?

8–7 THE EQUATION OF A STRAIGHT LINE

An understanding of many simple electronic devices depends on an understanding of their linear (straight-line) properties. Moreover, the ability to interpret complex curves that display performance data is based to a considerable extent on estimating intervals that may be considered as linear. For these and other reasons, we will take up the rudimentary study of the equations of a straight line (linear equations) in this section. Later, we will apply what we have learned here to the interpretation of graphs and graphical data.

The Slope of a Line

In order to begin our discussion of the straight line, we must first define a few simple terms. Every straight line that is not vertical has some amount of **slope.** That is, every line is inclined to some degree. Slope is simply a measure of the degree of inclination. Referring to Figure 8–27, we see that the line rises to the right. We may pick *any* two points on this line and use them to make the following definition.

Rule 8–6

$$\text{Slope} = m = \frac{\text{rise}}{\text{run}} = \frac{y_1 - y_2}{x_1 - x_2}$$

Note that the order of subtraction is unimportant provided we are consistent in both the numerator and denominator. In other words, the same answer will be obtained whether we write

$$m = \frac{y_1 - y_2}{x_1 - x_2} \quad \text{or} \quad m = \frac{y_2 - y_1}{x_2 - x_1}$$

For the line shown in Figure 8–27, the slope is always *positive,* and the line *rises to the right.* If the line *falls to the right,* the slope is *negative.* A straight line that is horizontal has zero slope ($m = 0$), since the rise is always zero. On the other hand, since the run of a vertical line is zero, the slope is undefined, since division by zero is not a permissible operation in mathematics. These four situations are summarized in Figure 8–28.

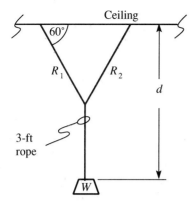

Figure 8–24

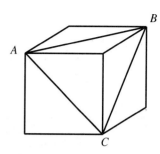

Figure 8–25

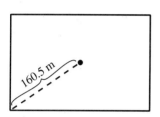

Figure 8–26

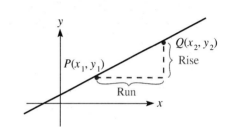

Figure 8-27

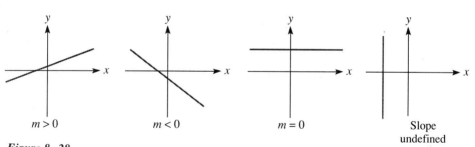

Figure 8–28

Equations of a Straight Line

Two common forms of a linear equation based on the definition of slope will now be developed. Recall that Rule 8–6 defined the slope (m) of a straight line as

$$m = \frac{y_1 - y_2}{x_1 - x_2}$$

If the slope and the coordinates of one point are known, we can rewrite Rule 8–6 as

$$m = \frac{y - y_1}{x - x_1}$$

Solving this last equation for y, we obtain the following.

Rule 8–7

$$y = m(x - x_1) + y_1$$

Since we used the value of the slope and the coordinates of a given point on the line, this form of the straight-line equation is called the **point-slope form.**

If we make use of the coordinates of the point where the line crosses the y-axis (the y-intercept), we can obtain another useful form of the equation of a straight line. Referring to Figure 8–29, we see that the coordinates of the y-intercept are $(0, b)$. Upon substituting these values for (x_1, y_1) in Rule 8–7, we obtain

$$m(x - 0) + b = y$$

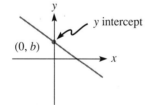

Figure 8–29

or

Rule 8–8

$$y = mx + b$$

Since we have made use of information concerning the slope and the y-intercept of a straight line in obtaining Rule 8–8, we call this form of the linear equation the **slope-intercept form.**

Graphs and Equations

Example 8-7

Find an equation of the line between points A and B of Figure 8-5 using both forms of the linear equation.

Solution The coordinates of the two points are $A(3, 3)$ and $B(-8, 6)$. The slope of this line is $m = (6 - 3)/(-8 - 3) = -3/11$. Using Rule 8-7, we can obtain the required equation using either point A or B. We will do both and then compare our answers.

Using Point A(3, 3)
$$y = m(x - x_1) + y_1$$
$$= -\frac{3}{11}(x - 3) + 3$$
$$= -\frac{3x}{11} + \frac{9}{11} + 3$$
$$y = -\frac{3x}{11} + \frac{42}{11}$$

Using Point B(−8, 6)
$$y = m(x - x_1) + y_1$$
$$= -\frac{3}{11}(x + 8) + 6$$
$$= -\frac{3x}{11} - \frac{24}{11} + 6$$
$$y = -\frac{3x}{11} + \frac{42}{11}$$

Note that the two equations agree and would be identical no matter which two points we select, provided they are on the line in question.

From either of the preceding equations, we see that when $x = 0$, $y = 42/11$. Therefore, the y-intercept coordinates are $(0, 42/11)$. Substituting this information and the value of the slope ($m = -3/11$) into Rule 8-8, we obtain

$$y = mx + b = -\frac{3x}{11} + \frac{42}{11}$$

We see, then, that both Rules 8-7 and 8-8 yield the same result in all cases.

Exercise Set 8-4

Questions

1. Define the terms *rise* and *run*.
2. What is meant by the slope of a line?
3. Why do lines rising to the right always have a positive slope?
4. Why do lines falling to the right always have a negative slope?
5. What is the point-slope form of a linear equation?
6. What is the slope-intercept form of a linear equation?
7. Why is the slope of a straight line parallel to the y-axis said to be undefined?
8. Why is a line perpendicular to the y-axis said to have zero slope?

Problems

1. What is the slope of a line inclined 45° to the x-axis and falling to the right? Make a drawing to support your answer.
2. Show that a line passing through the origin and inclined 60° to the x-axis has a slope of either $+\sqrt{3}$ or $-\sqrt{3}$.
3. What is an equation of the line passing through the origin and the point $S(-16, -2)$?
4. Demonstrate that the slope of any line parallel to the abscissa is always zero.

Advanced Topics in Algebra

5. In Figure 8–30, show whether the point R is on or not on the line connecting the points P and Q. (*Hint:* Derive equation of line thru P and Q; substitute coordinates of R in the result.)

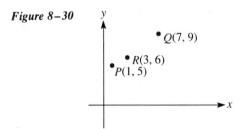

Figure 8–30

6. Derive the equation of the straight line passing through $M(-3, 10)$ and $N(5, 2)$.
7. The slope of a line is -3.5 and the y-intercept is $(0, -7)$. What is an equation of the line?
8. What is the equation of a line passing through the origin and having a slope of -1? What angle does this line make with the x-axis?
9. What are the coordinates of the point where the straight line $y = \frac{1}{4} - 3$ crosses the y-axis? At what point does this line cross the x-axis?
10. Two straight lines are parallel. The equation of one of the lines is $y = -2x + 8$. The y-intercept of the other line is $(6, 0)$. Find the equation of the other line.

8-8 GRAPHING LINEAR EQUATIONS

In the previous section, it was shown that linear equations are commonly (though not exclusively) expressed in either the point-slope form or the slope-intercept form. Given an equation in either form, we now consider how the straight line representing a given linear equation can be drawn (graphed) on a system of rectangular coordinates.

An equation given in the point-slope form is easily converted to the slope-intercept form, which is more easily graphed. For example, given the linear equation $y = -2(x - 2) + 4$, which is in the point-slope form, we can expand and simplify as follows:

$$y = -2(x - 2) + 4$$
$$= -2x + 4 + 4$$
$$y = -2x + 8$$

This last equation is in the slope-intercept form. The slope is -2; the y-intercept is the point $(0, 8)$.

We can now make use of a simple fact from plane geometry stating that one and only one straight line can be drawn through any two points. Using this fact, if we can find any two points on the line, then we can draw the line representing the equation through these two points. Which two points shall we use? Note that when $x = 0$, $y = 8$.

$$y = -2(0) + 8 = 8$$

This is simply the y-intercept. Moreover, when $y = 0$, $x = 4$.

$$0 = -2x + 8$$
$$-8 = -2x$$
$$x = 4$$

This is the x-intercept.

Graphs and Equations

We now have two points, (0, 8) and (4, 0), through which we can draw the graph representing the straight line $y = -2x + 8$. This is shown in Figure 8–31.

Figure 8–31

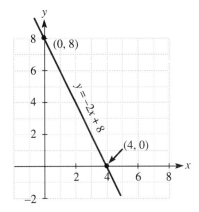

Observe that we can use *any other two points* and still be able to draw the line representing the equation. For example, if $x = 5$, $y = -2(5) + 8 = -2$. Also, if $y = -9$, then $-9 = -2x + 8$, so $-17 = -2x$, or $x = 17/2$. We now have the two points $(5, -2)$ and $(17/2, -9)$ through which to draw our straight line. You will probably agree, however, that using zero as one of the coordinates of each point makes the problem that much simpler. In this case, it is often a simple matter to determine the points mentally.

Example 8–8

Graph the equation represented by $25y = 100x - 250$.

Solution We can obtain a simpler form of this equation by dividing each side by 25, obtaining

$$y = 4x - 10$$

When $x = 0$, $y = -10$, and when $y = 0$, $x = 10/4$, or 2.5. We can now draw a straight line through the two points $(0, -10)$ and $(2.5, 0)$ to represent the equation. This is shown in Figure 8–32.

Figure 8–32

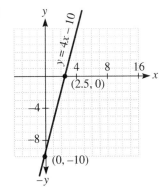

182 Advanced Topics in Algebra

Exercise Set 8-5

Problems

Draw the graph of each of the following linear equations.
1. $y = 3x - 7$
2. $-y = 24x + 48$
3. $\dfrac{2y}{3} = 10x$
4. $-1000y - 500x = 2500$
5. $y = -2(x + 16) + 5$
6. $y = -\tfrac{3}{2}x + 2$
7. $y = 2(x + 2) - 5$
8. $R = 5V + 288$
9. $T = 0.03k - 0.005$
10. $P = 5\dfrac{R}{8} + 22$
11. Write a BASIC program that will give the x- and y-intercepts of any linear equation in slope-intercept form.

8-9 INTERPRETATION OF GRAPHS

In this section we apply what we have learned to the interpretation of graphical data. Although many of the initial examples may seem trivial and perhaps even obvious, they will form the foundation for more complex situations to follow.

Figure 8–33 shows a voltage-versus-current graph for a 2 Ω resistor in a simple DC circuit. Suppose it is desired to determine the slope of the curve* shown. We may select *any two points on the curve* such as (6, 3) and (14, 7) and apply Rule 8–6 as follows.

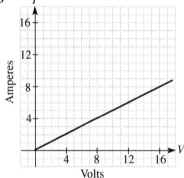

Figure 8–33

$$\text{Slope} = m = \frac{\text{rise}}{\text{run}} = \frac{y_1 - y_2}{x_1 - x_2} = \frac{3 - 7}{6 - 14} = \frac{-4}{-8} = \frac{1}{2}$$

We see that the slope is a positive number, as expected, since the curve rises to the right.

---

*Even though the line is straight, it has become common usage to refer to such a line as a *curve*.

Graphs and Equations

As another example, suppose we want to find the value of some current not on the graph. That is, suppose we need to know the value of *I* when *V* is 128 V. Since this is strictly an illustrative example, we will pretend that we cannot use Ohm's law to obtain our answer. Since the curve passes through the origin, it is perhaps simplest to use Rule 8–8 to derive the equation of the line, as follows:

$$y = mx + b = \frac{1}{2}x + 0$$

or

$$y = \frac{x}{2}$$

However, since we are dealing with volts (V), represented by *V*, and amperes (A), represented by *I*, we can relabel the variables to obtain the equation $I = V/2$. If we now substitute 128 for *V*, we obtain 64 A. Note that the equation we derived from $y = mx + b$ is the familiar $I = V/R$ of Ohm's law for $R = 2\ \Omega$.

Whenever we mathematically extend a curve beyond its given range for the purpose of finding other values not shown on the curve, we say we are using **extrapolation** techniques. In the foregoing example, we can even extrapolate for negative values of *V* if it suits our needs.

As a further example, we may wish to determine the rate of change in *I* per unit change in *V*. That is, we may want to know how fast *I* is changing with *V*. If you are especially observant, you will realize that the slope of the curve gives precisely this information. In the preceding example, the current is changing at the rate of 0.5 A/V, since $m = \frac{1}{2}$. We conclude, then, that the slope of a straight line is a measure of how fast one variable is changing per unit change in the other.

In Figure 8–34 we see a curve that is not linear. Nonetheless, we can use ideas from linear equations to answer some important questions about nonlinear curves. For example, at which point *A*, *B*, or *C* is the value of *y* changing the fastest per unit change in *x*? Here, we are actually talking about the **instantaneous rate of change** at a point, not over an interval. Although an exact answer to this question involves calculus, we can obtain a fairly close estimate, as shown in Figure 8–35. An approximate tangent line* is drawn at each of these three points, and we conclude from what we know about the slopes of straight lines that the fastest change is occurring at *A*, that the curve is not changing at all at *B*, and that *C* indicates a point at which the curve is changing more slowly than at *A*. These conclusions are apparent since the steeper the line, the faster the change in *y* per unit change in *x*.

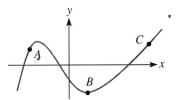

Figure 8–34

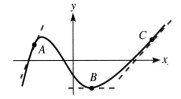

Figure 8–35

---

*A **tangent** line is a straight line that touches the curve at one and only one point.

Advanced Topics in Algebra

Consider the graph in Figure 8–36, which shows the power dissipated in a 5 Ω resistor. What is the *average* rate of change in the power as the current changes from 2 A to 5 A? Since the graph of Figure 8–36 is nonlinear, we can speak only of the *average* change over the interval in question. This should be apparent, since a tangent line at M indicates that the graph is changing slowest there. Moreover, a tangent line at N shows the graph changing fastest at that point. The secant line,* however, defines a sort of average rate of change over the entire interval between M and N.

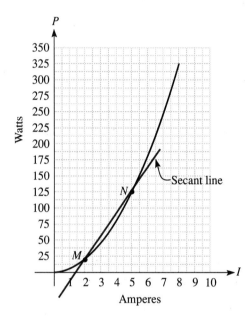

Figure 8–36

In higher mathematics it is common practice to indicate the change over an interval with the uppercase Greek letter delta, Δ. Accordingly, we can write the rise shown in Figure 8–37 as ΔP and the run as ΔI. The slope, then, is written as

$$m = \frac{\Delta P}{\Delta I} = \frac{106.25}{3} \approx 35.42$$

We conclude that the average rate of change in the power per unit change in the current over the interval specified is 35.42 W/A.

In addition to the graphical method for obtaining the average rate of change over an interval, there is a more precise method that we will now discuss. Referring again to Figure 8–37, we see that the ordinate of point M lies about three-fourths of the way between 0 and 25, or at 18.75. Moreover, we observe that the ordinate of point N is 125. Therefore, we conclude that $\Delta P = 125 - 18.75 = 106.25$. The value of ΔI is $5 - 2 = 3$. Consequently, the slope of the secant line (i.e., the average rate of change) can be determined to be $106.25/3 \approx 35.42$ W/A. In point of fact, however, this number somewhat overestimates the actual change, since we were not entirely certain of the exact value of M's ordinate.

---

*A **secant** line is a straight line passing through two points.

Graphs and Equations

Figure 8-37

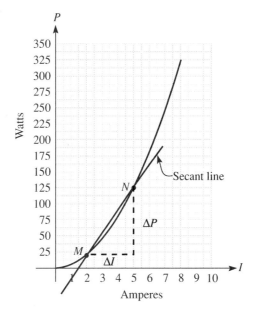

In order to gain an accurate quantification of the change, we resort to a nongraphical solution that does not involve guesses about the coordinates of the points based strictly on uncertain observation. This method is sometimes referred to as the **delta method.**

The equation of the curve shown in Figure 8-37 is given by $P = I^2R$, as you might have guessed. And since $R = 5$, as previously stated, we can write $P = I^2 5$. Knowing the equation of the graph, we can now determine the values of ΔP and ΔI precisely by mathematical means.

It is apparent that since the interval of interest ranges from $I = 2$ to $I = 5$, then $\Delta I = 5 - 2 = 3$. The value of ΔP is simply the value of P when $I = 5$ minus the value of P when $I = 2$. We note, however, that the value of P when $I = 5$ is simply $(I + \Delta I)^2 R$, as clearly seen from Figure 8-37. We can use this fact to write ΔP as

$$\Delta P = (I + \Delta I)^2 R - I^2 R$$

We conclude that

$$m = \frac{\Delta P}{\Delta I} = \frac{(I + \Delta I)^2 R - I^2 R}{\Delta I}$$

Upon expanding the numerator of this equation, we obtain

$$\frac{I^2 R + 2I\Delta I R + (\Delta I)^2 R - I^2 R}{\Delta I} = \frac{2I\Delta I R + (\Delta I)^2 R}{\Delta I}$$

Factoring out $\Delta I R$ from the numerator and dividing by ΔI, we obtain

$$m = \frac{\Delta I R(2I + \Delta I)}{\Delta I} = R(2I + \Delta I)$$

As stated before, the value of ΔI is $5 - 2 = 3$. Moreover, we let $I = 2$, since this was its value before the change began. Therefore,

$$m = R(2I + \Delta I) = 5(4 + 3) = 35 \text{ W/A}$$

Advanced Topics in Algebra

This is the precise average rate of change in power (W) per unit change in current (I) over the interval

$$2 \leq I \leq 5$$

Do not be fooled by the fact that this answer has fewer digits than our previous answer of 35.42 obtained by graphical methods. Remember that a graphical solution is only an estimate; its accuracy depends on how well we measured. The delta method, on the other hand, gives us a precise mathematical result independent of measurement errors. In fact, the answer obtained by the graphical method has an error of 1.2%.

Example 8–9 Use the delta method to find the average rate of change of the graph in Figure 8–33 between the interval $V = 4$ and $V = 12$. Figure 8–33 is redrawn as Figure 8–38.

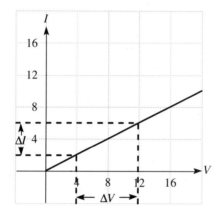

Figure 8–38

Solution The equation representing the graph is $I = \frac{1}{2}V$, since the resistance is 2 Ω.

$$m = \frac{\Delta I}{\Delta V} = \frac{\left(\frac{V + \Delta V}{2}\right) - \frac{V}{2}}{\Delta V} = \frac{\Delta V}{2} \times \frac{1}{\Delta V} = \frac{1}{2}$$

This is precisely the answer obtained earlier by graphical means.

Exercise Set 8–6

Questions

1. What is meant by the term *extrapolation*?
2. What is a tangent line?
3. What is the delta method?

Problems

1. In the drawing shown in Figure 8–39, the horizontal sweep speed is 0.8 ms/div and the vertical sensitivity is 20 V/div.

Graphs and Equations

a. What is the average rate of change (V/ms) between points *A* and *B*?
b. At which point (*P* or *Q*) is the waveform changing more slowly?
c. How long does it take for the pulse to change from *C* to *Q*?
d. What is the average rate of change from *C* to *Q*?

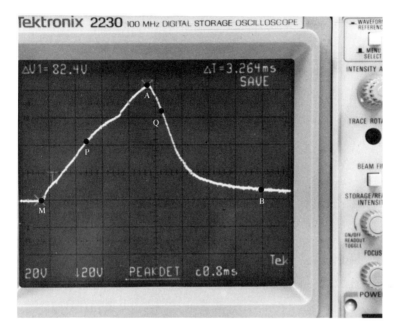

Figure 8–39
(Courtesy of Tektronix, Inc., Beaverton, OR)

2. In Figure 8–40 (page 188), assume that the vertical sensitivity is 2 mV/div.
   a. What is the average rate of change in millivolts per division between points *A* and *B*?
   b. What is the slope of the waveform between points *B* and *C*?
3. The risetime (t_r) of a pulse is defined as the time required for a waveform to increase in amplitude from 10% to 90% of its maximum value. For the pulse shown in Problem 2, what is the risetime if the horizontal sweep speed is set at 0.5 μs/div?
   Hint: Note that the waveform of the pulse has already been set to correspond with 0% and 100% of maximum as shown along the left edge of the graticule.
4. In Figure 8–41 (page 188), the vertical sensitivity is set at 2 V/div and the horizontal sweep is set at 5 μs/div. What is the approximate average rate of change between points *A* and *B* for the sinusoidal voltage displayed?
5. In Problem 1, if the waveform continued to rise at the same rate as it does between points *P* and *B*, by approximately how many volts would the amplitude change 4 mS after point *B* was reached?
6. In Figure 8–42 (page 189), in which curve (*Q* = 10, 50, or 100) does the amplitude vary most rapidly per unit change in the frequency?
7. For the filter circuit phase-response curve shown in Figure 8–43 (page 189), at what phase angle (or angles) is the frequency changing at the smallest rate?

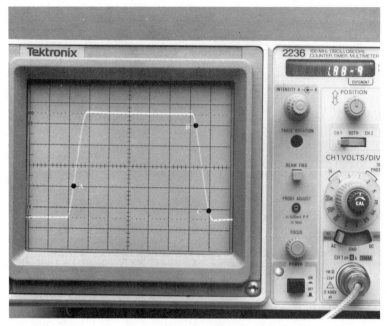

Figure 8–40
(Courtesy of Tektronix, Inc., Beaverton, OR)

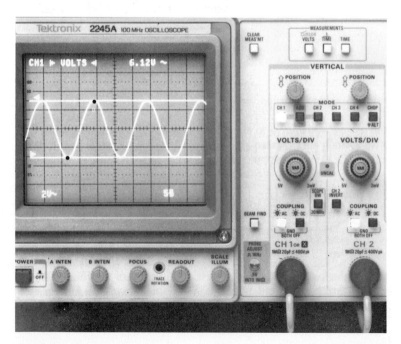

Figure 8–41
(Courtesy of Tektronix, Inc., Beaverton, OR)

Figure 8–42

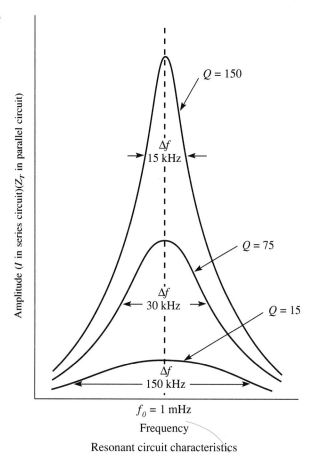

Figure 8–43

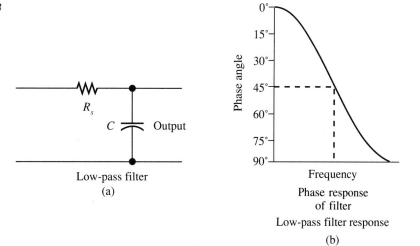

Advanced Topics in Algebra

8. In Figure 8–44, which sine wave (A, B, or C) has the fastest rate of change per unit change in the phase angle?

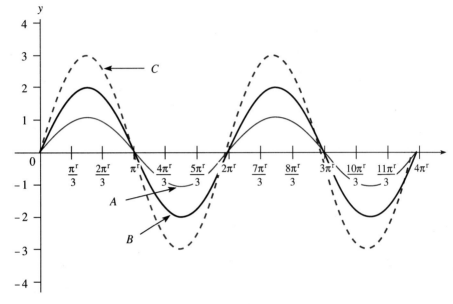

Figure 8–44

9. In the graph of $t = a^2 - 3a + 2$ shown in Figure 8–45, determine the average rate of change over the interval $-1 \leq a \leq 3$ using the graphical method.

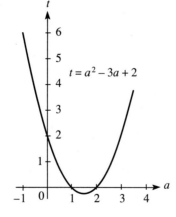

Figure 8–45

10. Using graphical techniques, find the average rate of change for the graph of $E = 12 - 4V - V^2$ over the interval $-5 \leq V \leq 0.5$. (See Figure 8–46.)

Graphs and Equations

Figure 8–46

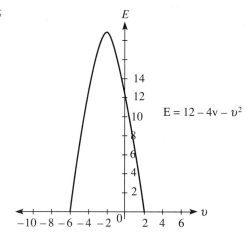

11. Use the delta method to find the exact average rate of change for the interval specified in Problem 9.
12. Find the precise average rate of change in Problem 10 using the delta method.

Key Terms

plane	coordinates	point-slope form
horizontal	quadrant	slope-intercept form
vertical	Pythagorean theorem	extrapolation
Cartesian coordinate system	hypotenuse	instantaneous rate of change
abscissa	45-45 triangle	tangent
ordinate	30-60 triangle	secant
origin	slope	delta method

Important Rules

The Pythagorean Theorem

Rule 8–1 In any right triangle (one that contains a 90° angle), the lengths of the three sides are related by the equation

$$c = a^2 + b^2$$

Rule 8–2

$$c = \sqrt{a^2 + b^2}$$

Distance Between Two Points

Rule 8–3

$$D = \sqrt{(x_1 - x_2)^2 + (y_1 - y_2)^2}$$

Advanced Topics in Algebra

The 45-45 Triangle

Rule 8–4

1. Both acute angles (those less than 90°) are the same and equal 45°.
2. Both the legs (sides) are equal.
3. The length of the hypotenuse is always $\sqrt{2} \approx 1.414$ times longer than either leg. (See Figure 8–47.)

Figure 8–47

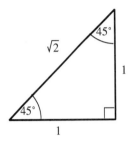

The 30-60 Triangle

Rule 8–5

1. The side opposite the 30° angle is half the hypotenuse.
2. The side opposite the 60° angle is $\sqrt{3} \approx 1.732$ times longer than the side opposite the 30° angle. (See Figure 8–48.)

Figure 8–48

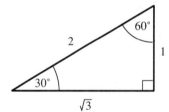

Slope of a Line

Rule 8–6

$$\text{Slope} = m = \frac{\text{rise}}{\text{run}} = \frac{(y_1 - y_2)}{(x_1 - x_2)}$$

Point-slope Form

Rule 8–7

$$y = m(x - x_1) + y_1$$

Slope-intercept Form

Rule 8–8

$$y = mx + b$$

9 Simultaneous Linear Equations

Upon completion of this chapter, you will have:
1. Learned how to solve simultaneous linear equations by the methods of
   a. Addition and subtraction
   b. Substitution
   c. Comparison
   d. Second- and third-order determinants
2. Learned how to use determinants in BASIC programs for solving systems of linear equations.
3. Learned how to interpret the graphical results of a simultaneous solution.

194   Advanced Topics in Algebra

9-1   INTRODUCTION

It frequently happens that more than one set of conditions affect the operation of a circuit at the same time. For example, many electronic networks have currents flowing that are influenced by more than one voltage source acting at the same instant. In such situations, it is often desirable to determine the effect that each source acting alone has on the total current. One way to do this is to solve a system of circuit-specific linear equations simultaneously.

In this chapter, we discuss several of the many methods available for solving simultaneous linear equations. In Chapter 11, we apply what we have learned to the solution of specific network problems.

9-2   THE SIMULTANEOUS SOLUTION

Figure 9-1 shows the graph of two linear equations that cross one another at the point $P(3, -1)$. If we substitute the value of the x-coordinate of P into the two equations, we obtain the following result:

$$y = -2x + 5 \qquad y = \frac{x}{3} - 2$$
$$= -2(3) + 5 \qquad = \frac{3}{3} - 2$$
$$= -1 \qquad\qquad = -1$$

Note that when $x = 3$ in *both* equations, $y = -1$ in *both* equations. This result demonstrates that the coordinates of P satisfy the conditions of *both* equations *at the same time*—that is, simultaneously. There is no other point—out of the infinite number available—that gives this result for the particular equations shown. Therefore, we say that $P(3, -1)$ is the **simultaneous solution** for the two equations in question.

Although it is not quite as apparent, an infinite number of linear equations have graphs that pass through point P and are satisfied simultaneously by the coordinates $(3, -1)$. Eight such equations are shown in Figure 9-2. It is left as an exercise for you to verify that each of these equations is indeed satisfied by the coordinates of $P(3, -1)$.

9-3   SOLVING SIMULTANEOUS LINEAR EQUATIONS

Figure 9-3 shows the simultaneous solution of two linear equations. Note that it is a simple matter to read the coordinates of the point where the two lines cross.

The usefulness of the graphical method of solution to linear systems of equations depends to a high degree on how accurately the coordinates of the point of intersection can be read from the Cartesian grid. In Figure 9-4, we observe that the point of intersection appears to lie quite close to the point $(1, 1.50)$ as was the case in Figure 9-3. However, it turns out that the actual coordinates are more accurately given by $(1.023, 1.517)$. Note that the only difference between the equations of Figure 9-3 and 9-4 is a slight decrease of 0.05 in the slope of one of the equations.

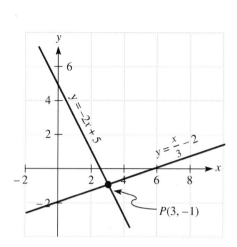

Figure 9-1

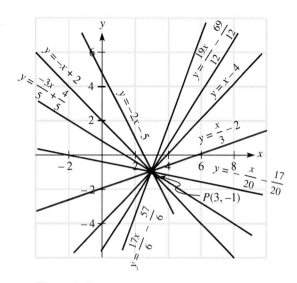

Figure 9-2

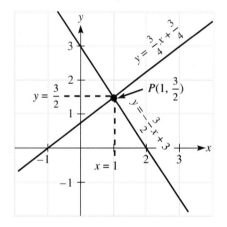

Figure 9-3

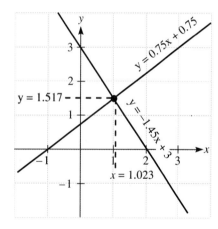

Figure 9-4

We must conclude that the usefulness of the graphical method is determined by how far away from the true solution we permit ourselves to be. If the application is critical, we may not be able to allow any error whatsoever. In these cases, the graphical solution is of little value. For this reason, we will not pursue the graphical solution further but will turn now to solutions that are purely mathematical and hence yield more accurate answers.

9-4 SOLUTION BY ADDITION AND SUBTRACTION

Note what happens when we add the two linear equations shown next.

$$x + 5y = 26$$
$$-x + 3y = 30$$
$$8y = 56$$

Since the coefficients of the x terms differed only in sign, adding these equations eliminated the x terms entirely, giving us a new linear equation in only one unknown. This new equation can now be solved to give $y = 7$. If we then substitute $y = 7$ back into *either* of the original equations and solve for x, we see that $x = -9$. Is the point given by the coordinates $(-9, 7)$ the simultaneous solution? Let us substitute these coordinates into *both* equations and see what happens.

$$-9 + 5(7) \stackrel{?}{=} 26 \qquad -(-9) + 3(7) \stackrel{?}{=} 30$$
$$-9 + 35 = 26 \qquad 9 + 21 = 30$$

Since both members in either equation are equal, we conclude that the coordinates of the point $(-9, 7)$ do, indeed, simultaneously satisfy the set of given equations.

The important point to note here is that we were able to derive the simultaneous solution by the following simple method.

> **Rule 9-1**
> 1. Eliminate one of the variables in both equations by adding or subtracting the two given equations or multiples of these equations.
> 2. Solve the resulting linear equation for the value of its variable.
> 3. Substitute this value back into *either* of the original equations, and solve for the remaining variable.

In implementing the first step (elimination of one variable), it is totally unimportant whether we subtract or add. All that is important is that one or the other of the variables is eliminated, giving us a new linear equation in only one unknown. The elimination by either addition or subtraction is the key idea to solving simultaneous linear equations by this method. We illustrate this technique further with a few more examples.

Example 9-1

Find the simultaneous solution of the equation set

$$3x - 4y = 13$$
$$5x + 6y = 9$$

Solution Unlike the equations given previously, neither of these equations has a variable with the same numerical coefficient. Consequently, it is not possible simply to add or subtract in order to eliminate either x or y. Note, however, that if we

Simultaneous Linear Equations

multiply both sides of the first equation by 3 and both sides of the second equation by 2, we obtain the new equivalent equations

$$9x - 12y = 39$$
$$10x + 12y = 18$$

Note that in these new equations we have the same numerical coefficient for both y terms. Adding now gives us

$$19x = 57$$

from which $x = 3$.

Substituting $x = 3$ into the first equation, for example, gives us $9(3) - 12y = 39$, or $y = -1$. Note that this same value of y is obtained if we substitute $x = 3$ into the second equivalent equation or either of the two original equations. You should verify this statement for yourself.

On checking our solution $(3, -1)$, we see that

$$3(3) - 4(-1) \stackrel{?}{=} 13$$
$$9 + 4 = 13$$

and

$$5(3) + 6(-1) \stackrel{?}{=} 9$$
$$15 - 6 = 9$$

You should always check the simultaneous solution by substituting the values obtained into the *original* set of equations just to be certain you haven't made any errors.

Example 9-2 Find the simultaneous solution for the equations

$$\frac{x}{3} + \frac{y}{5} = \frac{-1}{15}$$
$$\frac{x}{10} + \frac{7y}{30} = \frac{1}{2}$$

Solution When working with equations with fractional coefficients, it is usually best to clear the fractions before beginning. Note that if we multiply both sides of the first equation by 15, we obtain $5x + 3y = -1$. Moreover, multiplying both sides of the second equation by 30 gives $3x + 7y = 15$.

We can now work with this new set of simpler, equivalent equations.

$$5x + 3y = -1$$
$$3x + 7y = 15$$

Multiplying the first equation by -3 gives $-15x - 9y = 3$. Multiplying the second equation by 5 gives $15x + 35y = 75$. Since the coefficients of the x terms in both equations are equal in magnitude but opposite in sign, we can add these to obtain

$$26y = 78$$

from which $y = 3$.

Substituting $y = 3$ into the first equivalent equation, for example, gives $5x + 3(3) = -1$, or $x = -2$.

We now have the simultaneous solution $(-2, 3)$. The correctness of this answer can be checked by substituting these coordinates into the original equations. It is left as an exercise for you to verify the correctness of this assertion.

Example 9-3

Find the simultaneous solution to the equations

$$Ax + By = C$$
$$Px + Qy = R$$

Solution This equation set differs from those we have been dealing with only in that the coefficients are literal rather than numerical. Nonetheless, Rule 9-1 can be applied.

We begin by multiplying the first equation by P and the second by A.

$$APx + BPy = CP$$
$$APx + AQy = AR$$

Subtracting, we obtain

$$BPy - AQy = CP - AR$$

from which

$$y = \frac{CP - AR}{BP - AQ}$$

We now return to the original equations and eliminate the y variable to obtain x. We multiply the first equation by Q and the second by B, obtaining

$$AQx + BQy = CQ$$
$$BPx + BQy = BR$$

Subtracting gives us

$$AQx - BPx = CQ - BR$$

from which

$$x = \frac{CQ - BR}{AQ - BP}$$

It is left as an exercise for you to check the correctness of the solution by substituting the coordinates of the point

$$\left(\frac{CQ - BR}{AQ - BP}, \frac{CP - AR}{BP - AQ}\right)$$

into the original equation.

Example 9-4

Solve the three-equation system of linear equations shown.

$$x - 2y + z = 3 \qquad (9\text{-}1)$$

$$x + y + 2z = 1 \qquad (9\text{-}2)$$

$$2x - y + z = 2 \qquad (9\text{-}3)$$

Simultaneous Linear Equations

Solution We can apply essentially the same method we have been using to the solution of three simultaneous linear equations in three unknowns. We begin by eliminating any variable we choose from any two of the equations. We then eliminate the same variable from any other two equations. This results in two new equations in just two unknowns, which may be solved by our usual method. Finally, we substitute these two variables back into any of the original three equations and solve for the remaining variable.

Let us arbitrarily choose to eliminate x from Equations (9–1) and (9–2). We may begin by multiplying Equation (9–2) by -1 to obtain

$$x - 2y + z = 3 \qquad (9\text{–}1)$$

$$-x - y - 2z = -1 \qquad (9\text{–}2a)$$

Upon adding, we get

$$-3y - z = 2 \qquad (9\text{–}4)$$

We can now eliminate x between Equations (9–2) and (9–3), for example. We multiply Equation (9–2) by -2, obtaining

$$-2x - 2y - 4z = -2 \qquad (9\text{–}2b)$$

$$2x - y + z = 2 \qquad (9\text{–}3)$$

Adding, we get

$$-3y - 3z = 0 \qquad (9\text{–}5)$$

We now solve (9–4) and (9–5) simultaneously by our previous methods, obtaining $y = -1$ and $z = 1$. Upon substituting these in Equation (9–1), for example, we get $x - 2(-1) + 1 = 3$, from which $x = 0$.

We test these values of $x = 0$, $y = -1$, $z = 1$ by substituting into the original equations as follows:

$$0 - 2(-1) + 1 \stackrel{?}{=} 3 \qquad (9\text{–}1)$$

$$2 + 1 = 3$$

$$0 + -1 + 2 \stackrel{?}{=} 1 \qquad (9\text{–}2)$$

$$1 = 1$$

$$0 - (-1) + 1 \stackrel{?}{=} 2 \qquad (9\text{–}3)$$

$$1 + 1 = 2$$

We conclude that these values do satisfy the three equations simultaneously.

Example 9–5 Find the simultaneous solution for the equation set

$$4x + 2y = 8$$
$$2x + y = -3$$

Solution At first glance, we might be tempted to multiply the second equation by -2 in an attempt to eliminate the x terms. Note, however, what happens:

$$4x + 2y = 8$$
$$-4x - 2y = 6$$

Upon adding, we obtain 0 = 14, which is obviously not true. We see that both the x and y terms have been eliminated, leaving us with no solution. Similarly, if we tried eliminating the y terms, we would end up with the same situation. The problem is that we have a system of equations that does not have a solution. If we graph the equations, as shown in Figure 9–5, the situation becomes clear. Since the lines representing the equations are parallel, they cannot possibly intersect; hence, a simultaneous solution does not exist.

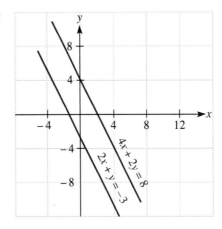

Figure 9–5

Example 9–6

Find the solution for the equations

$$2y - 4x = 2$$
$$y - 2x = 1$$

Solution If we multiply the second equation by -2, we obtain

$$-2y + 4x = -2$$

Upon adding, we have 0 = 0, which is certainly true. Unfortunately, however, all the variables have disappeared. Figure 9–6 clearly shows the situation. The two lines are identical, so there are an infinite number of solutions.

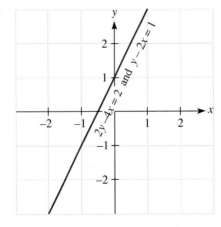

Figure 9–6

Simultaneous Linear Equations 201

From Examples 9–5 and 9–6, we see that not all systems of equations can be solved simultaneously. Thus there are three distinct cases when solving equations simultaneously.

> **Rule 9–2**
> 1. The lines cross. In this case, there is one unique solution.
> 2. The two lines are parallel. In this case, there is no simultaneous solution.
> 3. Both equations have the same graph. In this case, there is no unique solution; instead, an infinite number of solutions exist.

Exercise Set 9–1

Questions

1. What exactly is meant by the simultaneous solution of a set of linear equations?
2. What is the primary limitation to the graphical method of solution?
3. How many linear equations do you think are required to solve a linear equation in N unknowns?
4. Under what condition is a unique simultaneous solution not possible?
5. Under what condition will there be no simultaneous solution?

Problems

Solve simultaneously by addition or subtraction. Check each answer.

1. $x - 2y = -5$
   $3x + 4y = 20$

2. $6V - W = 15$
   $2V + 5W = 21$

3. $E + 2I = 26$
   $4E - I = 32$

4. $2P + Q = 9$
   $4P - Q = 6$

5. $2x - 3y = 6$
   $6x - 9y = 36$

6. $2x + y = 6$
   $x - y = 0$

7. $x + 5 = 0$
   $y = -2$

8. $8M - 10N = 16$
   $15N - 12M = -24$

9. $x + 2y = 0$
   $2x - y = 0$

10. $14V + 30E = -70$
    $15E + 7V = -35$

11. $x - 8y = 0$
    $x + y = 45$

12. $R + 2S = -2$
$15R - 4S = 106$

13. $6x + 2y = 7$
$3x - 2y = -1\frac{1}{2}$

14. $-3x + 4y = 9$
$3x + 2y = -18$

15. $3x + 5y = 15$
$6x + 10y = 30$

16. $x = -3$
$y - 4 = 0$

17. $\dfrac{P}{x} + \dfrac{Q}{y} = \dfrac{1}{xy}$
$\dfrac{R}{x} + \dfrac{S}{y} = \dfrac{1}{xy}$

18. $L - R + C = 0$
$L + R + C = 0$
$L + R - C = 2$

19. $2a + 3b + 5c = 0$
$6a - 2b - 3c = 3$
$8a - 5b - 6c = 1$

20. $X - 8 = Y$
$X - 2Z = -4$
$3Z - 3Y = 12$

9–5 SOLUTION BY SUBSTITUTION

Example 9–7 illustrates how substitution can be used to solve a system of equations.

Example 9–7

Consider the system of equations.

$$2x + 3y = -10$$
$$3x - 2y = -2$$

We can solve either of these equations for whichever variable we choose and then substitute that result in the other equation, forming a linear equation in only one variable. This equation can then be solved and the answer can be substituted into either of the original equations to solve for the remaining variable. For example, let us choose to solve the first equation for x in terms of y.

$$x = \frac{-10 - 3y}{2}$$

We now substitute this into the second equation, obtaining

$$3\frac{-10 - 3y}{2} - 2y = -2$$

Simultaneous Linear Equations

which may be solved for y, giving $y = -2$. Substituting this value of $y = -2$ into either original equation gives the value of x. For example, substituting $y = -2$ in the first equation gives

$$2x + 3(-2) = -10$$

Upon solving for x, we get $x = -2$.

We check our answer by the usual method of substituting the values of x and y into both original equations.

$$2x + 3y = -10 \qquad\qquad 3x - 2y = -2$$
$$2(-2) + 3(-2) \stackrel{?}{=} -10 \qquad 3(-2) - 2(-2) \stackrel{?}{=} -2$$
$$-4 - 6 = -10 \qquad\qquad -6 + 4 = -2$$

Therefore, we conclude that the point $(-2, -2)$ is the solution of the given linear system.

Rule 9–3 summarizes the method of solving by substitution.

Rule 9–3
1. Solve either equation for one of the variables in terms of the other.
2. Substitute the result from step 1 into the other original equation.
3. Solve the equation from step 2.
4. Substitute the solution from step 3 into either original equation and solve for the other variable.

Exercise Set 9–2

Problems

Solve the following problems by the method of substitution. Check each answer.

1. $V - 2I = 11$
   $3V + 5I = -11$
2. $2x - 5y = 4$
   $y = 3x + 7$
3. $2a + 6b = 3$
   $a - 4b = 2$
4. $2t - 13s = 45$
   $5t - 10s = 45$
5. $x + y = 2$
   $2y - 5x = -10$
6. $P = 3 - 2Q$
   $P + 2Q = 0$
7. $x - 2y = 2$
   $2x = 4y + 4$
8. $3i + v = 8$
   $2i - 3v = -13$
9. $4x - 14 = 2y$
   $2x - 3y = 5$
10. $5k - 4m = 5$
    $7k - 5m = 4$

9-6 SOLUTION BY COMPARISON

Often it is convenient to obtain a simultaneous solution by solving *both* equations for the *same* variable, equating the results, then solving for the single variable thus produced. This can then be substituted into either of the original equations, and the remaining variable can be found.

Example 9-8

Consider the two linear equations

$$x + 2y = 1$$
$$3x + y = -2$$

Solving both equations for x gives

$$x = 1 - 2y \quad \text{(from the first equation)}$$

and

$$x = \frac{-2 - y}{3} \quad \text{(from the second equation)}$$

We now equate both equations, obtaining

$$1 - 2y = \frac{-2 - y}{3}$$

We solve for y, finding $y = 1$.

Substituting $y = 1$ into either original equation gives the required value of x. For example, in the first equation

$$x + 2(1) = 1$$
$$x = -1$$

Checking our answer,

$$x + 2y = 1 \qquad\qquad 3x + y = -2$$
$$(-1) + 2(1) \stackrel{?}{=} 1 \qquad 3(-1) + 1 \stackrel{?}{=} -2$$
$$-1 + 2 = 1 \qquad\qquad -3 + 1 = -2$$

We conclude that $(-1, 1)$ satisfies both equations simultaneously.

Rule 9-4 summarizes this process.

Rule 9-4

1. Solve both equations for the same variable.
2. Equate both results from step 1.
3. Solve for the variable from step 2.
4. Substitute the solution in step 3 into either original equation and solve for the remaining variable.

Exercise Set 9-3

Problems

Solve the following problems by the method of comparison. Check each answer.

1. $2a - 8 = b$
   $2b + 3a = 5$

2. $7L - 2C = 40$
   $2L - 5C = 7$

3. $4x - 10y - 164 = 0$
   $3x - 2y - 68 = 0$

4. $k + t = 8$
   $k - t = 2$

5. $3x - 4y = -11$
   $2x + 3y = 9$

6. $2x + 15y = 35$
   $12x = 21 - 9y$

7. $3m - 8n + 20 = 0$
   $8m + 3n = 44$

8. $-9x + 4y + 16 = 0$
   $7x - 6y = -2$

9. $P + Q = 50$
   $2P + 4Q = 140$

10. $2s - 3t = 10$
    $5s + 6t = 29$

Second-Order Determinants

You will probably agree that the pencil-and-paper solution of simultaneous linear equations in two or three unknowns can be tedious. As the number of unknowns increases, so does the time spent in making boring, repetitious computations of the type so prone to simple arithmetical errors. And if you happen to make even a slight mistake that goes unnoticed, it gets carried through *all* subsequent calculations, often compounding the error.

However, the very fact that such solutions are monotonous has led to the development of algorithms (mathematical recipes), which are more or less mechanical and, once understood, are less likely to be a source of error. These recipes employ what are called *determinants*. Not only do they lessen the chance of error, since *no algebra is involved*, but they also greatly speed the solution. We will now review these algorithms and then see how we can apply them to the number-crunching capabilities of the personal computer. For it is here that the real advantages of determinants are best appreciated and applied.

Take a look at the two following equations:

$$a_1 x + b_1 y = c_1$$
$$a_2 x + b_2 y = c_2$$

If we multiply the first equation by b_2 and the second by b_1, we obtain

$$a_1 b_2 x + b_1 b_2 y = b_2 c_1$$
$$a_2 b_1 x + b_1 b_2 y = b_1 c_2$$

If we now subtract the second equation from the first, we can see that the $b_1 b_2 y$ terms drop out, leaving

$$a_1 b_2 x - a_2 b_1 x = b_2 c_1 - b_1 c_2$$

which can be rearranged, after factoring out x, to give

$$x = \frac{b_2 c_1 - b_1 c_2}{a_1 b_2 - a_2 b_1}$$

Similarly, we can show that

$$y = \frac{a_1 c_2 - a_2 c_1}{a_1 b_2 - a_2 b_1}$$

Note that the denominators in *both* equations are exactly alike. Also note that the numerator for the x variable makes use of the b and c coefficients, whereas the y variable uses the a and c coefficients. These facts, together with the symmetrical appearance of the terms in both fractions, suggests the possibility that there might be a simple, mechanical way of manipulating the constants to yield the values of x and y. Thus we ask this question: Is it possible to arrange the coefficients of *any* set of two linear equations in two unknowns in such a way that their simultaneous solution reveals itself in the physical arrangement of the terms? The answer is yes. Such a physical arrangement of four coefficients is known as a second-order **matrix**, or **array**; associated with each matrix is a **second-order determinant,** which has a numerical value.

In particular, the determinant for the common denominator of the last two equations is

$$\begin{vmatrix} a_1 & b_1 \\ a_2 & b_2 \end{vmatrix}$$

and the determinant has a value defined by

$$\begin{vmatrix} a_1 & b_1 \\ a_2 & b_2 \end{vmatrix} = a_1 b_2 - a_2 b_1$$

The numbers in the determinant are called its *elements;* the horizontal line of elements is called a *row*, and the vertical line is known as a *column.*

In defining the determinant's value, as shown by the previous equation, certain rules must be set up regarding the manner in which the elements are to be manipulated. Figure 9–7 illustrates the rules for evaluating *any* second-order determinant. The product diagonal running from upper left to lower right is called the **principal diagonal;** the other is known as the **secondary diagonal.** Therefore, strictly by definition, the principal-diagonal product ($a_1 b_2$) minus the secondary-diagonal product ($a_2 b_1$) equals the value of a second-order determinant.

Simultaneous Linear Equations

Figure 9-7

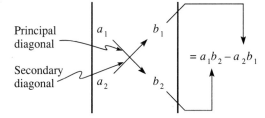

We now see that the numerators of the equations for x and y can also be written as second-order determinants:

$$\begin{vmatrix} c_1 & b_1 \\ c_2 & b_2 \end{vmatrix} = b_2 c_1 - b_1 c_2$$

$$\begin{vmatrix} a_1 & c_1 \\ a_2 & c_2 \end{vmatrix} = a_1 c_2 - a_2 c_1$$

At this point, we are in a position to write the equations for x and y using these determinants:

Rule 9-5

$$x = \frac{\begin{vmatrix} c_1 & b_1 \\ c_2 & b_2 \end{vmatrix}}{\begin{vmatrix} a_1 & b_1 \\ a_2 & b_2 \end{vmatrix}} = \frac{b_2 c_1 - b_1 c_2}{a_1 b_2 - a_2 b_1}$$

$$y = \frac{\begin{vmatrix} a_1 & c_1 \\ a_2 & c_2 \end{vmatrix}}{\begin{vmatrix} a_1 & b_1 \\ a_2 & b_2 \end{vmatrix}} = \frac{a_1 c_2 - a_2 c_1}{a_1 b_2 - a_2 b_1}$$

As with all simultaneous linear equations in two unknowns, we have three possible situations to consider:

1. If none of the determinants has a value of zero, there is only *one unique solution* to the system of equations.
2. If the denominator determinant is zero and the numerator determinants are *not* zero, there is *no solution* to the system of equations.
3. If *both* the numerator and denominator determinants are zero, there are *an infinite number of solutions*.

Here, then, are the steps for using determinants to solve any two simultaneous linear equations in two unknowns:

1. Write each equation in *standard form*: $a_n x + b_n y = c_n$.
2. Form the determinant of the denominator (which will be the same for both equations) and find its value.
3. Form the determinants of the two numerators and find the value of each.
4. Carry out the simple division of determinants in Rule 9-5.
5. Check the solution by substitution into the original equations.

Example 9-9

Find the simultaneous solution to $5x - 11 = -y$ and $2y - 8 = -3x$.

Step 1 Write each equation in standard form: $a_n x + b_n y = c_n$.
$$5x + y = 11$$
$$3x + 2y = 8$$

Step 2 Form the determinant of the denominator (which will be the same for both equations) and find its value.
$$\begin{vmatrix} a_1 & b_1 \\ a_2 & b_2 \end{vmatrix} = a_1 b_2 - a_2 b_1$$
$$\begin{vmatrix} 5 & 1 \\ 3 & 2 \end{vmatrix} = 10 - 3 = 7$$

Step 3 Form the determinants of the two numerators and find the value of each.
$$\begin{vmatrix} c_1 & b_1 \\ c_2 & b_2 \end{vmatrix} = b_2 c_1 - b_1 c_2$$
$$\begin{vmatrix} 11 & 1 \\ 8 & 2 \end{vmatrix} = 22 - 8 = 14$$
$$\begin{vmatrix} a_1 & c_1 \\ a_2 & c_2 \end{vmatrix} = a_1 c_2 - a_2 c_1$$
$$\begin{vmatrix} 5 & 11 \\ 3 & 8 \end{vmatrix} = 40 - 33 = 7$$

Step 4 Carry out the simple division of the determinants given by
$$x = \frac{\begin{vmatrix} c_1 & b_1 \\ c_2 & b_2 \end{vmatrix}}{\begin{vmatrix} a_1 & b_1 \\ a_2 & b_2 \end{vmatrix}} = \frac{14}{7} = 2$$

$$y = \frac{\begin{vmatrix} a_1 & c_1 \\ a_2 & c_2 \end{vmatrix}}{\begin{vmatrix} a_1 & b_1 \\ a_2 & b_2 \end{vmatrix}} = \frac{7}{7} = 1$$

Step 5 Check the solution set by substitution into the original equations.
$$5x + y = 11 \qquad 3x + 2y = 8$$
$$5(2) + 1 \stackrel{?}{=} 11 \qquad 3(2) + 2(1) \stackrel{?}{=} 8$$
$$10 + 1 = 11 \qquad 6 + 2 = 8$$

Determinants and Computer-Assisted Analysis

Now that we have begun to understand determinants, we will pause here to see how they can be used in computer-aided solutions. Then in the next section, we take up the topic of third-order determinants.

Figure 9–8a shows a BASIC program that utilizes the concept of the second-order determinant developed in the previous section. Line 120 is simply the expression of the denominator determinant (DD)—that is,

Simultaneous Linear Equations

$$\mathrm{DD} = \begin{vmatrix} a_1 & b_1 \\ a_2 & b_2 \end{vmatrix} = a_1 b_2 - a_2 b_1$$

which, in BASIC, is written as A1*B2−A2*B1.

Line 130 is the numerator determinant of the *x* variable (DX):

$$\mathrm{DX} = \begin{vmatrix} c_1 & b_1 \\ c_2 & b_2 \end{vmatrix} = b_2 c_1 - b_1 c_2$$

which can be written in BASIC as B2*C1−B1*C2.

Figure 9–8

```
LIST
10 PRINT"                    filename: DTRMNT-2
20 PRINT
30 PRINT"This program solves a system of two simultaneous linear
40 PRINT"equations in two unknowns whose coefficients are real numbers
50 PRINT
60 PRINT"The general form of the equations is:
70 PRINT"            a1X + b1Y = c1
80 PRINT"            a2X + b2Y = c2
90 PRINT
100 INPUT"Enter the values of a1, b1, and c1"; A1,B1,C1
110 INPUT"Enter the values of a2, b2, and c2";. A2,B2,C2
120 LET DD=A1*B2-A2*B1
130 LET DX=B2*C1-B1*C2
140 LET DY=A1*C2-A2*C1
150 PRINT"The value of x is ";DX/DD
160 PRINT"The value of y is ";DY/DD
170 END
Ok
```

a

```
RUN
                    filename: DTRMNT-2

This program solves a system of two simultaneous linear
equations in two unknowns whose coefficients are real numbers

The general form of the equations is:
            a1X + b1Y = c1
            a2X + b2Y = c2

Enter the values of a1, b1, and c1? 5,1,11
Enter the values of a2, b2, and c2? 3,2,8
The value of x is 2
The value of y is 1
Ok
```

b

Line 140 is the numerator determinant of the *y* variable (DY):

$$\mathrm{DY} = \begin{vmatrix} a_1 & c_1 \\ a_2 & c_2 \end{vmatrix} = a_1 c_2 - a_2 c_1$$

In BASIC, this is written as A1*C2−A2*C1.

Line 150 gives the value of the *x* variable as the quotient

$$x = \frac{\begin{vmatrix} c_1 & b_1 \\ c_2 & b_2 \end{vmatrix}}{\begin{vmatrix} a_1 & b_1 \\ a_2 & b_2 \end{vmatrix}} = \frac{\mathrm{DX}}{\mathrm{DD}}$$

Finally, line 160 gives the value of the y variable as the quotient

$$y = \frac{\begin{vmatrix} a_1 & c_1 \\ a_2 & c_2 \end{vmatrix}}{\begin{vmatrix} a_1 & b_1 \\ a_2 & b_2 \end{vmatrix}} = \frac{DY}{DD}$$

In Figure 9–8b, we have RUN the BASIC program for the simultaneous solution of the linear system

$$5x + y = 11$$
$$3x + 2y = 8$$

The results yield $x = 2$ and $y = 1$, as obtained earlier.

Third-Order Determinants

We have seen how second-order determinants simplify the solution of simultaneous linear equations. More importantly, we have learned how determinants can be applied in writing a BASIC computer program that allows us to find the general solution to *any* system of two linear equations in which the coefficients of the x and y variables are real numbers. We can quickly see the time-saving, error-free advantages of this approach!

We are now ready to begin our discussion of third-order determinants. In general, the three equations have the form

$$a_1x + b_1y + c_1z = d_1$$
$$a_2x + b_2y + c_2z = d_2$$
$$a_3x + b_3y + c_3z = d_3$$

By using the method of addition and subtraction, it can be shown that the solution to the system of equations is

$$x = \frac{b_2c_3d_1 + b_1c_2d_3 + b_3c_1d_2 - b_2c_1d_3 - b_3c_2d_1 - b_1c_3d_2}{a_1b_2c_3 + a_3b_1c_2 + a_2b_3c_1 - a_3b_2c_1 - a_1b_3c_2 - a_2b_1c_3}$$

$$y = \frac{a_1c_3d_2 + a_3c_2d_1 + a_2c_1d_3 - a_3c_1d_2 - a_1c_2d_3 - a_2c_3d_1}{a_1b_2c_3 + a_3b_1c_2 + a_2b_3c_1 - a_3b_2c_1 - a_1b_3c_2 - a_2b_1c_3}$$

$$z = \frac{a_1b_2d_3 + a_3b_1d_2 + a_2b_3d_1 - a_3b_2d_1 - a_1b_3d_2 - a_2b_1d_3}{a_1b_2c_3 + a_3b_1c_2 + a_2b_3c_1 - a_3b_2c_1 - a_1b_3c_2 - a_2b_1c_3}$$

A glance at these solutions reveals that the denominator is the same for each variable. Therefore, it seems appropriate to define a determinant for this expression. This **third-order determinant** is given by

$$\begin{vmatrix} a_1 & b_1 & c_1 \\ a_2 & b_2 & c_2 \\ a_3 & b_3 & c_3 \end{vmatrix} = a_1b_2c_3 + a_3b_1c_2 + a_2b_3c_1 - a_3b_2c_1 - a_1b_3c_2 - a_2b_1c_3$$

In making this definition of a third-order determinant, we are once again compelled to establish a rule of operation that represents the value of the determinant. Figure 9–9 shows the matrix with the first two columns rewritten to the right of the array. The columns to the right are not part of the determinant; instead they are part of a simple strategy devised to make the evaluation a purely mechanical process, bereft of algebra. As shown in Figure 9–9, the products of the elements in the principal diagonal as well as those of the elements in the two parallel diagonals to its right are *added* together to form

Simultaneous Linear Equations

the first three terms of the determinant. Also, the products of the elements in the secondary diagonal as well as those of the elements in the two parallel diagonals to its right are *subtracted* from the first three terms. This scheme makes a good mnemonic and is frequently used for evaluating third-order determinants.

Figure 9-9

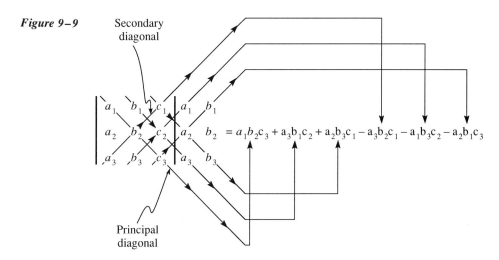

In a similar manner, we may define determinants for the numerators of the three variables. By using the results of all four determinants to write fractions, we obtain the general solution to a system of three linear equations in three unknowns.

Rule 9-6

$$x = \frac{\begin{vmatrix} d_1 & b_1 & c_1 \\ d_2 & b_2 & c_2 \\ d_3 & b_3 & c_3 \end{vmatrix}}{\begin{vmatrix} a_1 & b_1 & c_1 \\ a_2 & b_2 & c_2 \\ a_3 & b_3 & c_3 \end{vmatrix}} = \frac{b_2 c_3 d_1 + b_1 c_2 d_3 + b_3 c_1 d_2 - b_2 c_1 d_3 - b_3 c_2 d_1 - b_1 c_3 d_2}{a_1 b_2 c_3 + a_3 b_1 c_2 + a_2 b_3 c_1 - a_3 b_2 c_1 - a_1 b_3 c_2 - a_2 b_1 c_3}$$

$$y = \frac{\begin{vmatrix} a_1 & d_1 & c_1 \\ a_2 & d_2 & c_2 \\ a_3 & d_3 & c_3 \end{vmatrix}}{\begin{vmatrix} a_1 & b_1 & c_1 \\ a_2 & b_2 & c_2 \\ a_3 & b_3 & c_3 \end{vmatrix}} = \frac{a_1 c_3 d_2 + a_3 c_2 d_1 + a_2 c_1 d_3 - a_3 c_1 d_2 - a_1 c_2 d_3 - a_2 c_3 d_1}{a_1 b_2 c_3 + a_3 b_1 c_2 + a_2 b_3 c_1 - a_3 b_2 c_1 - a_1 b_3 c_2 - a_2 b_1 c_3}$$

$$z = \frac{\begin{vmatrix} a_1 & b_1 & d_1 \\ a_2 & b_2 & d_2 \\ a_3 & b_3 & d_3 \end{vmatrix}}{\begin{vmatrix} a_1 & b_1 & c_1 \\ a_2 & b_2 & c_2 \\ a_3 & b_3 & c_3 \end{vmatrix}} = \frac{a_1 b_2 d_3 + a_3 b_1 d_2 + a_2 b_3 d_1 - a_3 b_2 d_1 - a_1 b_3 d_2 - a_2 b_1 d_3}{a_1 b_2 c_3 + a_3 b_1 c_2 + a_2 b_3 c_1 - a_3 b_2 c_1 - a_1 b_3 c_2 - a_2 b_1 c_3}$$

Advanced Topics in Algebra

As was the case with second-order determinants, there are three possibilities to consider:

1. If none of the determinants has a value of zero, there is only *one unique solution* to the system of equations.
2. If the value of the denominator determinant is zero and the values of the numerator determinants are *not* zero, there is *no solution* to the system of equations.
3. If the value of both the numerator and denominator determinants is zero, there are *an infinite number of solutions* to the system of equations.

We may now summarize the process of solving three simultaneous linear equations in three unknowns with the following steps:

1. Write each equation in the *standard form*:

$$a_n x + b_n y + c_n z = d_n$$

2. Form the determinant of the denominator (which will be the same for all three equations) and find its value.
3. Form the determinants of the three numerators and find the value of each.
4. Carry out the simple division indicated by the equations in Rule 9–6.
5. Check the solutions by substitution into the original equations.

Example 9–10

Find the simultaneous solution to:

$$-S + 1 + P + 5R = 0$$
$$-3R - 3 + 2S = -P$$
$$2S - 4R = -2 - 3P$$

Step 1 Write each equation in standard form: $a_n x + b_n y + c_n z = d_n$.

$$P + 5R - S = -1$$
$$P - 3R + 2S = 3$$
$$3P - 4R + 2S = -2$$

Step 2 Form the determinant of the denominator (which will be the same for all three equations) and find its value. (See Figure 9–10.)

Figure 9–10

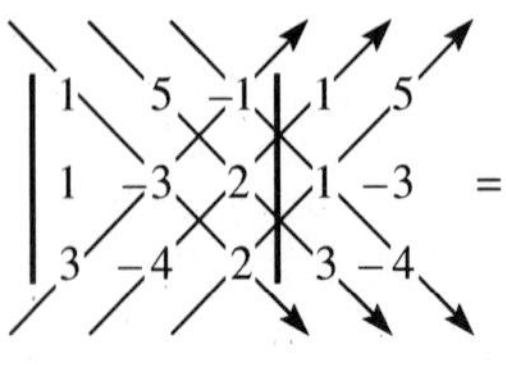

$$\begin{vmatrix} 1 & 5 & -1 \\ 1 & -3 & 2 \\ 3 & -4 & 2 \end{vmatrix} = (-6) + (30) + (4) - (9) - (-8) - (10)$$

$$= -6 + 30 + 4 - 9 + 8 - 10 = 17$$

Step 3 Form the determinants of the three numerators and find the value of each. (See Figure 9–11.)

Simultaneous Linear Equations

Figure 9–11

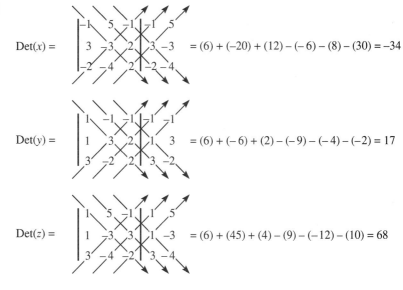

$$\text{Det}(x) = \begin{vmatrix} -1 & 5 & -1 \\ 3 & -3 & 2 \\ -2 & -4 & 2 \end{vmatrix} \begin{matrix} -1 & 5 \\ 3 & -3 \\ -2 & -4 \end{matrix} = (6) + (-20) + (12) - (-6) - (8) - (30) = -34$$

$$\text{Det}(y) = \begin{vmatrix} 1 & -1 & -1 \\ 1 & 3 & 2 \\ 3 & -2 & 2 \end{vmatrix} \begin{matrix} 1 & -1 \\ 1 & 3 \\ 3 & -2 \end{matrix} = (6) + (-6) + (2) - (-9) - (-4) - (-2) = 17$$

$$\text{Det}(z) = \begin{vmatrix} 1 & 5 & -1 \\ 1 & -3 & 3 \\ 3 & -4 & -2 \end{vmatrix} \begin{matrix} 1 & 5 \\ 1 & -3 \\ 3 & -4 \end{matrix} = (6) + (45) + (4) - (9) - (-12) - (10) = 68$$

Step 4 Carry out the simple division of the determinants.

$$x = P = \frac{-34}{17} = -2$$

$$y = R = \frac{17}{17} = 1$$

$$z = S = \frac{68}{17} = 4$$

Step 5 Check the solutions by substitution into the original equations.

$$\begin{array}{ll} P + 5R - S = -1 & (-2) + 5(1) - (4) = -2 + 5 - 4 = -1 \\ P - 3R + 2S = 3 & (-2) - 3(1) + 2(4) = -2 - 3 + 8 = 3 \\ 3P - 4R + 2S = -2 & 3(-2) - 4(1) + 2(4) = -6 - 4 + 8 = -2 \end{array}$$

Third-Order Determinants Applied to Computer Analysis

Figure 9–12a (page 214) shows a BASIC program that utilizes third-order determinants. Note that lines 140 through 170 give the determinant values of the denominator (DD); the x variable (DX); the y variable (DY); and the z variable (DZ). This is exactly what we did in Figure 9–8 for second-order determinants. Finally, lines 180 through 200 give the required quotients of determinants for the x, y, and z variables.

In Figure 9–12b (page 214), we have RUN the BASIC program for the simultaneous solution of the linear system

$$\begin{array}{l} P + 5R - S = -1 \\ P - 3R + 2S = 3 \\ 3P - 4R + 2S = -2 \end{array}$$

The results obtained verify our earlier solution for this system—that is, $P = -2$, $R = 1$, and $S = 4$.

214 Advanced Topics in Algebra

Figure 9–12

```
LIST
10 PRINT"                       filename: DTRMNT-3
20 PRINT
30 PRINT"This program solves a system of three simultaneous linear
40 PRINT"equations in three unknowns whose coefficients are real numbers
50 PRINT
60 PRINT"The general form of the equations is:
70 PRINT"            a1X + b1Y + c1Z = d1
80 PRINT"            a2X + b2Y + c2Z = d2
90 PRINT"            a3X + b3Y + c3Z = d3
100 PRINT
110 INPUT"Enter the values of a1, b1, c1 and d1"; A1,B1,C1,D1
120 INPUT"Enter the values of a2, b2, c2 and d2"; A2,B2,C2,D2
130 INPUT"Enter the values of a3, b3, c3 and d3"; A3,B3,C3,D3
140 LET DD=A1*B2*C3+A3*B1*C2+A2*B3*C1-A3*B2*C1-A1*B3*C2-A2*B1*C3
150 LET DX=B2*C3*D1+B1*C2*D3+B3*C1*D2-B2*C1*D3-B3*C2*D1-B1*C3*D2
160 LET DY=A1*C3*D2+A3*C2*D1+A2*C1*D3-A3*C1*D2-A1*C2*D3-A2*C3*D1
170 LET DZ=A1*B2*D3+A3*B1*D2+A2*B3*D1-A3*B2*D1-A1*B3*D2-A2*B1*D3
180 PRINT"The value of x is ";DX/DD
190 PRINT"The value of y is ";DY/DD
200 PRINT"The value of z is ";DZ/DD
210 END
Ok
```

a

```
RUN
                        filename: DTRMNT-3

This program solves a system of three simultaneous linear
equations in three unknowns whose coefficients are real numbers

The general form of the equations is:
         a1X + b1Y + c1Z = d1
         a2X + b2Y + c2Z = d2
         a3X + b3Y + c3Z = d3

Enter the values of a1, b1, c1 and d1? 1,5,-1,-1
Enter the values of a2, b2, c2 and d2? 1,-3,2,3
Enter the values of a3, b3, c3 and d3? 3,-4,2,-2
The value of x is -2
The value of y is 1
The value of z is 4
Ok
```

b

Exercise Set 9–4

Problems

1. Use second-order determinants to verify your solutions to Problems 1 through 17 in Exercise Set 9–1.
2. Use third-order determinants to verify your solutions to Problems 18 through 20 in Exercise Set 9–1.
3. Write a BASIC program similar to the ones in Figure 9–8 and 9–12 that can be used to solve *either* a two- or three-equation linear system.
4. Use the program developed in Problem 3 to verify the solution to Example 9–4 and the solutions to the problems in Exercise Sets 9–2 and 9–3.

Key Terms

simultaneous solution
matrix
array

second-order determinant
principal diagonal

secondary diagonal
third-order determinant

Simultaneous Linear Equations

Important Rules

Solution by Addition and Subtraction
Rule 9–1

1. Eliminate one of the variables in both equations by adding or subtracting the two given equations or multiples of these equations.
2. Solve the resulting linear equation for the value of its variable.
3. Substitute this value back into *either* of the original equations, and solve for the remaining variable.

Three Cases of the Simultaneous Solution
Rule 9–2

1. One unique solution
2. No solution
3. Infinite number of solutions

(Lines cross.)   (Lines are parallel.)   (Lines coincide.)

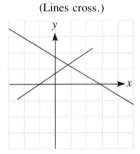 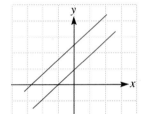 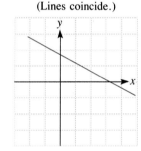

Figure 9–13   *Figure 9–14*   *Figure 9–15*

Solution by Substitution
Rule 9–3

1. Solve either equation for one of the variables in terms of the other.
2. Substitute the result from step 1 into the other original equation.
3. Solve the equation from step 2.
4. Substitute the solution from step 3 into either original equation and solve for the other variable.

Solution by Comparison
Rule 9–4

1. Solve both equations for the same variable.
2. Equate both results from step 1.
3. Solve for the variable from step 2.
4. Substitute the solution in step 3 into either original equation and solve for the remaining variable.

Solution with Determinants

To solve a system of two equations in two unknowns:

$$a_1 x + b_1 y = c_1$$
$$a_2 x + b_2 y = c_2$$

Rule 9-5

$$x = \frac{\begin{vmatrix} c_1 & b_1 \\ c_2 & b_2 \end{vmatrix}}{\begin{vmatrix} a_1 & b_1 \\ a_2 & b_2 \end{vmatrix}} = \frac{b_2 c_1 - b_1 c_2}{a_1 b_2 - a_2 b_1}$$

$$y = \frac{\begin{vmatrix} a_1 & c_1 \\ a_2 & c_2 \end{vmatrix}}{\begin{vmatrix} a_1 & b_1 \\ a_2 & b_2 \end{vmatrix}} = \frac{a_1 c_2 - a_2 c_1}{a_1 b_2 - a_2 b_1}$$

To solve a system of three equations in three unknowns:

$$a_1 x + b_1 y + c_1 z = d_1$$
$$a_2 x + b_2 y + c_2 z = d_2$$
$$a_3 x + b_3 y + c_3 z = d_3$$

Rule 9-6

$$x = \frac{\begin{vmatrix} d_1 & b_1 & c_1 \\ d_2 & b_2 & c_2 \\ d_3 & b_3 & c_3 \end{vmatrix}}{\begin{vmatrix} a_1 & b_1 & c_1 \\ a_2 & b_2 & c_2 \\ a_3 & b_3 & c_3 \end{vmatrix}} = \frac{b_2 c_3 d_1 + b_1 c_2 d_3 + b_3 c_1 d_2 - b_2 c_1 d_3 - b_3 c_2 d_1 - b_1 c_3 d_2}{a_1 b_2 c_3 + a_3 b_1 c_2 + a_2 b_3 c_1 - a_3 b_2 c_1 - a_1 b_3 c_2 - a_2 b_1 c_3}$$

$$y = \frac{\begin{vmatrix} a_1 & d_1 & c_1 \\ a_2 & d_2 & c_2 \\ a_3 & d_3 & c_3 \end{vmatrix}}{\begin{vmatrix} a_1 & b_1 & c_1 \\ a_2 & b_2 & c_2 \\ a_3 & b_3 & c_3 \end{vmatrix}} = \frac{a_1 c_3 d_2 + a_3 c_2 d_1 + a_2 c_1 d_3 - a_3 c_1 d_2 - a_1 c_2 d_3 - a_2 c_3 d_1}{a_1 b_2 c_3 + a_3 b_1 c_2 + a_2 b_3 c_1 - a_3 b_2 c_1 - a_1 b_3 c_2 - a_2 b_1 c_3}$$

$$z = \frac{\begin{vmatrix} a_1 & b_1 & d_1 \\ a_2 & b_2 & d_2 \\ a_3 & b_3 & d_3 \end{vmatrix}}{\begin{vmatrix} a_1 & b_1 & c_1 \\ a_2 & b_2 & c_2 \\ a_3 & b_3 & c_3 \end{vmatrix}} = \frac{a_1 b_2 d_3 + a_3 b_1 d_2 + a_2 b_3 d_1 - a_3 b_2 d_1 - a_1 b_3 d_2 - a_2 b_1 d_3}{a_1 b_2 c_3 + a_3 b_1 c_2 + a_2 b_3 c_1 - a_3 b_2 c_1 - a_1 b_3 c_2 - a_2 b_1 c_3}$$

10 Quadratic Equations

Upon completion of this chapter, you will have:
1. Learned how to graph a quadratic relationship.
2. Learned how to apply the quadratic formula to the solution of quadratic equations.
3. Learned how to interpret the nature of quadratic roots.
4. Learned how to find a quadratic equation given its roots.

10-1 INTRODUCTION

In Chapters 8 and 9, we studied the properties of linear relationships. Recall that the graph of a linear equation is a straight line, and the x term has an implied exponent of 1. In this chapter, we explore some of the common properties of another useful relation called a **quadratic relationship.** The graph of this relationship is not a straight line but instead is curved, and the exponent of the x term is 2.

In Figure 10-1, we see the graph of the power dissipated in a 5 Ω resistor. Here, the abscissa (x) has been relabeled I for current. The ordinate (y) has been relabeled P for power. Note that the I term is affected by an exponent of 2. We say, therefore, that this curve representing the power dissipated by current flowing through a 5 Ω resistor is a quadratic relationship.

The general form of a quadratic relationship is given as follows.

Rule 10-1

$$y = ax^2 + bx + c \quad (a \neq 0)$$

In the example shown by Figure 10-1, the coefficients given as b and c in Rule 10-1 are both equal to zero, and $a = 5$. As we shall see later, the coefficients b and c merely help to determine the general appearance of the quadratic curve. The shape of the curve given by Rule 10-1 is called a **parabola** (pa-RAB-o-la) and can be thought of as the shape obtained by taking a slice through a cone, as shown in Figure 10-2. Of course, the current-versus-power curve in Figure 10-1 is only half of a parabola, since we are not dealing with negative values of current. If we were to allow negative values, then the curve would appear as shown in Figure 10-3. All further discussion of parabolas in this chapter will permit negative values of x.

10-2 GRAPHING QUADRATIC RELATIONSHIPS

Any quadratic relationship given by Rule 10-1 can be graphed by using a few key points as reference. In Chapter 8, we observed that by knowing the coordinates of the y-intercept $(0, b)$ and the slope (m), we could draw a straight line using this information as a guide. Similarly, we will now show a method for graphing the general quadratic relationship.

Consider the quadratic relationship $y = x^2 + x - 6$, whose graph is shown in Figure 10-4. We begin by finding the y-intercept. Recall that the **y-intercept** is the point where the graph crosses the y-axis—that is, the point at which $x = 0$. If we substitute $x = 0$ into the given equation, we obtain

$$y = (0)^2 + (0) - 6$$

Note that the y-intercept is simply the value of c in Rule 10-1. In Figure 10-4, $c = -6$. Therefore, we may make the following definition.

Rule 10-2

The y-intercept of the equation $y = ax^2 + bx + c$ is the point $(0, c)$.

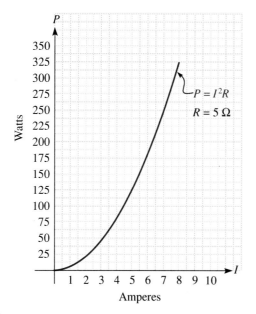

Figure 10–1

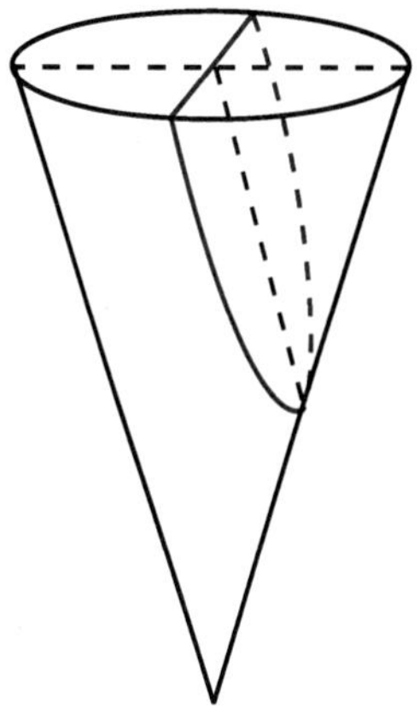

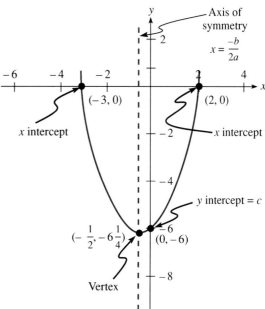

Figure 10–2

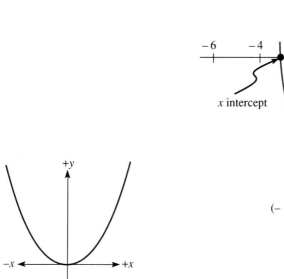

Figure 10–3

Figure 10–4

219

Obviously, we cannot tell much about the curve with just one point. It would be helpful if we could also find the points where the curve crosses the x-axis—that is, the **x-intercepts** (see Figure 10–4). At these points, the value of y must equal zero. Therefore, we write

$$y = 0 = x^2 + x - 6$$

From Chapter 5, we recall that expressions of this general form are often factorable. Indeed, this particular expression factors as $(x - 2)(x + 3)$. It is now obvious that there are only two values of x that will make $(x - 2)(x + 3)$ equal to zero: $x = 2$ and $x = -3$. We conclude that the x-intercepts are at the points $(2, 0)$ and $(-3, 0)$.

We are still not quite ready to sketch the graph. Some additional information is necessary. Notice that the graph of the parabola shown in Figure 10–4 is symmetrical about some vertical axis. In other words, the left and right halves are mirror images of each other. The vertical line about which these two halves exist is called the **axis of symmetry**. In Figure 10–3, the axis of symmetry happens to be the y-axis. This is *not* always the case. To find the axis of symmetry, then, we note that the x-intercepts themselves are distributed symmetrically about the axis. To find the x-coordinate of this axis, we need to determine the midpoint between the x-intercepts $(-3, 0)$ and $(2, 0)$. This is equivalent to finding the average value between two numbers. Recall that the average of two numbers is obtained by finding their sum and then dividing by 2. Therefore, we see that the average of -3 and 2 gives us the required axis of symmetry.

$$\text{Axis of symmetry:} \quad x = \frac{x_2 + x_1}{2} = \frac{-3 + 2}{2} = \frac{-1}{2}$$

Later, we prove that the axis of symmetry is given simply as follows:

Rule 10–3

$$x = \frac{-b}{2a}$$

In Figure 10–4, $b = 1$, and $a = 1$. Therefore, by Rule 10–3, the axis of symmetry is $x = -\frac{1}{2}$, which is consistent with the equation obtained previously from the average-value method.

Having obtained the equation of the axis of symmetry, we can determine the lowest point of the parabola through which this axis passes. This point is called the **vertex**, and its ordinate is simply the value of y when $x = -b/2a = -\frac{1}{2}$. Therefore, we see that

$$y = x^2 + x - 6 = \left(-\frac{1}{2}\right)^2 + \left(-\frac{1}{2}\right) - 6$$

$$= \frac{1}{4} - \frac{1}{2} - 6 = -\frac{25}{4}, \text{ or } -6\frac{1}{4}$$

The vertex, then, is at the point $(-\frac{1}{2}, -6\frac{1}{4})$.

We are finally in a position to sketch the graph of the quadratic relationship given by $y = x^2 + x - 6$ by drawing a smooth curve between the four key points just found. If greater accuracy is required, we can simply plug in particular values of x that interest us and solve for the associated values of y. We can use as many additional points as we wish. The key points serve as a guide. We summarize these points as follows.

Quadratic Equations

> **Rule 10–4**
> 1. y-intercept: $y = c$.
> 2. x-intercepts: Set $y = 0$ and solve the resulting equation for x.
> 3. Axis of symmetry: $x = {-b}/{2a}$.
> 4. Vertex: $x = {-b}/{2a}$. The y-value is the corresponding value obtained for this x.

Unfortunately, not all expressions of the form given by Rule 10–1 are factorable by ordinary means. This makes finding the x-intercepts a little more difficult. Also, not all graphs of parabolas cross the x-axis. In this latter case, we say the parabola has imaginary roots. Finally, the vertex may be a maximum rather than a minimum, as shown in Figure 10–4, and the parabola then opens downward rather than upward. We will see the implications of these variations in subsequent sections.

Exercise Set 10–1

Questions

Define the following terms in your own words.
1. Quadratic relationship
2. Parabola
3. Conic section
4. y-intercept
5. x-intercepts
6. Axis of symmetry
7. Vertex

Problems

Graph the following quadratic relationships. On each graph, label the vertical-axis intercept (y-intercept), the horizontal-axis intercepts (x-intercepts), axis of symmetry, and vertex.
1. $y = x^2 - x - 2$
2. $y = -3x^2 - 2x + 5$
3. $P = I^2 - 10I + 16$
4. $-3x - 4x^2 = M - 27$
5. $y = 2x^2 + 4x - 6$
6. $y = 27x^2 + 24x - 20$
7. $V = 2e^2 - 3e + 1$
8. $p = 144t^2 - 24t - 63$
9. $y = -12x^2 - 11x + 1$
10. $y = 25x^2 + 10x + 1$

10–3 THE QUADRATIC FORMULA

In the previous section, we were able to graph quadratic relationships with the help of the x-intercepts, which were found by factoring quadratic equations of the form $ax^2 + bx + c = 0$. These intercepts are more commonly referred to as the **roots** of the quadratic equation. It frequently happens, however, that the roots cannot be found by our usual factoring techniques discussed in Chapter 5, and we are forced to seek other methods.

Advanced Topics in Algebra

Consider the equation $x^2 - 4x + 1 = 0$. We see almost instantly that this expression cannot be factored into the form $(x + m)(x + n) = 0$. We can, however, force an equation of this type into a form that lends itself to an easy solution. The key to this strategy rests in transforming part of the equation into a perfect trinomial square that can, as we discovered in Chapter 5, be factored in the form of $(x + k)^2$.

Recall that a perfect trinomial square such as $x^2 - 4x + 4$ can be factored as $(x - 2)^2$ and has the characteristic that the middle term is twice the product of the square roots of the end terms. Note that this perfect trinomial is the same as the one we tried to factor earlier ($x^2 - 4x + 1$) except for the last term. Let us now try to change this expression so it becomes a perfect trinomial square.

We begin by rewriting the expression as

$$x^2 - 4x = -1$$

At this point we have at least part of a perfect trinomial square in the left member. The question now, however, is how do we convert the entire left member to a perfect square? Recall that in a perfect square the middle term is twice the product of the square roots of the end terms. Therefore, we need some term N^2 in the left member that fulfills the requirement

$$-4x = 2xN$$

Solving for N, we obtain $N = -2$, or $N^2 = 4$.

We can now add this N^2 term to both sides of our equation, obtaining

$$x^2 - 4x + 4 = -1 + 4$$
$$(x - 2)^2 = 3$$

On taking the square root of both sides (Axiom 4–7), we obtain

$$x - 2 = \pm\sqrt{3}$$

Hence,

$$x = 2 \pm \sqrt{3}$$

On checking our answer in the original equation, we get

$$(2 + \sqrt{3})^2 - 4(2 + \sqrt{3}) + 1 \stackrel{?}{=} 0$$
$$4 + 4\sqrt{3} + 3 - 8 - 4\sqrt{3} + 1 \stackrel{?}{=} 0$$
$$4 + 3 - 8 + 1 \stackrel{?}{=} 0$$
$$8 - 8 = 0$$

It is left as an exercise for you to show that the other root $(2 - \sqrt{3})$ also satisfies this equation.

The foregoing procedure is often referred to as the method of **completing the square.** It is presented here for reasons other than mere mathematical novelty. For if $ax^2 + bx + c = y$ is any quadratic relationship, then completing the square when this expression is set equal to zero will give us the roots for any quadratic equation. The outcome of this procedure, then, is a general formula for finding quadratic roots.

We begin by writing $ax^2 + bx + c = 0$ as

$$ax^2 + bx = -c$$

We then divide through by the coefficient a, giving

$$x^2 + \frac{b}{a}x = \frac{-c}{a}$$

Quadratic Equations

We now seek a number N^2 that can be added to both sides, which fulfills the requirement

$$\frac{b}{a}x = 2xN$$

On solving for N, we obtain

$$N = \frac{b}{2a}$$

Adding N^2 to each side gives us

$$x^2 + \frac{b}{a}x + \frac{b^2}{4a^2} = \frac{-c}{a} + \frac{b^2}{4a^2}$$

which can be written as

$$\left(x + \frac{b}{2a}\right)^2 = \frac{b^2 - 4ac}{4a^2}$$

Taking the square root of each side gives

$$x + \frac{b}{2a} = \pm\sqrt{\frac{b^2 - 4ac}{4a^2}}$$

from which

$$x = -\frac{b}{2a} \pm \frac{\sqrt{b^2 - 4ac}}{2a}$$

or

$$x = \frac{-b \pm \sqrt{b^2 - 4ac}}{2a}$$

This last result, called the **quadratic formula,** can be used to determine the roots of any quadratic equation.

Rule 10-5 The Quadratic Formula

$$x = \frac{-b \pm \sqrt{b^2 - 4ac}}{2a}$$

In order to apply the quadratic formula to an equation, the equation must be in the form $ax^2 + bx + c = 0$. For example, the quadratic equation $\frac{1}{4}x^2 = 1 - x$ is not in the standard form. However, multiplying through by 4 and rearranging, we obtain

$$x^2 + 4x - 4 = 0$$

which is in the required form. In this case, $a = 1$, $b = 4$, and $c = -4$. On applying Rule 10-5, we get

$$x = \frac{-b \pm \sqrt{b^2 - 4ac}}{2a} = \frac{-4 \pm \sqrt{16 - 4(1)(-4)}}{2(1)}$$

$$= \frac{-4 \pm \sqrt{32}}{2} = \frac{-4 \pm 4\sqrt{2}}{2} = -2 \pm 2\sqrt{2}$$

In graphing the relationship given by $y = x^2 + 4x - 4$, we can use the approximate values (0.828, 0) and (−4.828, 0) for the roots, as shown in Figure 10–5.

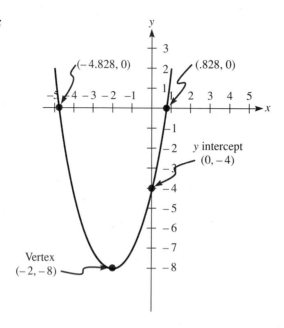

Figure 10–5

Now that we have this powerful tool in our possession, we can verify an assertion made in an earlier section that the axis of symmetry is given by $x = -b/2a$. Since the axis lies midway between the two roots, the x-coordinate of this midpoint is simply the average value of the roots. And because the roots are given by Rule 10–5, we obtain the average by adding the roots and dividing by 2 as follows:

$$\text{Root 1 plus root 2} = \frac{(-b + \sqrt{b^2 - 4ac}) + (-b - \sqrt{b^2 - 4ac})}{2a}$$

$$= \frac{-2b}{2a} = \frac{-b}{a}$$

Dividing by 2, we obtain the average value $x = -b/2a$, as stated previously.

Example 10–1

Use the quadratic formula in graphing the quadratic relationship $y = -x^2 - 4x + 2$.

Solution

$$x = \frac{4 \pm \sqrt{(-4)^2 - 4(-1)(2)}}{2(-1)} = -2 \pm \sqrt{6}$$

The y-intercept is (0, 2). The x-coordinate of the vertex is $x = -b/2a = -(-4)/2(-1) = -2$. Therefore, $y = -(-2)^2 - 4(-2) + 2 = 6$. The coordinates of the vertex are (−2, 6).

Quadratic Equations

The graph is shown in Figure 10–6. Note that the graph opens downward rather than upward. This is always the case when the coefficient of the x^2 term is negative.

Figure 10–6

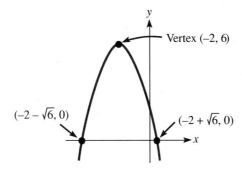

Exercise Set 10–2

Questions

1. What is the relationship between the x-intercepts of the graph of a quadratic equation and the roots of the equation?
2. Explain the logic behind the method of completing the square.
3. It is not always necessary to use the quadratic formula. For what situations is it necessary?

Problems

Use the quadratic formula to find the roots of the following quadratic relationships. Graph each relationship showing the vertex, axis of symmetry, and vertical-axis intercept (y-intercept).

1. $y = -5x + 6 + x^2$
2. $P - 16 = -10Q + Q^2$
3. $y = -3x + 27 - 4x^2$
4. $E = v^2 - 4v - 5$
5. $y - x^2 = -6x + 5$
6. $y = t^2 + 9 - 6t$
7. $y = -2x^2$
8. $y = x^2 - 4x$
9. $Z = -2p^2 + 4p + 6$
10. $y = 8x + 2x^2 + 3$
11. The following BASIC program calculates the real roots of any quadratic equation.

```
10 REM   The following program computes the real roots
20 REM   of the quadratic equation ax^2 + bx + c = 0.
30 REM
40 PRINT"Enter the value of a."
50 INPUT A
60 PRINT"Enter the value of b."
70 INPUT B
80 PRINT"Enter the value of c."
90 INPUT C
100 LET X1=(-B+(SQR((B^2)-(4*A*C))))/(2*A)
110 PRINT"The value of X1 is: ";X1
120 LET X2=(-B-(SQR((B^2)-(4*A*C))))/(2*A)
130 PRINT"The value of X2 is: ";X2
```

Extend this program to give the y-intercept, the axis of symmetry, and the vertex. Verify your answers to Problems 1 through 10 using this program.

10-4 THE NATURE OF QUADRATIC ROOTS

The portion of the quadratic formula under the radical sign ($b^2 - 4ac$) is called the **discriminant** and can be used to reveal the nature of the roots without actually having to solve the quadratic equation. Knowing what to expect of the roots prior to solution is often an aid to graphing and to solving quadratic problems in general.

There are three general situations associated with the value of the discriminant.

Rule 10-6

1. If $b^2 - 4ac = 0$, the roots are real and equal.
   For example, in the equation $9x^2 - 6x + 1 = 0$, the value of the discriminant is $b^2 - 4ac = (-6)^2 - 4(9)(1) = 36 - 36 = 0$. Therefore, we would expect the roots to be real and equal. As shown in Figure 10-7, the graph touches the x-axis at (1/3, 0).

   Figure 10-7

   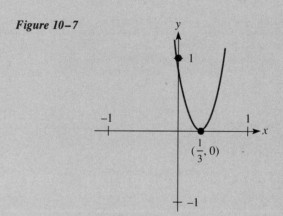

2. If $b^2 - 4ac > 0$, the roots are real and unequal.
   For example, in the equation $2x^2 + 5x - 12 = 0$, the value of the discriminant is $b^2 - 4ac = 25 - 4(2)(-12) = 121$, which is greater than zero. Therefore, the roots are real and unequal, as shown in Figure 10-8.

3. If $b^2 - 4ac < 0$, the roots are **imaginary roots**.
   In practical terms, this means that the graph of the equation does not touch or cross the x-axis. We discuss imaginary numbers briefly in Chapter 16. In the equation given by $x^2 - 6x + 11 = 0$, the discriminant is $b^2 - 4ac = 36 - 44 < 0$. Therefore, the roots are imaginary, and the graph does not touch the x-axis, as shown in Figure 10-9.

Quadratic Equations

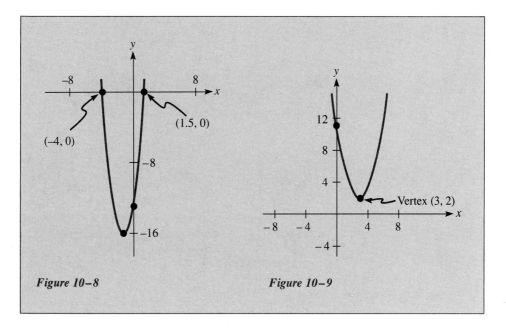

Figure 10-8

Figure 10-9

Exercise Set 10-3

Questions

1. For what purpose is the discriminant used?
2. What three general situations are described by the discriminant?
3. In terms of graphing, what is the implication of imaginary roots?

Problems

In each of the following problems, determine the nature of the roots without actually solving the problem.

1. $x^2 - x - 12 = 0$
2. $-4x + x^2 = 0$
3. $w^2 + 25 = 10w$
4. $-2 = -6x + 9x^2$
5. $2x^2 + 6x + 5 = 0$
6. $3E^2 + 12E = -12$
7. $M^2 - 1 = 2M$
8. $2t^2 + 4t + 5 = 0$
9. $10p^2 - 11p = 5$
10. $-5x^2 + 2x - 1 = 0$
11. Write a BASIC program that will evaluate the discriminant of a quadratic equation.

10-5 USING ROOTS TO FIND AN EQUATION

Often it is useful to be able to determine an equation given its roots. We may use the factoring method in reverse to accomplish this task. For example, given the roots 5 and -3, we know that since $x = 5$, then $x - 5 = 0$. Moreover, $x + 3 = 0$. Therefore, the two factors of the desired quadratic equation must be $x - 5$ and $x + 3$. Consequently, we write

$$(x - 5)(x + 3) = 0$$
$$x^2 + 3x - 5x - 15 = 0$$
$$x^2 - 2x - 15 = 0$$

As a further example, consider the roots 0 and ¼. We see that the factors are x and $x - ¼$. Therefore, we write

$$x\left(x - \frac{1}{4}\right) = x^2 - \frac{x}{4}$$

or

$$4x^2 - x = 0$$

In both examples, note that the derived equations are not the only ones with the given roots. In $x^2 - 2x - 15 = 0$, if we were to multiply through by 3, for example, we would obtain the equation $3x^2 - 6x - 45 = 0$. Obviously, this equation has the same roots (5 and -3) as the given equation. In fact, there are an infinite number of such equations. Note, however, that $x^2 - 2x - 15 = 0$ is the simplest form of the required equation, having been stripped of all common factors.

We do not consider the situation here when the roots are imaginary.

Exercise Set 10-4

Problems

Find the simplest quadratic equation for each pair of roots given.
1. 3, 2
2. −5, 1
3. ⅒, 6
4. −1, ¹⁄₁₂
5. −⅔, 1
6. Write a BASIC program that will write out the simplest quadratic equation in the form of $y = ax^2 + bx + c$ when the roots are real integers only. Note that since all common factors have been removed, the coefficient of x^2 is $+1$.

Key Terms

quadratic relationship
parabola
y-intercept
x-intercepts

axis of symmetry
vertex
roots
completing the square

quadratic formula
discriminant
imaginary roots

Important Rules

The General Quadratic Relationship

Rule 10–1 $y = ax^2 + bx + c$ $(a \neq 0)$

Rule 10–2 The y-intercept of the equation $y = ax^2 + bx + c$ is the point $(0, c)$.

The Axis of Symmetry

Rule 10–3 $x = \dfrac{-b}{2a}$

Quadratic Equations

Key Points Used in Graphing Quadratic Relationships
Rule 10-4

1. y-intercept: $y = c$.
2. x-intercepts: Set $y = 0$ and solve the resulting equation for x.
3. Axis of symmetry: $x = -b/2a$.
4. Vertex: $x = -b/2a$. The y-value is the corresponding value obtained for this x.

The Quadratic Formula

Rule 10-5 $\quad x = \dfrac{-b \pm \sqrt{b^2 - 4ac}}{2a}$

The Discriminant

Rule 10-6

1. If $b^2 - 4ac = 0$, the roots are real and equal.
2. If $b^2 - 4ac > 0$, the roots are real and unequal.
3. If $b^2 - 4ac < 0$, the roots are imaginary roots (the graph does not cross the x-axis).

PART FIVE
Applications of Algebra to Electronics

11
Applications in Network Analysis

11 Applications in Network Analysis

Upon completion of this chapter, you will have learned how to apply simultaneous linear equations to the solution of electrical networks.

11–1 INTRODUCTION

In the analysis of electrical networks, a technician with only a knowledge of Ohm's law is seriously handicapped. In fact, many networks are impossible to solve without use of more sophisticated mathematics. For example, what is the current flowing through resistor R_3 in Figure 11–1? After a bit of head scratching, you may conclude that something more than Ohm's law is required to solve this problem. In this section, we present a variety of networks and solve them using the mathematical methods developed in earlier chapters.

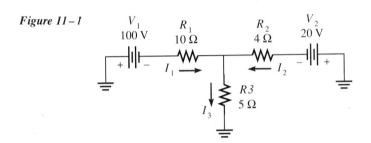

Figure 11–1

We should point out here that mesh and nodal analysis methods have been intentionally omitted from our discussion, since they are extensions of Kirchhoff's laws and require no additional mathematical concepts beyond simultaneous linear equations. Likewise, the Thevenin circuit-reduction method has been omitted because no new mathematical concepts are required in applying this procedure. Many fine reference texts in basic electrical circuit analysis are available for the student interested in these shortcut methods.* Therefore, since our primary concern here is in mathematical (rather than electrical) techniques, we limit our discussion of networks to the more general methods of Kirchhoff.

11–2 KIRCHHOFF'S LAWS

The nineteenth-century German physicist Gustav Kirchhoff stated two very essential and axiomatic laws whose importance to electrical circuit theory remain paramount. His first law, the voltage law, states that the algebraic sum of all the voltage drops around any closed circuit must be zero. We see the simple clarity of this statement in Figure 11–2.

As seen in Figure 11–2, the sum of the individual voltage drops around the closed circuit must equal the supply voltage—that is, $V_s = V_1 + V_2 + V_3 + V_4$. Another way of stating this, of course, is that the algebraic sum of all the voltages is zero—that is, $V_s - V_1 - V_2 - V_3 - V_4 = 0$. In obtaining this last expression, we simply subtracted each voltage from both sides of the original equation until we had zero in the right member.

---

*For example,
Boctor, S. *Electric Circuit Analysis*. Englewood Cliffs, N.J.: Prentice Hall, 1987.
Floyd, T. *Principles of Electric Circuits*. Columbus, Oh.: Merrill, 1989.
Suprynowicz, V. *Electrical and Electronics Fundamentals*. St. Paul, Minn.: West, 1987.

Applications in Network Analysis

Kirchhoff's second law, his current law, states that at any point in a circuit, the algebraic sum of the currents flowing into and out of that point is zero. Again, the simple truth of this statement can be seen from Figure 11–3.

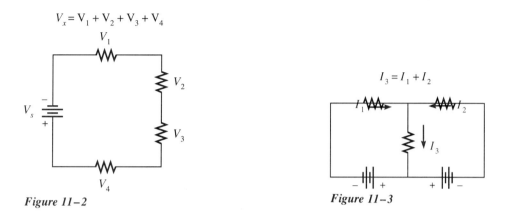

Figure 11–2

Figure 11–3

In Figure 11–3, we observe that $I_3 = I_1 + I_2$. Or, stated in the language of Kirchhoff's current law, $I_3 - I_1 - I_2 = 0$.

We will now see how to apply Kirchhoff's voltage law (KVL) and Kirchhoff's current law (KCL) to practical circuits. Kirchhoff's laws are fundamental and can be applied to any circuit in solving for any required parameters.

Returning to Figure 11–1, we see that since there are two voltage sources acting on the network, there are several possibilities as to the actual path taken by the various currents. See Figure 11–4.

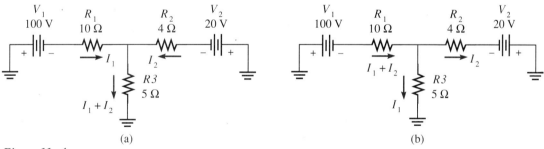

Figure 11–4

In Figure 11–4a, it seems logical—given the polarities of the two voltage sources—that I_1 and I_2 would flow in the directions indicated and combine into $I_1 + I_2$ through R_3. Further observation, however, reveals that since V_1 is much greater than V_2, the current paths shown in Figure 11–4b are equally likely. In order to answer the question posed at the beginning of the chapter—that is, to find the current through R_3—it would seem necessary for us to make a choice at this point. Surprisingly, the use of

Applications of Algebra to Electronics

Kirchhoff's laws makes such a choice unnecessary. Furthermore, the correct answer will be obtained no matter which choice we make. Although this statement probably runs counter to your intuition, we present proof here that either choice provides the same correct answer.

Since we have two unknown currents flowing in the circuit, it seems logical to expect that we could make use of two simultaneous linear equations in solving for I_1 and I_2. What are the required equations? Suppose we arbitrarily choose the current paths shown in Figure 11–4a to guide our thinking. As I_1 flows through R_1, a voltage drop of $10I_1$ appears across R_1. Similarly, as I_2 flows through R_2, a voltage drop of $4I_2$ appears across R_2. Finally, as I_1 and I_2 combine in R_3, a voltage drop appears across R_3 equal to $5(I_1 + I_2)$. These voltage drops are obvious from the simple application of Ohm's law to each resistor. Therefore, according to KVL, the voltage drops around the entire V_1-R_1-R_3 circuit must be

$$100 - 10I_1 - 5(I_1 + I_2) = 0$$

And the voltage drops around the complete V_2-R_2-R_3 loop must be

$$20 - 4I_2 - 5(I_1 + I_2) = 0$$

Upon simplifying these expressions by combining like terms and removing common factors, we obtain our two required simultaneous linear equations in two unknowns:

$$20 - 3I_1 - I_2 = 0$$
$$20 - 5I_1 - 9I_2 = 0$$

On solving these simultaneously, we obtain $I_1 \approx 7.273$ A and $I_2 \approx -1.818$. Finally, the current through R_3 is

$$I_1 + I_2 = 7.273 + (-1.818) = 5.455 \text{ A}$$

We now show that if we had chosen the current paths in Figure 11–4b, we would have obtained exactly the same value of current through R_3.

The voltage drop across R_1 is $10(I_1 + I_2)$ and that across R_3 is $5I_1$. Therefore, the V_1-R_1-R_3 loop equation is

$$100 - 10(I_1 + I_2) - 5I_1 = 0$$

which simplifies to

$$100 - 15I_1 - 10I_2 = 0$$

The voltage drops around the V_1-R_1-R_2-V_2 circuit give us the equation

$$100 - 10(I_1 + I_2) - 4I_2 - 20 = 0$$

which simplifies to

$$80 - 10I_1 - 14I_2 = 0$$

Solving these equations simultaneously gives us $I_1 \approx 5.455$, which is exactly what we obtained in the previous solution. Note that $I_2 \approx +1.818$ A rather than -1.818, as obtained earlier. The reason for this is that Figure 11–4a actually gives the wrong direction for I_2. It should be obvious, though, that the actual current direction does not alter the outcome. This fact is one of the principal advantages to using Kirchhoff's laws. If you get a negative current value, you know you guessed wrong about the direction, but you still get the right answer! Few things in life are as forgiving as Kirchhoff's laws.

Applications in Network Analysis

As a further example, consider the circuit in Figure 11–5. Part (a) shows the assumed direction of the three currents, whereas part (b) shows the three loops involved in the formation of the three required simultaneous linear equations.

Figure 11–5

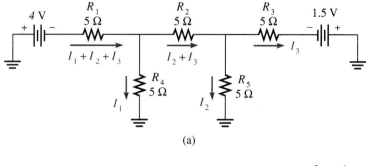

(a)

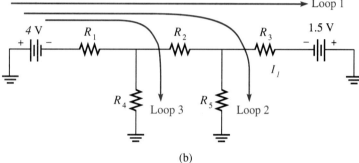

(b)

In writing the three equations, we note the following:

1. Loop 1 involves both voltage sources and the voltage drops across R_1, R_2, and R_3.
$$4 - 5(I_1 + I_2 + I_3) - 5(I_2 + I_3) - 5I_3 - 1.5 = 0$$
which simplifies to
$$2.5 - 5I_1 - 10I_2 - 15I_3 = 0 \qquad (11\text{–}1)$$

2. Loop 2 involves only the 4 V source and the voltage drops across R_1, R_2, and R_5.
$$4 - 5(I_1 + I_2 + I_3) - 5(I_2 + I_3) - 5I_2 = 0$$
which simplifies to
$$4 - 5I_1 - 15I_2 - 10I_3 = 0 \qquad (11\text{–}2)$$

3. Loop 3 involves the 4 V source, R_1 and R_4.
$$4 - 5(I_1 + I_2 + I_3) - 5I_1 = 0$$
which simplifies to
$$4 - 10I_1 - 5I_2 - 5I_3 = 0 \qquad (11\text{–}3)$$

According to the solution algorithm in Chapter 10, we can choose to eliminate I_3, for example, between loop Equations 11–2 and 11–3.

$$\begin{aligned}
(11\text{–}2): &\quad 4 - 5I_1 - 15I_2 - 10I_3 = 0 \\
\text{Multiply } (11\text{–}3) \text{ by } -2: &\quad \underline{-8 + 20I_1 + 10I_2 + 10I_3 = 0} \\
&\quad -4 + 15I_1 - 5I_2 = 0
\end{aligned}$$

We can now eliminate I_3 from Equations 11–1 and 11–2.

$$\begin{array}{rl} (11-1): & 2.5 - 5\,I_1 - 10\,I_2 - 15I_3 = 0 \\ \text{Multiply } (11-2) \text{ by } -1.5: & \underline{-6\ \ + 7.5I_1 + 22.5I_2 + 15I_3 = 0} \\ & -3.5 + 2.5I_1 + 12.5I_2 \ \ \ \ \ \ \ \ \ \ = 0 \end{array}$$

Solving simultaneously these last two new equations in two unknowns, we get $I_1 \approx 0.338$ A and $I_2 \approx 0.213$ A. Substituting these values into Equation 11–1, we have $I_3 = -0.088$ A. The negative sign tells us that I_3 actually flows in a direction opposite to that shown in Figure 11–5.

It is now possible to write the various currents that flow in each section of the network.

$$\begin{aligned} I_1 + I_2 + I_3 &= 463 \text{ mA} \\ I_2 + I_3 &= 125 \text{ mA} \\ I_3 &= 88 \text{ mA} \\ I_2 &= 213 \text{ mA} \\ I_1 &= 338 \text{ mA} \end{aligned}$$

At this point, you are encouraged to calculate the voltage drops across the individual resistors and then verify that each loop equation sums to zero. Remember that the polarity across R_5 is opposite to that originally assumed.

11–3 NETWORK ANALYSIS

The following problems are further examples of the application of simultaneous linear equations to the solution of electrical networks.

Example 11–1

Find the currents flowing in the various parts of the network shown in Figure 11–6. What is the voltage drop across R_4? Across R_5?

Figure 11–6

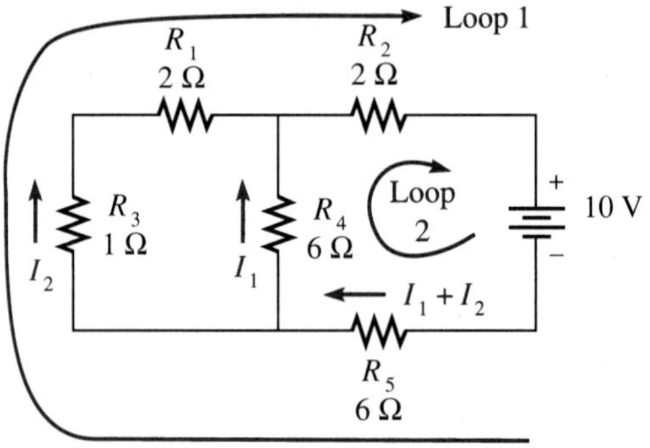

Solution   *Loop 1:* $10 - 6(I_1 + I_2) - I_2 - 2I_2 - 2(I_1 + I_2) = 0$, which simplifies to

$$10 - 8I_1 - 11I_2 = 0$$

Loop 2: $10 - 6(I_1 + I_2) - 6I_1 - 2(I_1 + I_2) = 0$, which simplifies to

$$10 - 14I_1 - 8I_2 = 0$$

Applications in Network Analysis

Solving the two simplified equations simultaneously gives

$$I_1 \approx 0.333 \text{ A}, I_2 \approx 0.667 \text{ A, and } I_1 + I_2 = 1.00 \text{ A}$$

The voltage drop across R_4 is 1.998 V, and the voltage drop across R_5 is 6 V.

Example 11-2

Figure 11-7 shows a bridge network. Find the various branch currents, and determine the voltage drop across R_3.

Figure 11-7

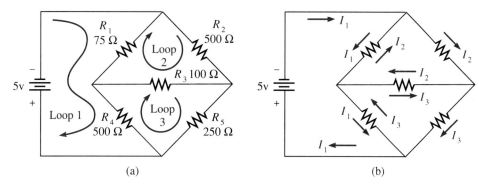

(a) (b)

Solution In setting up the loop equations for all the previous problems, we observe that the source voltage was included in each loop. In the present situation, however, the source appears only in loop 1. Loops 2 and 3 do not include the voltage source. Nonetheless, the sum of the individual voltage drops around each loop must still be zero. Therefore, paying close attention to algebraic signs, we write the following loop equations.

Loop 1: $5 - 75(I_1 - I_2) - 500(I_1 - I_3) = 0$, which simplifies to

$$575I_1 - 75I_2 - 500I_3 = 5$$

Loop 2: $75(I_2 - I_1) + 500I_2 + 100(I_2 - I_3) = 0$, which simplifies to

$$75I_1 - 675I_2 + 100I_3 = 0$$

Loop 3: $500(I_3 - I_1) + 100(I_3 - I_2) + 250I_3 = 0$, which simplifies to

$$500I_1 + 100I_2 - 850I_3 = 0$$

On solving the system simultaneously, we obtain the following currents:

$$I_1 \approx 20 \text{ mA}$$
$$I_2 \approx 4 \text{ mA}$$
$$I_3 \approx 12 \text{ mA}$$
$$I_3 - I_2 \approx 12 - 4 = 8 \text{ mA}$$

Therefore, the voltage drop across R_3 is $0.008 \times 100 \, \Omega = 0.8$ V.
You should solve the loop equations yourself in order to verify the results.

Example 11–3

Figure 11–8 shows a *resistor cube* whose sides are 100 Ω resistors. The ohmmeter indicates 83.29 Ω, which is close to the actual value of 83.33 Ω. How could you apply Kirchhoff's laws to determine this value?

Solution Assume a 1 A current entering node (vertex) A in Figure 11–9. Since this current "sees" the same opposition in every leg, it splits equally into three smaller ⅓ A currents, which split again (at node B, for example) into two ⅙ A currents. At node C, two ⅙ A currents reunite, forming ⅓ A. Finally, each ⅓ A current combines at node D, producing a 1 A output current. All this, of course, was predicted by Kirchhoff's current law.

Now, applying Ohm's law to R_1, R_2, and R_3, we can find the total voltage required to sustain this assumed 1 A current: $V_1 = IR_1 = (⅓)100 = 100/3$ V, $V_2 = IR_2 = (⅙)100 = 100/6$ V, and finally, $V_3 = IR_3 = (⅓)100 = 100/3$ V. Adding these voltages, we get

$$V_t = \frac{100}{3} + \frac{100}{6} + \frac{100}{3} = \frac{200 + 100 + 200}{6}$$

$$= \frac{500}{6} \text{ V}$$

Notice that the sum of the voltage drops across any three legs in any direction is the same, namely, 500/6 V. The total resistance, then, is

$$R = \frac{V}{I} = \frac{500/6 \text{ V}}{1 \text{ A}} \approx 83.33 \text{ Ω}$$

Example 11–4

What voltage is read between points X and Y in Figure 11–10?

Solution In answering this question, we make use of the superposition theorem. Here, we seek to evaluate the effect of each source acting alone. Our final answer is the algebraic sum of these separate influences. Figure 11–11 shows the effect of each source acting alone. In Figure 11–11a we see the effect of eliminating the 55 V source and treating the network as if it were powered solely by the +20 V supply. The current can be calculated as

$$I = \frac{V}{R} = \frac{20}{175} \approx 114 \text{ mA}$$

Therefore, the voltage between point X and ground is

$$V = IR = 0.114 \times 100 = 11.4 \text{ V}$$

In Figure 11–11b we see the −55 V source acting alone. Here, the current is

$$I = \frac{V}{R} = \frac{-55}{175} \approx -314 \text{ mA}$$

and the voltage at X is

$$V = IR = -0.314 \times 75 = -23.55 \text{ V}$$

On combining these two results, we see that the total voltage between point X and point Y (ground) is $-23.55 + 11.4 \approx -12.2$ V.

Figure 11–8
(Courtesy of Juan Jorge Arteaga)

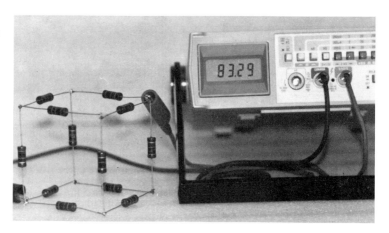

Figure 11–9

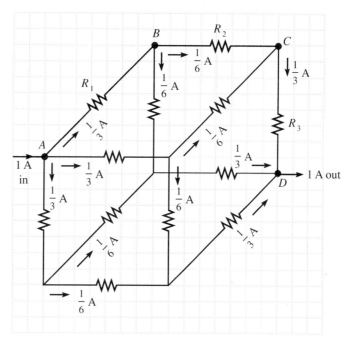

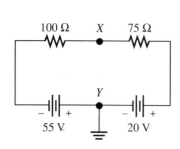

Figure 11–10

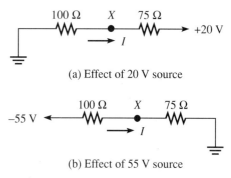

(a) Effect of 20 V source

(b) Effect of 55 V source

Figure 11–11

Exercise Set 11-1

Questions

1. In your own words, state Kirchhoff's current and voltage laws.
2. In the application of Kirchhoff's laws, what meaning is to be attached to a negative current?
3. What is a network *loop*?

Problems

Solve each network for the required value(s).

1. Using Figure 11–12, find the voltage drop across R_3.

Figure 11–12

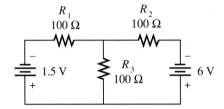

2. Using Figure 11–13, find the voltage drop across each resistor.

Figure 11–13

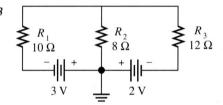

3. Find the voltage drop across each resistor. All resistors are 100 Ω. (See Figure 11–14.)

Figure 11–14

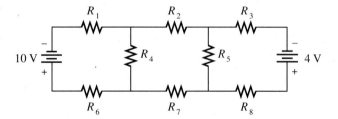

4. Find the current flowing through R_4 and the voltage drop across R_5. (See Figure 11–15.)

Figure 11–15

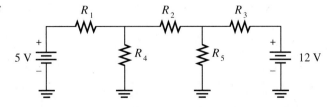

Applications in Network Analysis

5. What is the voltage drop across the 18 Ω resistor? (See Figure 11–16.)

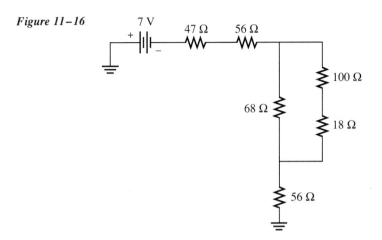

Figure 11–16

6. What current is read on the ammeter? (See Figure 11–17.)

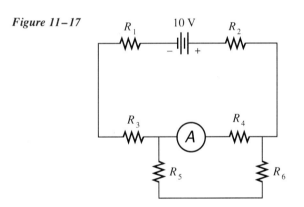

Figure 11–17

7. Using Figure 11–18, find the voltage drop across R_3.

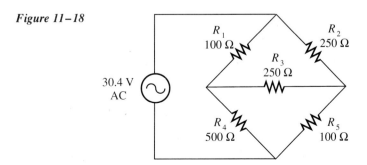

Figure 11–18

Applications of Algebra to Electronics

8. What is the voltage drop across R_5? (See Figure 11–19.)

Figure 11–19

R_1 1 kΩ
R_2 2.2 kΩ
10 V
R_3 1.2 kΩ
R_4 3.3 kΩ
R_5 4.7 kΩ

9. If each resistor is 36 Ω, what is the total resistance between points A and B? (See Figure 11–20.)
10. What is the voltage from P to ground? (See Figure 11–21.)

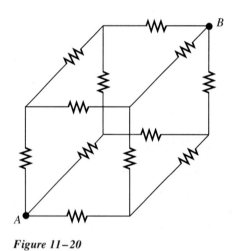

Figure 11–20

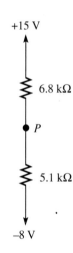

Figure 11–21

Summary

In this chapter, we are able to combine Kirchhoff's circuit laws with the methods of simultaneous linear equations in solving network problems. The general procedure is as follows:

1. Identify the number of loops in the network.
2. Assume an arbitrary direction of current flow, either clockwise or counterclockwise.
3. Write an equation that sets the algebraic sum of all voltage drops in an individual loop equal to zero.
4. Write as many equations as there are loops.
5. Solve the system of equations simultaneously.

In applying the superposition theorem, we treated each source as though it acted alone and then combined the results algebraically.

PART SIX
Advanced Topics in Electronics Mathematics

12
Elementary Trigonometry

13
Exponents and Logarithms

14
Introduction to Vectors

Upon completion of this chapter, you will have:
1. Learned how to measure and specify angles.
2. Learned how the trigonometric relationships are derived.
3. Learned how the methods of trigonometry can be applied to problems involving the right triangle.
4. Learned how trigonometric ideas apply to the generation of sinusoidal AC voltage.
5. Learned how to sketch basic and composite trigonometric relationships.

12 Elementary Trigonometry

248 Advanced Topics in Electronics Mathematics

12-1 INTRODUCTION

The word *trigonometry* is derived from two Greek words, *trigonon,* meaning a triangle, and *metron,* measure. In the most literal sense, then, trigonometry means to measure a triangle. And, indeed, our basic purpose in this chapter is to derive the relationships that exist among the various sides and angles of a triangle. Subsequently, we show how trigonometric methods are applied to electronic circuits.

Before we begin our detailed study of circuit applications, however, it is necessary to define many new terms and concepts. Accordingly, this first section deals with the nomenclature and definitions used in the language of trigonometry. You are encouraged to learn these new words and to employ them often in your conversations about angles and triangles. By so doing, they will become a routine part of your technical language; this will enable you better to communicate ideas and questions that are important elements to learning trigonometry in general and electronics in particular.

12-2 ANGLES

Technically speaking, an **angle** is an *amount of rotation*. The drawing shown in Figure 12-1, therefore, is a picture of an amount of rotation—a convenient way of representing an angle using straight lines.

Figure 12-1

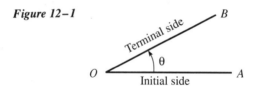

In the figure, the point about which the rotation occurs (*O*) is called the **vertex.** The angle itself is often named using the three letters describing the endpoints of the intersecting straight lines showing the extent of the rotation. In Figure 12-1, the angle can be named angle *AOB*. The symbol ∠ is frequently used as a convenient substitute for the word angle. Therefore, we can write angle *AOB* as ∠*AOB*. The Greek letter θ (theta) is often used to indicate the amount of rotation in degrees, or it can actually refer to the name of the angle—that is, ∠θ.

Straight lines are almost always used to indicate the beginning and the end of the rotation. Accordingly, in Figure 12-1 the beginning of rotation is shown by a line called the **initial side,** and the end of the rotation is marked by a straight line called the **terminal side.** The lengths of the lines have nothing to do with the size of the angle.

Another convention used in trigonometry is to refer to angles as *positive* when they are generated by *counterclockwise* rotation of the terminal side. Conversely, a *negative* angle is formed when the terminal side moves *clockwise*. Figure 12-2a shows a positive angle; 12-2b is the picture of a negative angle.

An angle is said to be in **standard position** when its initial side extends along the positive *x*-axis of a rectangular coordinate system and its vertex is at the origin. Using this convention, Figure 12-3 shows a positive 105° angle and a negative 85° angle in standard position.

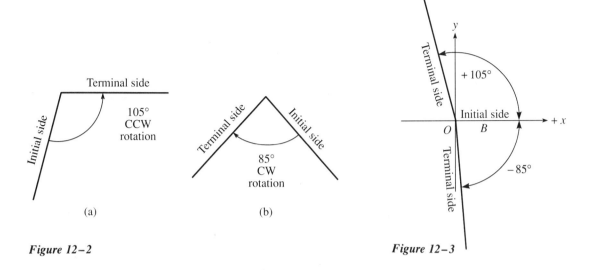

Figure 12–2

Figure 12–3

12–3 MEASUREMENT OF ANGLES

In Figure 12–3, if the terminal side of the positive angle, for example, were to continue to rotate in a counterclockwise direction, it would eventually come back to the positive x-axis. We would say that the terminal side had completed one revolution. Imagine, then, a red light at the tip of the terminal side. What shape would the light describe as it rotates? The answer, of course, is a circle. And circles are said to contain 360°. Therefore, a full revolution also contains 360°. Buy why 360? Why not some other number? Is there some natural quantity, whose value is 360, associated with making a full revolution? The answer is no. The number 360 is not sacred and it certainly is not based on any natural phenomenon. More than likely, the number 360 comes from the ancient Babylonian astronomers, who calculated (incorrectly) that their year—the time for the earth to make one revolution about the sun—contained 360 days. The idea stuck and became the basis for measuring angles from that time forth. For example, an angle produced by rotating the terminal side ⅕ revolution was said to contain ⅕ of 360°, or 72°.

There are, however, other units used to measure angles. For example, consider the circle shown in Figure 12–4. If the length of arc *AB* is equal to the radius *r* of the circle, we define ∠*AOB* as being equal to 1 **radian** (rad). There are two questions you might ask immediately. First, how many arc lengths equal to the radius are there in a circle? And second, how many degrees are there in 1 rad?

Figure 12–4

It has been known for several thousand years that whenever a circle's circumference (distance around the circle) is divided by its diameter, the same number (about 3.14) is always obtained no matter how large or small the circle is. This particular number is a natural phenomenon and does not depend on ancient Babylonian beliefs (or anybody else's). The same result occurs every time the circumference (C) of *any* circle is divided by its diameter (D). In other words, C/D is always this same constant. Because this number is a constant, it has been given a special symbol. We denote the value C/D by the Greek letter pi (π) and write $\pi \approx 3.14. . .$, where the dots indicate that the number never ends and there is no repeating pattern to its digits. In the language of mathematics, we say the number is irrational—that is, it cannot be obtained as the quotient of two integers.

Observe, now, that since $\pi = C/D$ and the diameter is simply twice the radius (r), we may write

$$\pi = \frac{C}{2r}$$

or

$$2\pi = \frac{C}{r}$$

This is simply another way of saying that the radius (r) can be laid out along the circumference (C) 2π, or about 6.28, times. It is apparent, then, that there must be 2π (about 6.28) rad in a complete circle.

We further note that since there are 360° in every circle and also 2π rad in every circle, then there must be 360° in 2π rad—that is,

$$\frac{360°}{2\pi \text{ rad}}$$

Performing the actual division, we see that there must be about 57.3° per radian. Similarly, we may write

$$\frac{2\pi \text{ rad}}{360°}$$

which yields about 0.017 rad per degree. The following summarizes these relationships.

$$180° = \pi \text{ rad}$$
$$1 \text{ rad} \approx 57.29577951°$$
$$1° \approx 0.017453292 \text{ rad}$$

In electronics, radian measures are probably more useful than degree measures.

There is yet a third system of angular measurement called the *grad system,* wherein the circle is divided into 400 parts, each part being called a **grad**. A grad, then, is 1/400 of a complete revolution. Since there are 400 grads in 360°,

$$\frac{400 \text{ grads}}{360°} \approx 1.11 \text{ grads/degree}$$

and

$$\frac{360°}{400 \text{ grads}} = 0.9°/\text{grad}$$

Elementary Trigonometry

It remains as an exercise for you to show (using the method of analysis of units in Chapter 7) that the following conversion factors are correct:

$$1 \text{ rad} \approx 0.0157 \text{ grad}$$
$$63.66 \text{ grads} \approx 1 \text{ rad}$$

Exercise Set 12-1

Questions

1. Define the following terms in your own words.
   a. Angle
   b. Vertex
   c. Initial side
   d. Terminal side
   e. Positive (negative) angle
   f. Standard position
   g. Radian
   h. Grad
2. What is the relationship between the circumference of a circle, its diameter, and the constant π?

Problems

1. Using a protractor (a device for measuring angles), draw the following angles in standard position.
   a. 30°
   b. 2.2 rad
   c. −300 grads
   d. −175°
   e. $\pi/2$ rad
   f. $-3\pi/4$ rad
   g. 58 grads
   h. −390°
2. The following BASIC program converts the degree measure of any angle to radians and grads.

   ```
   10 PRINT"Input the number of degrees D"
   20 INPUT D
   30 LET R=D*0.017453292
   40 PRINT"The number of radians in D degrees is:";R
   50 LET G=D*1.111111111
   60 PRINT"The number of grads in D degrees is:";G
   70 END
   ```

   Write a BASIC program that will convert the radian measure of any angle to degrees and grads.

12-4 THE TRIGONOMETRIC RELATIONSHIPS

Consider the circle shown in Figure 12–5, whose radius is one unit long. The actual length of the unit is unimportant. It could be 1 in., 1 ft, or 1 mi. All that is important is that we have a radius of one unit length on the coordinate system from O to A. Note that the radius makes an angle θ with the abscissa and that θ is in standard position.

Consider, now, the effect of dropping a line from the tip of the radius arrow (point A) to the abscissa (point B), as shown in Figure 12–6. We see that the points AOB define a right triangle.

Advanced Topics in Electronics Mathematics

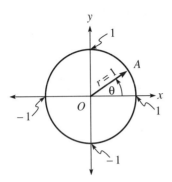

Figure 12–5

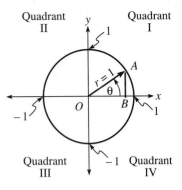

Figure 12–6

As seen from Figure 12–7, as θ rotates counterclockwise, the side labeled *AB* gets larger, whereas the line labeled *OB* becomes smaller. As θ passes the 90° point, as shown in Figure 12–8, we see that *OB* starts to get longer in the negative *x* direction, whereas *AB* becomes smaller. This situation in quadrant II is just the reverse of what was happening when θ was in quadrant I.

Figure 12–7

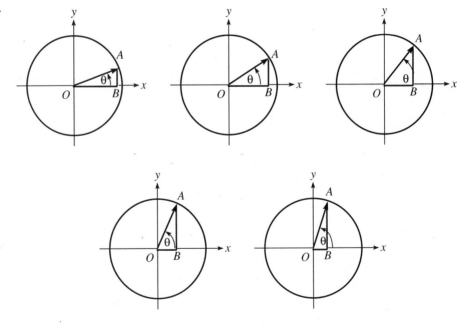

It should be easy for us to visualize what happens as θ continues to rotate counterclockwise past the 180° point and into quadrant III. Side *OB* once again starts getting smaller, and *AB* grows larger in the negative *y* direction. As θ continues to rotate and enters quadrant IV, side *AB* once again becomes smaller and *OB* grows larger in the positive *x* direction (see Figure 12–9).

Elementary Trigonometry

Figure 12-8

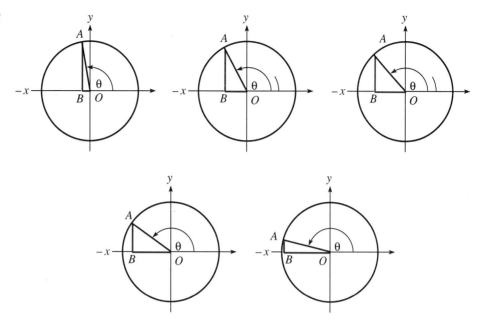

Figure 12-9

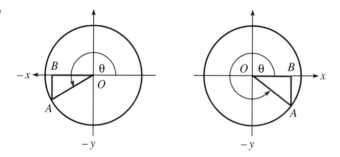

We begin to see, even without a great deal of thought, that perhaps there is some relationship between angle θ and the sides *OA* and *AB* of triangle *AOB*. As a matter of fact, the definitions of the trigonometric relationships are based on the lengths *AB* and *OB* as θ rotates. Specifically, the length *AB* is defined as the **sine** of θ, and length *OB* is defined as the **cosine** of θ. For example, when θ = 29°, length *AB* = 0.4848 and length *OB* = 0.8746. Therefore, 0.4848 is said to be the sine of 29°, and 0.8746 is said to be the cosine of 29°. We generally abbreviate this and write

$$\sin 29° = 0.4848$$

and

$$\cos 29° = 0.8746$$

If you were to draw a unit circle very carefully, you could actually measure the lengths of *AB* and *OB* for each angle θ and construct a table of trigonometric values for yourself. Fortunately, tables of such values have already been prepared by mathemati-

Advanced Topics in Electronics Mathematics

cians; a part of one such table is show in Figure 12–10, where the sine and cosine of 29° have been highlighted.

Angle	Radians	Sin	Cos	Tan	Cot	Sec	Csc
27°00′	0.47122	0.4540	0.8910	0.5095	1.963	1.122	2.203
10′	0.4741	0.4566	0.8897	0.5132	1.949	1.124	2.190
20′	0.4771	0.4592	0.8884	0.5169	1.935	1.126	2.178
30′	0.4800	0.4617	0.8370	0.5206	1.921	1.127	2.166
40′	0.4829	0.4643	0.8857	0.5243	1.907	1.129	2.154
50′	0.4858	0.4669	0.8843	0.5280	1.894	1.131	2.142
28°00′	0.4887	0.4695	0.8829	0.5317	1.881	1.133	2.130
10′	0.4916	0.4720	0.8816	0.5354	1.868	1.134	2.118
20′	0.4945	0.4746	0.8802	0.5392	1.855	1.136	2.107
30′	0.4974	0.4772	0.8788	0.5430	1.842	1.138	2.096
40′	0.5003	0.4797	0.8774	0.5467	1.829	1.140	2.085
50′	0.5032	0.4823	0.8760	0.5505	1.816	1.142	2.074
29°00′	0.5061	0.4848	0.8746	0.5543	1.804	1.143	2.063
10′	0.5091	0.4874	0.8732	0.5581	1.792	1.145	2.052
20′	0.5120	0.4899	0.8718	0.5619	1.780	1.147	2.041
30′	0.5149	0.4924	0.8704	0.5658	1.767	1.149	2.031
40′	0.5178	0.4950	0.8689	0.5696	1.756	1.151	2.020
50′	0.5207	0.4975	0.8675	0.5735	1.744	1.153	2.010
30°00′	0.5236	0.5000	0.8660	0.5774	1.732	1.155	2.000
10′	0.5265	0.5025	0.8646	0.5812	1.720	1.157	1.990
20′	0.5294	0.5050	0.8631	0.5851	1.709	1.159	1.980
30′	0.5323	0.5075	0.8616	0.5890	1.698	1.161	1.970
40′	0.5352	0.5100	0.8601	0.5930	1.686	1.163	1.961
50′	0.5381	0.5125	0.8587	0.5969	1.675	1.165	1.951
31°00′	0.5411	0.5150	0.8572	0.6009	1.664	1.167	1.942
10′	0.5440	0.5175	0.8557	0.6048	1.653	1.169	1.932
20′	0.5469	0.5200	0.8542	0.6088	1.643	1.171	1.923
30′	0.5498	0.5225	0.8526	0.6128	1.632	1.173	1.914
40′	0.5527	0.5250	0.8511	0.6168	1.621	1.175	1.905
50′	0.5556	0.5275	0.8496	0.6208	1.611	1.177	1.89⌐
32°00′	0.5585	0.5299	0.8480	0.6249	1.600	1.179	
10′	0.5614	0.5324	0.8465	0.6289	1.590	1.181	
20′	0.5643	0.5348	0.8450	0.6330	1.580	1 1⌐	
30′	0.5672	0.5373	0.8434	0.6371	1.570		
40′	0.5701	0.5398	0.8418	0.6412	1 ⌐⌐		
50′	0.5730	0.5422	0.8403	0.6453			
33°00′	0.5760	0.5446	0.8387	⌐ ⌐⌐⌐			
10′	0.5789	0.5471	0 ⌐⌐⌐				
20′	0.5818	0.5495					
30′	0.5847	0.5519					
40′	0.5876	0.5⌐					
50′	0.5905	⌐					
34°00′	0.5⌐⌐⌐						
10′							
20′							

Figure 12–10

Perhaps even more important, trigonometric tables have been built into many hand-held calculators. For example, if you want to know the sine of 29°, you merely enter the following key strokes:

[2] [9] [SIN]

Elementary Trigonometry

and read 0.48480962 on the display. For cosine of 29°, you key in

[2] [9] [COS]

and read 0.874619707. When entering measurements in degrees, be sure your calculator is in degree mode.

Referring back to Figure 12–9a, it appears that θ is about 210°. Using your calculator, key in [2], [1], [0], [SIN] and read −0.5 on the display. The negative sign is precisely what we would have expected, since AB is in the negative y direction. Note, too, that cos 210° = −0.8660, since OB lies along the negative x-axis.

In Figure 12–9b, the angle θ appears to be about 325°. Before you key this into your calculator, see if you can predict the signs (polarities) of sin 325° and cos 325°. Sine 325° is negative (−0.5736), and cosine 325° is positive (+0.8192).

You might have noticed that the values of sin θ are between −1 and +1 no matter in which quadrant θ lies. Moreover, the values of cos θ are between −1 and +1, regardless of how big θ is. Try this exercise: Think of an angle, positive or negative, no matter how large or how small, and try to find a sine or cosine that is smaller than −1 or bigger than +1. Try as you might, you cannot find such an angle. For example, if you want to find the sine of 10,000°, you must first realize that 360° divides into 10,000° twenty-seven times with a remainder of 280°. Therefore, what you're really after is the sine of 280°, which is −0.9848. Note that your calculator might give an error message (-E-) if your angle is much above 1000°. Note, too, that if you were to subtract 280° from 360°, you would obtain 80°, the sine of which is 0.9848. Note that, except for the polarity, sin 80°, sin 280°, and sin 10,000° all have the same magnitude, 0.9848. In looking back at Figures 12–7 through 12–9, this is exactly what you would expect, since *any* angle may be reduced to an angle between 0 and 90°. Moreover, except for the polarity, the values of sine and cosine are always between 0 and 1. You can always supply the missing sign by remembering the drawing in Figure 12–6. For example, find cos 453°. From Figure 12–11, we see that 453° is a second-quadrant angle; therefore, its cosine should be negative. Subtracting 360° results in 93°, and 180° − 93° is 87°. Therefore, we can key in [8] [7] [COS] and read 0.05233. But since 453° actually lies in quadrant II, we must supply the missing sign. Therefore, we write

$$\cos 453° = -0.05233$$

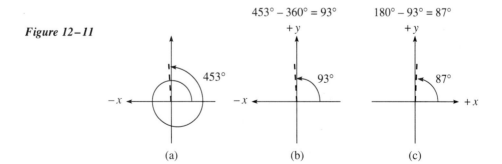

Figure 12–11

Note that if we key cos 453° directly into the calculator, we get −0.05233.

12-5 TRIGONOMETRY AND THE RIGHT TRIANGLE

Figure 12–12 shows the unit circle with triangle AOB, in which side AB equals $\sin \theta$ and side OB equals $\cos \theta$, as defined in the previous section. The question now before us is, Since triangles MON and POQ are similar to AOB, are the lengths of their appropriate sides also related to the sine and cosine of θ? Clearly, side MN is smaller than AB and PQ is larger. Accordingly, side MN understates the sine of θ, whereas side PQ exceeds $\sin \theta$. The same may be said for sides ON and OQ in regard to the cosine of θ. Note, however, that the ratio of MN to OM is the same as the ratio of AB to OA. Moreover, the ratio of PQ to OQ is also the same as the ratio of AB to OA. We see, therefore, that in any right triangle, $\sin \theta$ may also be defined as the ratio of the side opposite the acute angle θ to the hypotenuse. Of course, the same logic applies to the cosine of θ. Therefore, we make the following additional definitions of the sine and cosine of an angle in any right triangle:

$$\sin \theta = \frac{\text{opposite side}}{\text{hypotenuse}}$$

$$\cos \theta = \frac{\text{adjacent side}}{\text{hypotenuse}}$$

Figure 12–12

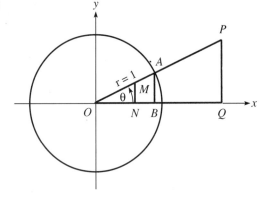

These ideas are summarized in Figure 12–13, where the arrow points from the numerator to the denominator of the appropriate relationship.

Figure 12–13

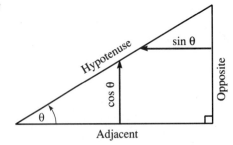

12-6 THE TANGENT OF θ

In Figure 12-14, an additional element has been added that was not shown earlier in Figure 12-6. We see that line *CD* has been drawn tangent to the circle at *D*. Recall from Figure 8-35 that a tangent was defined as a straight line that touches a curve at one and only one point. We define the length of *CD* in Figure 12-14 as the **tangent** of θ. From Figure 12-15, it is clear that tan θ increases rather rapidly and becomes infinitely large as θ approaches 90° (along the +*y*-axis) or 270° (along the −*y*-axis).

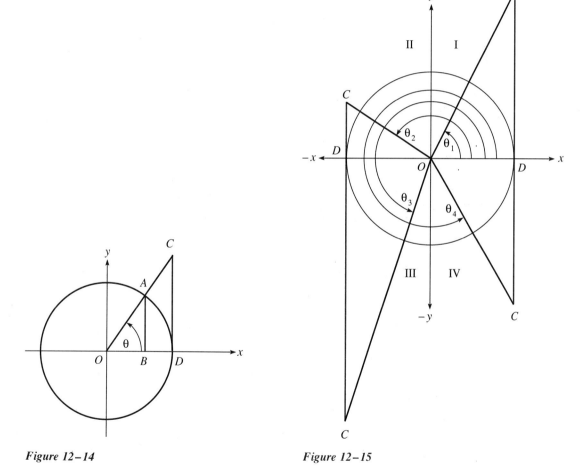

Figure 12-14   *Figure 12-15*

Referring to Figure 12-16, we can now ask a question similar to that posed earlier: Since triangles *OAB*, *OCD*, and *OJK* are all similar, how are the lengths of sides *AB* and *JK* related to the tangent of θ?

Again, we see that although the actual lengths of *AB* and *JK* do not accurately represent tan θ, it is apparent that the ratios *AB*/*OB*, *CD*/*OD*, and *JK*/*OK* are all identical.

Figure 12-16

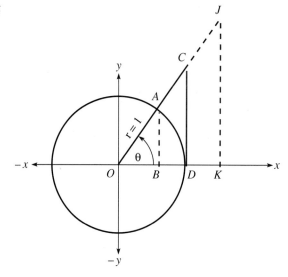

Therefore, we may extend our definition of the trigonometric relationships in any right triangle to include the tangent of θ.

$$\tan \theta = \frac{\text{opposite side}}{\text{adjacent side}}$$

Referring again to Figure 12–15 and using the ratio just given, it should be clear that tan θ is positive in quadrants I and III and negative in quadrants II and IV.

In Figure 12–17, we have extended the definitions shown in Figure 12–13 to include the tangent of θ. The arrow points from the numerator of the ratio to the denominator.

Figure 12-17

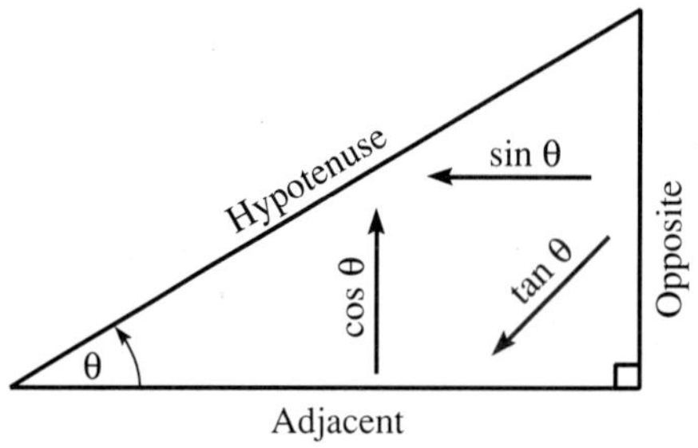

Exercise Set 12-2

Questions

1. Using a drawing of a right triangle, show the ratios existing among the various sides that define the three basic trigonometric relationships.

Elementary Trigonometry

2. In the triangle shown in Figure 12–18, what are the ratios that define the trigonometric relationships of angle θ?

Figure 12–18

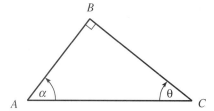

3. In Figure 12–18, what ratios define the trigonometric relationships for angle α (alpha)?

Problems

1. Carefully draw a unit circle about 3 in. (7⅝ cm) in diameter. Using a protractor, draw a positive 150° angle in standard position. Measure the length of the tangent line, and determine tan θ using the proper ratio. How does your answer compare with that given in Figure 12–10? (*Remember:* 180° − 150° = 30°.) How does it compare with the answer given on your calculator? (Key in [1] [5] [0] [TAN].) Why is tan 150° negative?

2. In BASIC programming, the following commands print the sine, cosine, and tangent of an angle A in radians.

```
10 PRINT SIN(A)
20 PRINT COS(A)
30 PRINT TAN(A)
```

For example, SIN(5) = −0.958924274, where the angle is 5 rad (286.479°).

Write a BASIC program that will display all three trigonometric relationships for any angle in degrees. (Remember, 1 rad ≈ 57.29577951°.)

3. Three other trigonometric relationships are defined using the elementary relationships previously given. These are called the *reciprocal trigonometric relationships* and are given by

$$\text{cosecant } \theta = \csc \theta = \frac{1}{\sin \theta} = \frac{\text{hypotenuse}}{\text{opposite side}}$$

$$\text{secant } \theta = \sec \theta = \frac{1}{\cos \theta} = \frac{\text{hypotenuse}}{\text{adjacent side}}$$

$$\text{cotangent } \theta = \cot \theta = \frac{1}{\tan \theta} = \frac{\text{adjacent side}}{\text{opposite side}}$$

Write a BASIC program that will compute the reciprocal relationships of any angle in degrees.

12–7 THE INVERSE TRIGONOMETRIC RELATIONSHIPS

Suppose we have a value of one of the three trigonometric relationships and want to find the angle associated with it. How would we proceed? For example, suppose we know sin θ = 0.5736 and wish to know the value of θ. We may key our calculator as follows:*

[.] [5] [7] [3] [6] [INV] [SIN]

and read 35.00164818°, or about 35.00°, on the display. We may write this relationship as $\theta = \sin^{-1} 0.5736$, which is read, "theta is equal to the angle whose sine is 0.5736."

---

*Some calculators require different key strokes in order to obtain inverse functions. Read your operator's manual carefully.

We write $\sin^{-1}$ to denote the **inverse** of sine θ and read $\sin^{-1}$ as, "the angle whose sine is . . . ."* Similarly, $\cos^{-1}$ and $\tan^{-1}$ refer to the angle whose cosine or tangent is given. When inverse notation is used on your calculator, there is a restriction placed on the range of values that θ may assume. These restrictions have to do with the singularity of values that a mathematical function must have.** The range restrictions for each inverse function are listed in Table 12–1.

Table 12–1
Restrictions on the Range of θ for an Inverse Function

Inverse Function	Range of θ
sine	−90° to +90°
cosine	0 to +180°
tangent	−90° to +90°

For example, suppose you want to obtain sin 240°. If you key in 240°, your calculator will return an answer of −0.866. If, now, you key in $\sin^{-1}(-0.866)$, you might be surprised when you read the displayed answer of −60°! A glance at Figure 12–19 should clear up the confusion. From Figure 12–19, we see that 240° − 180° = 60°, and sin 60° is +0.866. However, in order for the sine to be *negative* 0.866, the angle must be −60° in order to comply with the range restrictions given in Table 12–1. To prevent confusion, you should get into the habit of drawing a sketch of the problem before you start pressing calculator keys.

Figure 12–19

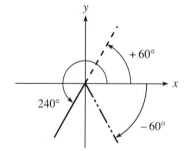

12–8 SOLUTIONS OF RIGHT TRIANGLES

The previous sections contained information that you need in order to apply trigonometry to practical situations in electronics. Before we undertake the direct application of these ideas to alternating current in Chapters 15 and 16, however, we pause here to examine the

---

*The abbreviation arcsin is used interchangeably with $\sin^{-1}$ to denote the angle whose sine is given. Do not confuse $\sin^{-1}$ with $1/\sin θ$, which is csc θ.

**We have intentionally avoided the concept of *function* in order to eliminate much of the formal discussion surrounding this fascinating topic. The interested student is referred to A. Taylor, *Calculus with Analytic Geometry*, Englewood Cliffs, N.J.: Prentice Hall, 1959.

Elementary Trigonometry

application of trigonometry to problems in the more visible physical world. In this way, the transition to the abstract world of electronics will be much easier. For example, consider Figure 12–20, which shows the shadow cast by an industrial smokestack. In the redrawing of Figure 12–20 as Figure 12–21, we see that the situation is basically a right triangle with an acute angle of 34° and a side (the adjacent side) of 104 ft.

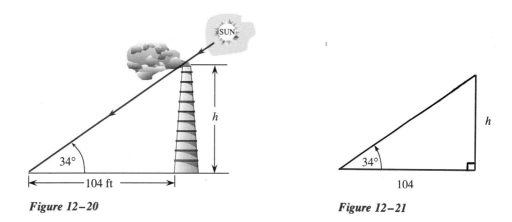

Figure 12–20 *Figure 12–21*

In order to determine the height (h) of the smokestack, we may employ the tangent function, since $\tan 34° = h/104$ (see Figure 12–17). If we multiply each side of this equation by 104, we obtain $104 \times \tan 34° = h$, or $h = 104 \times 0.6745 \approx 70.15$ ft.

The following additional examples give you some idea of the extent to which trigonometry can be applied to practical situations.

Example 12-1

How many inches apart must the marks be placed on the side of the tapered vat in order to indicate the depth of the liquid? (See Figure 12–22.)

Solution An enlarged section of the vat shown in Figure 12–23 reveals that the problem lends itself to solution by a right triangle. From the definition of cos θ, we may

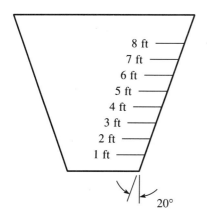

Figure 12–22

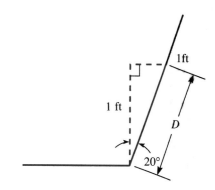

Figure 12–23

write

$$\cos 20° = \frac{12 \text{ in.}}{D}$$

from which

$$D = \frac{12 \text{ in.}}{\cos 20°} = \frac{12}{0.9397}$$
$$\approx 12.77 \text{ in.}$$

Thus the marks must be placed about 12.77 in. apart in order to indicate the depth of the liquid.

Example 12–2

An airplane flies due west at 320 mi/h with a 50 mi/h crosswind from the south. How many degrees off course is the actual direction of travel?

Solution From Figure 12–24, we see that the solution of the right triangle gives us the required angle. We write $\theta = \tan^{-1} \frac{50}{320}$, from which $\theta \approx 8.88°$.

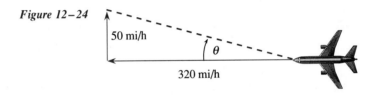

Figure 12–24

Example 12–3

A sailor at point A (Figure 12–25) observes that the angle made with the water to the top of the suspension-bridge support is 53°. At point B, 20 ft from point A, she notices that the angle has changed to 56°. How high is the support above the water?

Solution The situation has been redrawn in Figure 12–26, with the height of the support labeled. Note that we do not know the sides of any of the right triangles thus

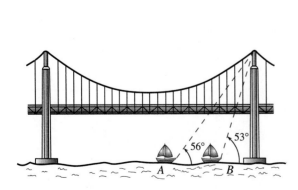

Figure 12–25

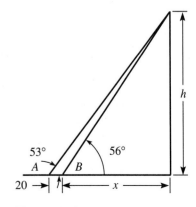

Figure 12–26

Elementary Trigonometry

formed. We can proceed as follows:

$$\tan 53° = \frac{h}{20 + x} \qquad \tan 56° = \frac{h}{x}$$

$$20 + x = \frac{h}{\tan 53°} \qquad x = \frac{h}{\tan 56°}$$

$$x = \frac{h}{\tan 53°} - 20$$

Therefore,

$$\frac{h}{\tan 53°} - 20 = \frac{h}{\tan 56°}$$

$$\frac{h}{\tan 53°} - \frac{h}{\tan 56°} = 20$$

$$\frac{1}{\tan 53°} - \frac{1}{\tan 56°} = \frac{20}{h}$$

$$h = \frac{20}{\frac{1}{\tan 53°} - \frac{1}{\tan 56°}}$$

$$= \frac{20}{\frac{1}{1.327} - \frac{1}{1.483}} \approx \frac{20}{0.7536 - 0.6743}$$

$$= \frac{20}{0.0793} \approx 252$$

Thus the support is about 252 ft above the water.

Example 12–4 The hour hand on the clock of the Homewood Bank Building is 3.5 ft long. The minute hand is 4 ft long. When the clock strikes 8:30, what will be the distance between the tips of the hands? (See Figure 12–27.)

Figure 12–27

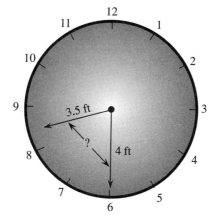

264   Advanced Topics in Electronics Mathematics

Solution Figure 12–28 defines the important distances that will give rise to the correct solution.

The 3 h between 6 and 9 divide the right angle *FBD* into three equal angles of 30° each. At 8:30, the hour hand divides the 30° angle between 8 and 9 in half, which gives an angle of 15°.

Since lines *FB* and *EC* are parallel, $\angle BAC$ is also 15°, from the laws of plane geometry.

The value of y can be found by using $\sin 15° = y/3.5$, from which $y = 3.5 \times \sin 15° \approx 0.9059$. Moreover, x can be found using $\cos 15° = x/3.5$, from which $x = 3.5 \times \cos 15° \approx 3.3807$.

Therefore, in the right triangle *ACD*, side $DC = 4 - y = 3.0941$ and side $AC = x = 3.3807$. Applying the Pythagorean theorem, $AD = \sqrt{(AC)^2 + (DC)^2} = \sqrt{3.3807^2 + 3.0941^2} \approx 4.583$ ft.

Example 12–5 Figure 12–29 shows a schematic diagram of a three-phase alternator (AC generator) connected in a "star," or Y, configuration such that the phase difference between each winding is 120° electrical. In electronics, the phase relationship between the voltage in each winding as well as its amplitude is often diagrammed using phasors. Phasor is another word for *vector*, which is a straight line used to show the magnitude and phase angle of a voltage. (We will study phasors in Chapter 14.) A phasor diagram for the Y-connected generator is shown in Figure 12–30, where the amplitude of each voltage is represented by a line 120 units long, separated from the other lines by 120°.

The diagram shows that each voltage is 120° out of phase with any other voltage and that the amplitude of each voltage is 120 V. Show that a voltmeter connected between any two output terminals will measure 208 V.

Solution By drawing any two of the phasors on a rectangular coordinate system, we can solve the resulting triangles for the total voltage, as shown in Figure 12–31.

$$\sin 30° = \frac{OC}{120} = \frac{y}{120}$$
$$y = 120 \sin 30°$$
$$= 60$$

$$\cos 30° = \frac{CB}{120} = \frac{x}{120}$$
$$x = 120 \cos 30°$$
$$x \approx 103.92$$

$$AC = 120 + y = 180$$
$$CB \approx 103.92$$

Therefore, from the Pythagorean theorem:

$$AB = \sqrt{(AC)^2 + (CB)^2}$$
$$AB = \sqrt{180^2 + 103.92^2}$$
$$\approx 208 \text{ V}$$

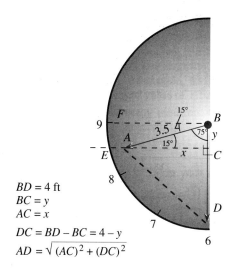

$BD = 4$ ft
$BC = y$
$AC = x$
$DC = BD - BC = 4 - y$
$AD = \sqrt{(AC)^2 + (DC)^2}$

Figure 12–28

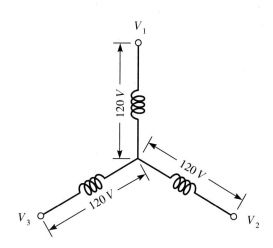

Figure 12–29

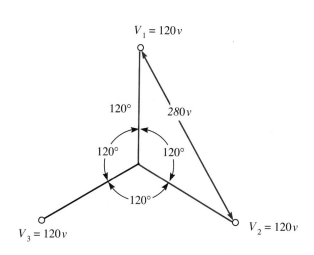

Figure 12–30

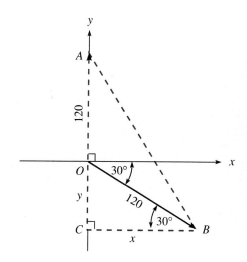

Figure 12–31

266 Advanced Topics in Electronics Mathematics

Example 12-6 In Example 12-5, show that the voltage V_t between any two output terminals of a Y-connected alternator whose phase differences are 120° and whose individual voltages have an amplitude of V_P volts is given by $V_t = V_P\sqrt{3}$.

Solution Using Figure 12-31, we relabel the phasors as shown in Figure 12-32.

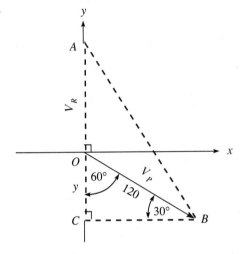

Figure 12-32

From our study of special triangles in Chapter 8, we see that the 30-60 triangle can be relabeled as shown in Figure 12-33.

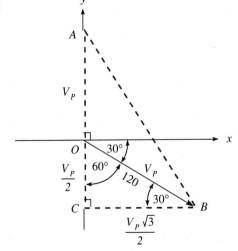

Figure 12-33

Elementary Trigonometry

$$AC = V_P + \frac{V_P}{2}$$

and

$$CB = \frac{V_P\sqrt{3}}{2}$$

From the Pythagorean theorem, we write:

$$V_T = AB = \sqrt{\left(V_P + \frac{V_P}{2}\right)^2 + \left(\frac{V_P\sqrt{3}}{2}\right)^2}$$

$$= \sqrt{V_P^2 + V_P^2 + \frac{V_P^2}{4} + \frac{3V_P^2}{4}}$$

$$= \frac{\sqrt{12V_P^2}}{2}$$

$$= \frac{2V_P\sqrt{3}}{2}$$

$$= V_P\sqrt{3}$$

$$\doteq V_T$$

The foregoing examples should give you some idea of the variety of situations that can be resolved using some basic algebra, the Pythagorean theorem, and the three basic trigonometric relationships derived in this chapter.

Exercise Set 12-3

Questions

1. What is meant by an inverse trigonometric function?
2. What notation symbolizes an inverse trigonometric function?
3. What are the meanings of $\sin^{-1} \theta$ and $1/\sin \theta$? How are they different?

Problems

1. An engineer uses a computer-controlled optical laser to direct a train of pulses at a mirror erected atop the Leaning Tower of Pisa. (See Figure 12–34.) The computer determines that the time required for a pulse to make the one-way return trip is 2 μs. The angle of elevation of the laser is 5.1°. What is the distance D from the top of the tower to the ground?

   Hint: The laser pulses travel at the speed of light, which is 300×10^6 m/s.

Figure 12–34

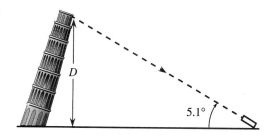

2. Two searchlights, 750 ft apart, illuminate the same spot underneath a cloud, as shown in Figure 12–35. What is the height (*h*) of the cloud?

Figure 12–35

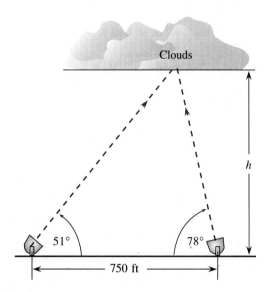

3. A 1.23 MHz Marconi antenna is secured with a network of guys, as shown in Figure 12–36. There are three such networks spaced 120° apart around the antenna. Ideally, how much wire is required to support the antenna in a vertical position?

Figure 12–36

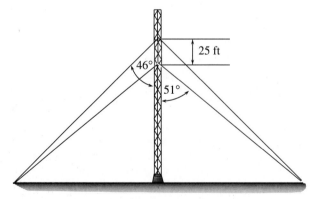

4. The distance between adjacent teeth on a 30-tooth sprocket is measured as 0.393 in. What is the overall diameter of the sprocket? (See Figure 12–37.)
5. A wooden plank rests against a 3 ft–diameter cylinder, as shown in Figure 12–38. What is angle θ?
6. A 1.7 ft rigid arm is attached to a 2.5 ft rigid rod and is free to pivot at points *A*, *B*, and *D* as the 2 ft–diameter wheel rotates about point *C*. Points *A* and *C* are separated by 3 ft. What is the total angle through which the arm moves as the wheel rotates? (See Figure 12–39.)

Figure 12–37

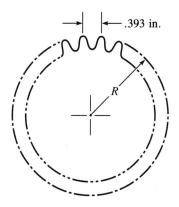

Figure 12–38

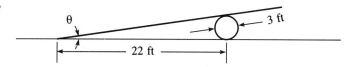

Figure 12–39 I.

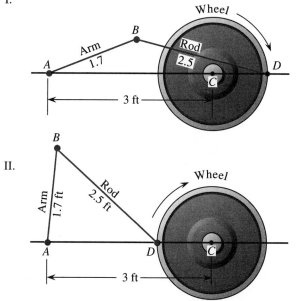

II.

7. An octagonal wire fence with eight equal sides is to be built around a 100 kVA high-voltage transformer to keep out unauthorized persons. The fence is 18 ft across, as shown in Figure 12–40. What is the total length of fencing material required?

Figure 12–40

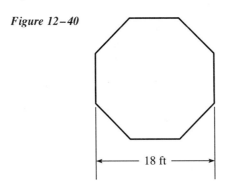

8. A regular pentagon (five equal sides) measures 20 cm on each side. What is the length of a straight line drawn between A and B? (See Figure 12–41.)

Figure 12–41

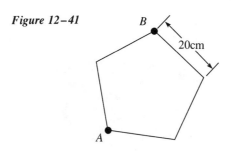

9. A radar installation detects a sailboat on a bearing of 108° at a range of 265 yd. A buoy is also detected on a bearing of 162° at a distance of 309 yd. How far apart are the sailboat and the buoy? (See Figure 12–42.)
10. Write a BASIC program that will solve Problem 9 as the sailboat moves relative to the buoy.

12–9 GRAPHS OF THE TRIGONOMETRIC RELATIONSHIPS

Imagine the visual effect produced by a red light located at point A on the unit circle in Figure 12–43 as θ rotates counterclockwise with uniform speed. Obviously, the pattern traced by the light as θ rotates is a circle. Imagine, now, the effect of simultaneously moving Figure 12–43 in its entirety along a straight line to the right with uniform speed as θ continues to rotate, as shown in Figure 12–44. What is the shape of the resultant light pattern?

To answer this question, we imagine taking a series of stop-action photos at various distances corresponding to θ in radians. For example, at $\theta = 90°$ ($\pi/2$ rad), the distance moved to the right is $\pi/2$ units, or about 1.57 units. Similarly, at $\theta = 180°$ (π rad), the distance moved to the right is another 1.57 units, or 3.14 units from zero. We continue in this manner until all five points shown in Figure 12–45 have been taken.

Figure 12-42

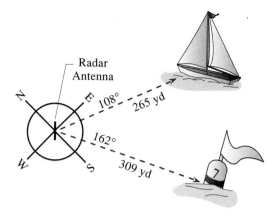

Figure 12-43

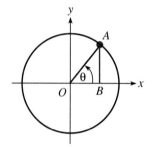

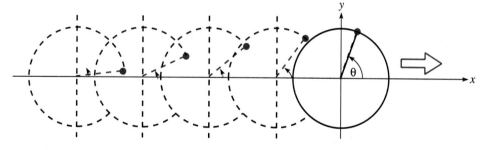

Figure 12-44

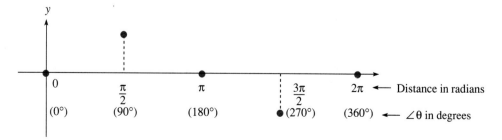

Figure 12-45

By filling in many more points with stop-action photos, we obtain a graph like that shown in Figure 12–46. As we fill in all the remaining spaces, the final shape shown in Figure 12–47 emerges. This shape is called a sine curve, or **sine wave,** since the height of the curve above (or below) the horizontal axis at any value of θ (in degrees or radians) is equal to the value of $\sin \theta$ at that point. Recall from Figure 12–6 that the length of line AB coincides with $\sin \theta$. Therefore, Figure 12–47 represents the graph of $y = \sin \theta$.

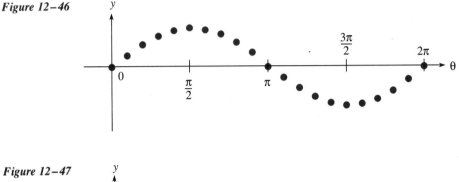

Figure 12–46

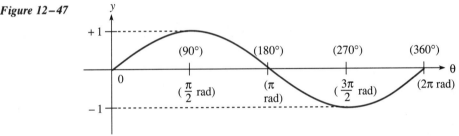

Figure 12–47

Figure 12–48 shows a single loop of copper wire rotating with uniform speed in the field of a permanent magnet. As the conductor cuts across the magnetic field, a voltage is induced in the conductor that is proportional to the *rate* at which the conductor cuts through the magnetic field. This rate, in turn, depends on where in the magnetic field the conductor is located at a particular moment. This is simply another way of saying that the rate is determined by the angle through which the conductor has rotated relative to an initial point where $V = 0$. For example, in part (a) of the figure, the conductor is moving parallel to the magnetic field at $\theta = 0°$; hence, the rate of cutting magnetic lines is zero and the induced voltage is zero. In part (b), the conductor is moving through the field at right angles to the field ($\theta = 90°$) and is, therefore, cutting magnetic lines at the maximum rate. Here, the voltage is maximum. It can be shown that the induced voltage is proportional to $\sin \theta$. Mathematically, we can write $V = k \sin \theta$, where k is the constant of proportionality determined by a variety of factors, among which is the strength of the magnetic field.

The voltage produced by our simple alternator (AC generator) is called a **sinusoidal** (si-nu-SOI-dal) voltage, or sine-wave voltage, and is the same type that is commonly available at the electrical outlets in a modern house for supplying 120 V, 60 Hz power. An oscilloscope connected across the terminals of such an outlet will display a sinusoidal voltage, as shown in Figure 12–49 on page 274.

By relabeling the axes of Figure 12–43 and rotating the entire figure 90° counterclockwise, we may place our imaginary red light at point B and thus draw the

Figure 12–48

(a) $\theta = 0°; V = 0v$

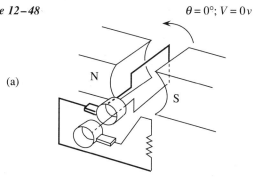 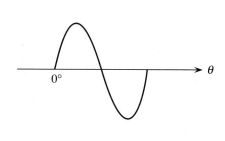

(b) $\theta = 90°; V = +V_{max}$

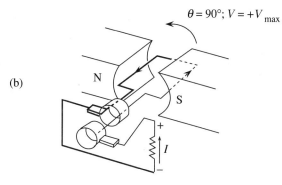 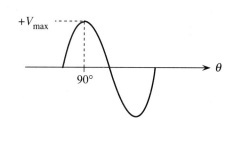

(c) $\theta = 180°; V = 0\ v$

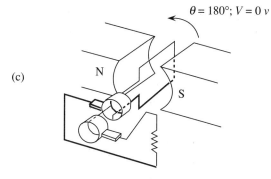 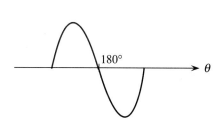

(d) $\theta = 270°; V = -V_{max}$

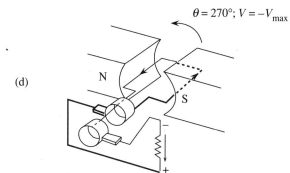 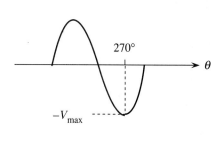

273

Figure 12-49

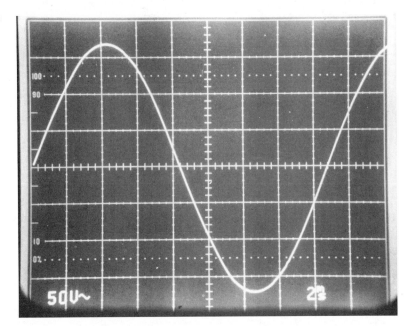

graph of cos θ, as shown in Figure 12-50. The shape of this graph is called a **cosine curve**, since the height of OB above or below the horizontal axis at any value of θ is equal to the value of cos θ at that point. Note that the graph of $y = \cos \theta$ is the same as $y = \sin \theta$ shifted 90° ($\pi/2$ rad) to the left.

Figure 12-14 has been redrawn as Figure 12-51 with our imaginary red light located at point C in order to represent a portion of the graph of $y = \tan \theta$. Note that the graph is undefined at $\theta = 90°$ ($\pi/2$ rad), since the length of CD approaches infinity as θ approaches 90°. However, observe that we may allow θ to get as close to 90° as we wish (e.g., 89.999999°) provided it does not actually equal 90°. In this case, we say that the graph of $y = \tan \theta$ approaches the vertical line at $\pi/2$ in an asymptotic (a-sim(p)-TOT-ic) manner. That is, the distance between the graph and the vertical line may become infinitely small but may never actually reach zero. In other words, the graph does not touch the line.

The complete graph of $y = \tan \theta$ (for $0° \leq \theta \leq 360°$) is shown in Figure 12-52. Note that the values of $y = \tan \theta$ are negative between 90° and 180° and positive between 180° and 360°. Careful inspection of Figure 12-52 reveals that this situation is exactly opposite of what our red-light graph would have looked like but is entirely consistent with the definition of $y = \tan \theta$ developed earlier as the ratio of signed numbers. Nonetheless, the basic shape of the red-light graph is correct except for its inversion between 90° and 270°.

12-10 GRAPHS OF $y = A \sin(Bx - C)$ AND $A \cos(Bx - C)$

In the previous section we developed the graphs of $y = \sin \theta$ and $y = \cos \theta$. We saw that one complete cycle of the graph is obtained each time the angle θ completes one revolution (360°, or 2π rad). During the second revolution, these cycles repeat again, yielding another portion of a graph exactly like the previous one. We say, then, that these two trigonometric relationships have a **period** of 360°, or 2π rad. Furthermore, we see

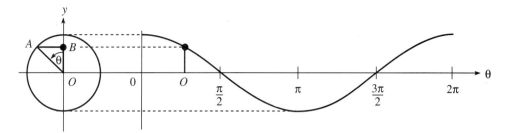

Figure 12–50

Figure 12–51

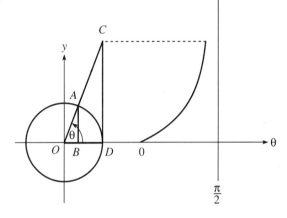

Figure 12–52

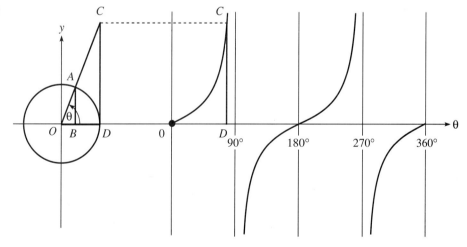

that both sin θ and cos θ have a maximum absolute value of 1. The maximum absolute value that either of these two waveforms attains is called the **amplitude** of the graph. It is possible to create an amplitude greater than 1.0 by multiplying the sine or cosine by some number. For example, the graph of $y = 3 \sin θ$ has an amplitude 3 times as great as that of $y = \sin θ$. The numeral 3 specifies the amplitude of the relationship. In general, $y = A \sin θ$ and $y = A \cos θ$ have amplitudes equal to $|A|$. For comparison, the graphs of $y = \sin θ$ and $y = 3 \sin θ$ are drawn in Figure 12–53.

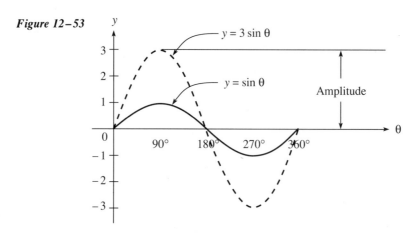

Figure 12–53

The trigonometric graphs can be altered in terms of their periodicity as well as their amplitudes. For example, consider the graph of $y = \sin 2x$. Here, θ has been replaced by the expression $2x$. We know from our previous discussion that $0 \le θ \le 2π$. Replacing θ by $2x$, we obtain $0 \le 2x \le 2π$. Finally, dividing each term by 2, we get $0 \le x \le π$. Thus the period of $y = \sin 2x$ is π rad (180°) rather than 2π rad (360°), as shown in Figure 12–54.

Note that the graph of $y = \sin 2x$ is twice as fast, so to speak, as the graph of $y = \sin x$. This is another way of saying that $y = \sin 2x$ has a **frequency** twice that of $y = \sin x$. Indeed, if we replace the angle (x) with time (t), we may use the familiar relationship $f = 1/t$, where f is the frequency, or number of complete cycles in 1 s. In Figure 12–55, we have redrawn Figure 12–54 with time in arbitrary values of milliseconds (ms) along the abscissa instead of θ in degrees.

We see that the frequency of sin x is given by $f = 1/t = 1/0.004 = 250$ cycles per second (hertz, abbreviated Hz), and the frequency of $y = \sin 2x$ is $f = 1/t = 1/0.002 = 500$ Hz.

In the expressions $y = \sin Bx$ and $y = \cos Bx$, the period is given by $2π/B$ radians, or $360/B$ degrees, where B is the periodicity factor. In the preceding example, $y = \sin 2x$, the period is $2π/2 = π$ rad, or $360/2 = 180°$.

We see, then, that the factors A and B affect the basic trigonometric relationships in terms of both amplitude as well as periodicity. In $y = A \sin Bx$, for example, it is a simple thing to see that the amplitude is A and the period is $2π/B$ radians (or $360/B$ degrees). There is, however, an additional third factor that determines where on the Cartesian coordinate system we begin to draw the graph of sin θ or cos θ.

Consider the expression $y = \cos(x - C)$. Replacing θ by $x - C$, we may write $0 \le x - C \le 2π$. Adding C to each term, we obtain $C \le x \le 2π + C$. This last result tells us that the starting point of the graph of $y = \cos(x - C)$ has been shifted along the x-axis by an amount equal to C. For example, Figure 12–56a shows the graph of $y = \sin(x - 45°)$, which has been shifted 45° to the right. For comparison, the graph of $y = \cos x$ is shown in part (b) of the figure, and the two graphs have been superimposed in part (c).

Figure 12–54

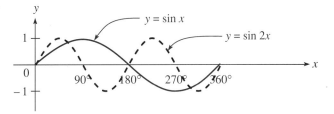

Figure 12–55

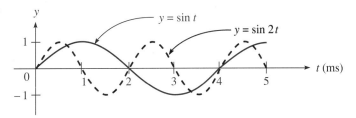

Figure 12–56

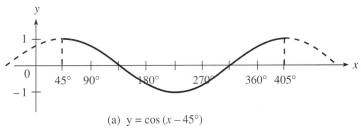

(a) $y = \cos(x - 45°)$

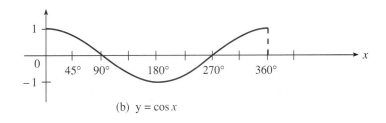

(b) $y = \cos x$

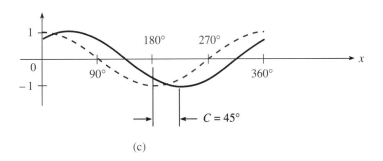

(c)

278   Advanced Topics in Electronics Mathematics

In the more general expressions $y = \sin(Bx - C)$ and $y = \cos(Bx - C)$, it may be intuitively obvious to you that since the period is affected by B, the amount of shift in the curve along the axis is also affected by B. This is correct and may be seen by replacing θ by $Bx - C$. This allows us to write $0 \leq Bx - C \leq 2\pi$. Adding C to each term, we get $C \leq Bx \leq 2\pi + C$. Finally, dividing through by B, we have

$$\frac{C}{B} \leq x \leq \frac{2\pi}{B} + \frac{C}{B}$$

Therefore, $-C/B$, called the **phase shift,** is simply the amount a given curve is shifted along the x-axis relative to its normal starting position.

Rule 12–1 summarizes the properties of the trigonometric relationships $y = A \sin(Bx - C)$ and $y = A \cos(Bx - C)$.

Rule 12–1

In $y = A \sin(Bx - C)$ and $y = A \cos(Bx - C)$,

$$\text{Amplitude} = A$$
$$\text{Period} = \frac{2\pi}{B} \text{ radians}$$
$$\text{Phase shift} = \frac{C}{B} \text{ radians}$$

Phase shift is sometimes symbolized by the Greek letter ϕ (phi).

The following examples should serve to clarify the foregoing concepts.

Example 12–7   Draw the graph of $y = 4 \sin(x - \frac{1}{3}\pi)$ (x is in radians).

Solution   Amplitude = 4; period = 2π; phase shift = $-C/B = \pi/3$. See Figure 12–57.

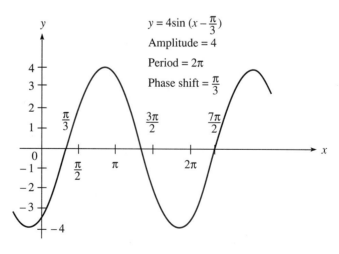

Figure 12–57

Elementary Trigonometry

Example 12-8

Draw the graph of $y = \sin(x + \frac{1}{4}\pi)$ (x is in radians).

Solution Amplitude = 1; period = 2π; phase shift = $-C/B = -\pi/4$. (Substituting $x + \pi/4$ for θ, we obtain $-\pi/4 \leq x \leq 2\pi - \pi/4$, from which we see that the phase shift is $-\pi/4$.) See Figure 12-58.

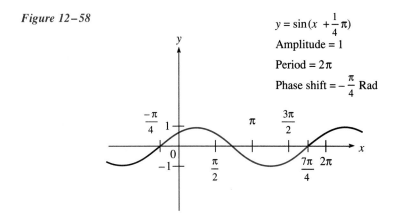

Figure 12-58

Example 12-9

Draw the graph of $y = \sin(-x + 6)$ (x is in radians).

Solution Amplitude = 1; period = 2π; phase shift = 6. Since $\sin(-x) = -\sin x$, we can write the problem as $y = \sin[-(x - 6)] = -\sin(x - 6)$. This gives us a positive value for B and simplifies the algebra. Therefore, $0 \leq x - 6 \leq 2\pi$, and $6 \leq x \leq 2\pi + 6$. Similarly, since $\cos(-x) = \cos(x)$, we may write $y = \cos(-x + c)$ as $\cos[-(x - c)] = \cos(x - c)$. See Figure 12-59. Although the graph may be sketched freehand by plotting a few selected points computed with the aid

Figure 12-59

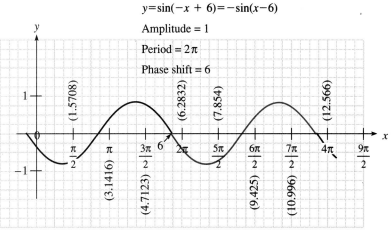

of a hand-held calculator, the BASIC program in Figure 12–60 provides a printed array of as many values of *y* as desired for *x* in radians. By slight modification, the program can be made to run in degrees as well. The program will generate a list of *y*-values for every 0.25 rad from 0 to 14, as shown in Figure 12–61.

```
Ok
LIST
10 PRINT TAB(1)"X" TAB(15)"SIN(-X+6)"
20 FOR X = 0 to 14 STEP .25
30 LET Y=SIN(-X+6)
40 PRINT TAB(1)X TAB(15)Y
50 NEXT X
60 END
Ok
```

Figure 12–60
(Courtesy of Kee Hyuk Yoon.)

Figure 12–61

```
Ok
RUN
X              SIN(-X+6)
 0             -.2794155
  .25          -.5082792
  .5           -.7055404
  .75          -.8589345
 1             -.9589243
 1.25          -.9992929
 1.5           -.9775302
 1.75          -.8949894
 2             -.7568026
 2.25          -.5715613
 2.5           -.3507832
 2.75          -.1081951
 3              .14112
 3.25           .381661
 3.5            .5984722
 3.75           .7780732
 4              .9092975
 4.25           .9839859
 4.5            .9974951
 4.75           .9489847
 5              .841471
 5.25           .6816388
 5.5            .4794255
 5.75           .247404
 6              0
 6.25          -.247404
 6.5           -.4794255
 6.75          -.6816388
```

Elementary Trigonometry

7	−.841471
7.25	−.9489847
7.5	−.9974951
7.75	−.9839859
8	−.9092975
8.25	−.7780732
8.5	−.5984722
8.75	−.381661
9	−.14112
9.25	.1081951
9.5	.3507832
9.75	.5715613
10	.7568026
10.25	.8949894
10.5	.9775302
10.75	.9992929
11	.9589243
11.25	.8589345
11.5	.7055404
11.75	.5082792
12	.2794155
12.25	3.317922E−02
12.5	−.21512
12.75	−.4500441
13	−.6569866
13.25	−.8230809
13.5	−.9380001
13.75	−.9945988
14	−.9893582

Example 12–10   Draw the graph of $y = \sin \theta + \sin 2\theta$ (θ in degrees).

Solution   Figure 12–62 shows the graphs of $y = \sin \theta$, $y = \sin 2\theta$, and $y = \sin \theta + \sin 2\theta$. The composite graph is obtained by taking the algebraic sum of the two component graphs at every point.

Figure 12–62

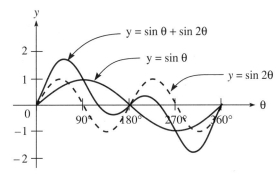

Although the composite graph can be obtained using only a calculator, the BASIC program presented in Figure 12–63 will greatly reduce the labor and help improve the accuracy of the curves. The program gives the y-component value for each of the three graphs every 5°. A partial list of values obtained from the program is shown in Figure 12–64 for illustrative purposes only.

The photo in Figure 12–65 was obtained from an oscilloscope display of $y_1 = \sin \theta$ and $y_2 = \sin 2\theta$. The frequency of y_1 is 60 Hz; the frequency of y_2 is 120 Hz. Note that the photo and composite graph are identical as to shape.

Exercise Set 12–4

Questions

1. Define the following terms in your own words.
   a. Amplitude
   b. Period
   c. Phase shift
2. Why is the amplitude of the graphs of the basic sine and cosine relationships always 1?
3. Why is the period of the graphs of the basic sine and cosine relationships always 2π rad (360°)?
4. Why is the phase shift of the graphs of the basic sine and cosine relationships always 0?
5. Draw a circle of unit radius that could be used to generate the graph of $y = \sin x$ with a phase shift of 45° (i.e., $y = \sin(x - 45°)$.
6. Draw a circle that could be used to generate the graph of $y = 2 \cos \theta$.

Problems

Draw the graph of each of the following trig relationships. Label the amplitude, period, and phase shift (x is in radians).

1. $y = -2 \sin x$
2. $y = \sin \dfrac{x}{2}$
3. $y = \cos\left(\dfrac{2x}{3} - \dfrac{3}{2}\right)$
4. $y = -4 \sin\left(\dfrac{x}{2} + 1\right)$
5. $y = 2 \sin(x + 1)$
6. $y = \sin\left(x - \dfrac{\pi}{3}\right)$
7. $y = \cos\left(x - \dfrac{1}{2}\right)$
8. $y = 5 \cos(-3x + 3)$
9. $y = \cos(\pi - x)$
10. $y = \dfrac{1}{2} \sin\left(\pi x + \dfrac{\pi}{2}\right)$

11. Write a BASIC program that will provide a table of arrayed data points for each of the above problems. (Use the table in Example 12–9 as a guide.)
12. Draw the graph of $y = 5 + \sin x$. In what way is it similar to the graph of $y = \sin x$? In what way is it different?

```
Ok
LIST
10 PRINT TAB(1)"DEGREES" TAB(15)"SIN X" TAB(30)"SIN 2X" TAB(50)"Y1+Y2"
20 FOR A = 0 to 360 STEP 5
30 LET Y1=SIN(A*.017453292#)
40 LET Y2=SIN(2*A*.017453292#)
50 LET Y=Y1+Y2
60 PRINT TAB(1)A TAB(15)Y1 TAB(30)Y2 TAB(50)Y
70 NEXT A
80 END
Ok
```

Figure 12–63

Figure 12–64

```
Ok
RUN
```

DEGREES	SIN X	SIN 2X	Y1+Y2
0	0	0	0
5	8.715574E-02	.1736482	.260804
10	.1736482	.3420202	.5156684
15	.2588191	.5	.758819
20	.3420202	.6427876	.9848078
25	.4226183	.7660444	1.188663
30	.5	.8660254	1.366025
35	.5735765	.9396928	1.513269
40	.6427876	.9848078	1.627595
45	.7071068	1	1.707107
50	.7660444	.9848079	1.750852
55	.8191521	.9396928	1.758845
60	.8660254		1.732051
	.9063078		1.672352

Figure 12–65

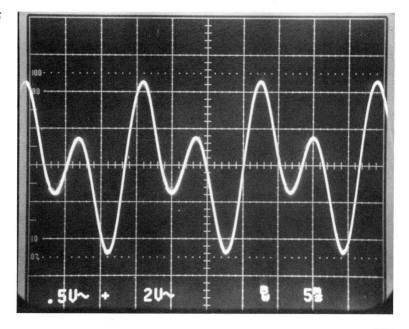

Advanced Topics in Electronics Mathematics

13. With the aid of a calculator, a student is trying to sketch the graph of $y = \cos x$ by plotting points every 5°. The following partial table of data is obtained.

x	$\cos x$
0	1
5	0.2837
10	−0.8391
15	−0.7597
20	0.4081
25	0.9912
30	-E-
35	-E-

What error has the student made?

14. Draw the graph of $y = \cos \theta + 2 \cos 3\theta$. What is the amplitude of the composite graph? The period? The phase shift?
15. Write a BASIC program that will assist in drawing the composite graph in Problem 14. (Use the program in Example 12–10 as a guide.)

Key Terms

angle
vertex
initial side
terminal side
standard position
radian

grad
sine
cosine
tangent
inverse
sine wave

sinusoidal
cosine curve
period
amplitude
frequency
phase shift

Important Rules

Radian Measure

$$180° = \pi \text{ rad}$$
$$1 \text{ rad} \approx 57.29577951°$$
$$1° \approx 0.017453292 \text{ rad}$$

Trigonometric Relationships in a Right Triangle

Figure 12–66

$$\sin \theta = \frac{\text{opposite side}}{\text{hypotenuse}}$$

$$\cos \theta = \frac{\text{adjacent side}}{\text{hypotenuse}}$$

$$\tan \theta = \frac{\text{opposite side}}{\text{adjacent side}}$$

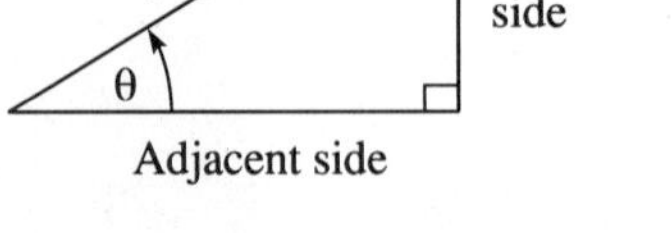

Elementary Trigonometry

Graphs of the Trigonometric Relationships

Figure 12–67

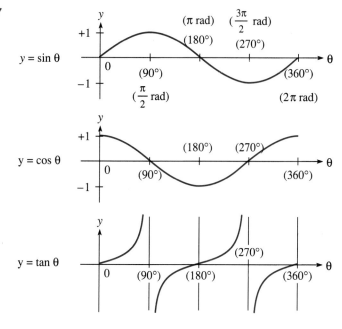

Properties of $y = A \sin(Bx - C)$ and $y = A \cos(Bx - C)$

Rule 12–1

Amplitude = A
Period = $2\pi/B$ rad
Phase Shift = $-C/B$ rad

Upon completion of this chapter, you will have:
1. Learned how to solve exponential and logarithmic equations.
2. Learned how to perform change-of-base operations.
3. Learned how logarithms are used to specify electrical device characteristics in decibels.

13 Exponents and Logarithms

13-1 INTRODUCTION

Historically, logarithms were used extensively to simplify complex engineering and scientific calculations. With the advent of the hand-held digital calculator, however, logarithms are no longer used to any real extent for this purpose. Rather, the study of logarithms and exponents is important because they describe so much of what we experience in the real world. For example, humans and most animals hear in a logarithmic manner. The intensity of earthquakes is measured on a logarithmic scale. Bacterial population growth occurs at an exponential rate. The money in our savings account increases exponentially. The growth and decay of current through various electrical components obeys exponential laws. Finally, gain and attenuation of electronic devices is frequently specified in units of decibels based on logarithms. It is necessary, then, to thoroughly understand the properties of exponential and logarithmic relationships.

13-2 EXPONENTIAL RELATIONSHIPS

Recall from Chapter 6 that an exponent was defined as a number written to the upper right of another number, called the base, that tells how many times the base is to be taken as a factor. For example, in the expression $2^5 = 32$, the exponent is 5, and the base is 2. Taking 2 as a factor 5 times, we obtain $2 \times 2 \times 2 \times 2 \times 2 = 32$.

The first question that might come to your mind is, What if we let the exponent of the base 2 take on any range of numbers and then graph the results? What would the graph look like? In other words, what shape does the graph of $y = 2^x$ have? We can answer this question by making a partial list of values for x and y, as shown in Table 13-1.* In the

---

*The table was generated from a BASIC computer program similar to the ones used in Chapter 12. The format xxxxxxE-xx is BASIC's way of indicating scientific notation. For example, 9.536743E-07 means 9.536743×10^{-7}.

Figure 13-1

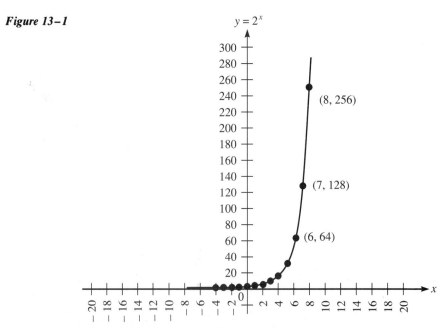

Exponents and Logarithms

Table 13–1

x	$y = 2^x$
−20	9.536743E−07
−19	1.907349E−06
−18	3.814697E−06
−17	7.629395E−06
−16	1.525879E−05
−15	3.051758E−05
−14	6.103516E−05
−13	1.220703E−04
−12	2.441406E−04
−11	4.882813E−04
−10	9.765625E−04
−9	1.953125E−03
−8	3.90625E−03
−7	.0078125
−6	.015625
−5	.03125
−4	.0625
−3	.125
−2	.25
−1	.5
0	1
1	2
2	4
3	8
4	16
5	32
6	64
7	128
8	256
9	512
10	1024
11	2048
12	4096
13	8192
14	16384
15	32768
16	65536
17	131072
18	262144
19	524288
20	1048576

table, we have allowed x to take on the range of numbers between −20 and +20 in one-step intervals. Figure 13–1 shows the results of graphing the **exponential relationship** $y = 2^x$ using the data in Table 13–1.

From the graph in Figure 13–1, we observe several important characteristics. First, it is apparent that the graph of $y = 2^x$ grows *very* rapidly. Even with considerable compression of the y-axis scale used in the figure, only a few points are actually capable of being plotted. The next integral point on the graph, (9, 512), is completely off the top of the figure! The second thing we notice is that as the exponent gets smaller, approaching −20, the graph comes very close to the x-axis without actually touching it. In fact, at

$x = -20$, $y = 0.000\ 000\ 954$. For the value $x = -100$, the y-value is 0.000 000 000 000 000 000 000 000 000 000 788 9. We may conclude, then, that no matter how small x becomes, the graph never touches the x-axis. We say that the x-axis is the asymptote of the curve. You may recall from Chapter 12 that the graph of $y = \tan \theta$ had vertical asymptotes at $\theta = \pi/2$ and $3\pi/2$ rad, among others.

The graph of $y = 2^x$ in Figure 13–1 has a general form that is typical of all exponential relationships having a base equal to any positive number other than 1 (since 1 raised to any power is always 1). Figure 13–2 shows the graphs of $y = 2^x$, $y = 3.5^x$, and $y = 10^x$. Note that all these curves pass through the point (0, 1). This is to be expected, since we know from Chapter 3 that $b^0 = 1$. Moreover, each graph has the x-axis as an asymptote.

13-3 LOGARITHMIC RELATIONSHIPS

Figure 13–3 shows the graph of $y = 2^x$ and the diagonal line $y = x$. The line of $y = x$ has an interesting mirror property in that any graph "reflected" in it becomes the *inverse* of the given relationship.

In Figure 13–4, we have reflected $y = 2^x$ through the line $y = x$ to obtain a graph that represents a new type of relationship called the **logarithmic relationship,** given as $y = \log_2 x$, which is read "y equals the log of x to the base 2." Observe how much more slowly a logarithmic curve grows in contrast to its exponential counterpart.

In the figure, note that every point on $y = 2^x$ is reflected through the line $y = x$ onto the graph of $y = \log_2 x$. The equation $y = 2^x$ is said to be in *exponential form,* and the equation $y = \log_2 x$ is said to be in *logarithmic form.*

A partial list of values of x and y is shown in Table 13–2. As you probably already expected, since one relationship is the inverse of the other, the values of x and y are merely interchanged.

Study of Table 13–2 will lead you to the conclusion that a logarithm is simply another name for an exponent. Since this is the case, it follows that we ought to be able to write the same information in two different forms, namely, exponential and

Table 13–2

Exponential Form $y = 2^x$		Logarithmic Form $y = \log_2 x$	
x	y	x	y
-4	0.0625	0.0625	-4
-3	0.125	0.125	-3
-2	0.25	0.25	-2
-1	0.5	0.5	-1
0	1	1	0
1	2	2	1
2	4	4	2
3	8	8	3
4	16	16	4

Figure 13–2

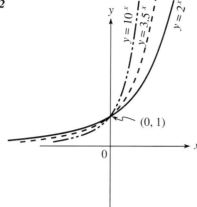

Figure 13–3

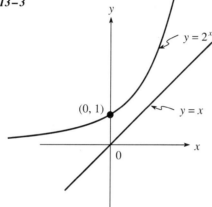

Figure 13–4

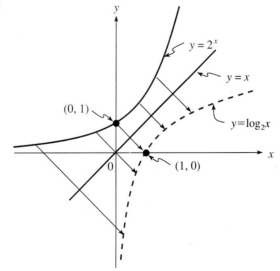

logarithmic. Table 13–3 shows how this is done. The last entry in the table is important, since it reveals the general connection between exponential and logarithmic forms. And since $B^x = N$ and $x = \log_B N$, it is apparent that we do, in fact, get the same information in either form.

Table 13–3

Exponential Form	Logarithmic Form
$2^3 = 8$	$3 = \log_2 8$
$5^2 = 25$	$2 = \log_5 25$
$10^4 = 10{,}000$	$4 = \log_{10} 10{,}000$
$B^x = N$	$x = \log_B N$
$B^{\log_B N} = N$	$N = \log_B B^N$

13–4 OPERATIONS WITH LOGARITHMS

Since a logarithm is an exponent, it must follow the laws of exponents. The laws of exponents of primary interest to us here are listed for reference:

$$b^p b^q = b^{p+q}$$
$$\frac{b^p}{b^q} = b^{p-q} \quad (b \neq 0)$$
$$(b^p)^q = b^{pq}$$

We now show that these laws can be used to establish many useful properties for logarithms.

Let $p = \log_b M$ and $q = \log_b N$. Then $b^p = M$ and $b^q = N$. Hence $b^p b^q = b^{p+q} = MN$. Therefore, $\log_b MN = \log_b b^{p+q} = p + q = \log_b M + \log_b N$.

Rule 13–1 Log of a Product

$$\log_b MN = \log_b M + \log_b N$$

In words, this says that the log of a product is the sum of the logs. For example,

$$\log_{10}(50) = \log_{10} 5 + \log_{10} 10$$
$$\approx 0.699 + 1 = 1.699$$

Let $p = \log_b M$ and $q = \log_b N$. Then $b^p = M$ and $b^q = N$. Thus $b^p/b^q = b^{p-q} = M/N$. Therefore $\log_b M/N = \log_b b^{p-q} = p - q = \log_b M - \log_b N$.

Rule 13–2 Log of a Quotient

$$\log_b \frac{M}{N} = \log_b M - \log_b N$$

Exponents and Logarithms

In words, this states that the log of a quotient is the difference of the logs. For example,

$$\log_{10}\frac{75}{5} = \log_{10}75 - \log_{10}5$$
$$\approx 1.875 - 0.699 = 1.176$$

Let $p = \log_b M$. Then $b^p = M$. Raising both sides to the qth power, we obtain

$$b^{pq} = M^q$$

Writing this in logarithmic form, we have

$$\log_b M^q = pq$$

On substituting for p, we get

$$\log_b M^q = q \log_b M$$

Rule 13–3 Log of a Power

$$\log_b M^q = q \log_b M$$

In words, this states that the logarithm of the qth power of a number is equal to q times the log of that number. For example,

$$\log_{10}3^5 = 5 \log_{10}3 \approx 5 \times 0.477 = 2.385$$

As a corollary to Rule 13–3, we see that $\log_b b^p = p$. The truth of this assertion becomes obvious when we use Rule 13–3 to write

$$\log_b b^p = p \log_b b = p \times 1 = p$$

because $b^1 = b$.

Corollary to Rule 13–3

$$\log_b b^p = p$$

For example,

$$\log_{10}10^3 = 3 \times \log_{10}10 = 3 \times 1 = 3$$

In Figure 13–4, it is apparent that logarithms are valid for positive numbers only. Therefore, M and N in the preceding rules must be positive numbers.

Example 13–1

Express $\log_2 64$ in three different ways using the given rules.

Solution By Rule 13–1, $\log_2 64 = \log_2(16 \times 4) = \log_2 16 + \log_2 4$.

By Rule 13–2, $\log_2 64 = \log_2\left(\frac{192}{3}\right) = \log_2 192 - \log_2 3$.

By Rule 13–3, $\log_2 64 = \log_2 4^3 = 3 \log_2 4$.

Advanced Topics in Electronics Mathematics

Example 13-2 Note that each of the following expressions may be written as a single logarithmic quantity using the rules of logarithms:

$$\log_{10} 17 + \log_{10} x = \log_{10} 17x \qquad \text{Rule 13-1}$$

$$\log_3 k - \log_3 5 = \log_3\left(\frac{k}{5}\right) \qquad \text{Rule 13-2}$$

$$\log_2 12 + 3 \log_2 x - \log_2 5 = \log_2\left(\frac{12x^3}{5}\right) \qquad \text{Rules 13-1, 13-2, and 13-3}$$

Example 13-3 Find $\log_3\left(\frac{1}{9}\right)$.

Solution We may use either of two equally appropriate methods:

First method: $\log_3\left(\frac{1}{9}\right) = \log_3 1 - \log_3 3^2$
$= 0 - 2 \log_3 3 = -2$

Second method: $\log_3(\frac{1}{9}) = \log_3 3^{-2} = -2 \log_3 3 = -2$

Example 13-4 $\log_2 \sqrt{15} = \log_2 15^{1/2} = \frac{1}{2} \log_2 15.$

Exercise Set 13-1

Questions

1. In what major ways do the exponential and logarithmic functions differ?
2. Which axis is the asymptote of the exponential relationship? Of the logarithmic relationship?
3. What is significant about the points (0, 1) and (1, 0) in the graphs of exponential and logarithmic relationships?
4. How is the logarithmic relationship obtained from the exponential relationship?
5. If M is raised to the Nth power, the result is F. Write this fact in exponential as well as logarithmic form.
6. State the rules for operating with logarithms.
7. Show that $y = 2^x$ and $y = x^2$ are two entirely different relationships. Which is logarithmic? What name is given to the other?

Problems

Using the rules for operating on logarithms, write each of the following expressions in two or more parts.

1. $\log_3 15$
2. $\log_2 6t$
3. $\log_2 4x$
4. $\log_5 50$
5. $\log_8\left(\frac{5}{x}\right)$
6. $\log_2 \frac{10}{m}$
7. $\log_4\left(\frac{16}{h}\right)$

Exponents and Logarithms

8. $2\log_2 5x$
9. $\log_7\left(\dfrac{4x^2}{3}\right)$
10. $\log_4\left(\dfrac{16t^3}{4}\right)$

Express each of the following expressions as the logarithm of a single quantity.
11. $\log_5 9 - \log_5 3$
12. $\log_b 12 - \log_b 3$
13. $\log_b x + \log_b y$
14. $\log_2 3 + \log_2 t$
15. $\log_4 3^3 + \log_4 9$
16. $\log_5 243 + \log_5 3$
17. $\frac{1}{2}\log_b x - 2\log_b 7$
18. $\frac{1}{2}\log_2 t - 4$
19. $\log_b M^2 - \log_b \sqrt{M}$
20. $\log_a x^2 - \log_a \sqrt{x^3}$
21. The following BASIC program generates a list of ordered pairs (x, y) for graphing the exponential relationship $y = 4^x$.

```
10 PRINT TAB(1)"X" TAB(6) "4^X"
20 FOR X=-20 TO 20 STEP 1
30 LET Y=4^X
40 PRINT TAB(1)X TAB(6) Y
50 NEXT X
```

Write a program for obtaining the data points used in sketching the graph of $y = 10^x$, and then draw the graph. In what way does this graph differ from that of $y = 4^x$?

13-5 COMMON AND NATURAL LOGARITHMS

The only restriction concerning logarithms is that the base must be some positive number greater than 1. Certain bases, however, have become more popular than others due to their usefulness in engineering and science. As a result, two systems of logarithms have emerged. One, the **common log,** or *Brigg's* system, uses 10 as its base; the other is the **natural log,** or *Naperian* system, whose base is the irrational number denoted by e, which is approximately equal to 2.71828.* The common (base 10) log at one time was used extensively for computational purposes. With the advent of the hand-held digital calculator, however, such use has been largely discontinued. Today, the common logarithm remains useful as the basis for specifying gain or loss in all kinds of electronic systems and components. On the other hand, the natural (base e) system finds many uses in quantifying phenomena in the natural sciences and engineering. Both systems have so many everyday applications that scientific calculators have separate keys for each one. In this section, we take a closer look at each of these logarithmic systems.

---

*The invention of logarithms is commonly credited to Lord Napier of Scotland in 1614, but logarithms were subsequently improved by Henry Briggs, a professor of geometry at Oxford, in 1624.

The Common Log

The common log has become so common that we ordinarily omit the base 10 in writing logarithmic expressions. For example, $\log_{10} 35$ is usually written simply as log 35 without the base 10 being specified. On your calculator,* you may determine the common log of 35 by keying in ③ ⑤ (LOG) and reading the display 1.544068044. Conversely, if you know the exponent (log) of a number to the base 10 and want to find the number, you use the exponential keys (INV) (LOG) or (10^x). A little reflection will demonstrate the logic of this procedure, since — as mentioned in an earlier section — the two types of operations are inverse relationships. For example, suppose you knew that the common logarithm of some number N is 5. That is, you knew that $\log N = 5$. Therefore, $10^5 = N$, and $N = 100{,}000$. In the language of logarithms, N is said to be the **antilog**, or *inverse log* of 5. The symbol $\log^{-1}$ is occasionally used to denote the inverse log, or antilog, of a number. In other words, antilog 5, inverse log 5, 10^5, and $\log^{-1} 5$ all have the same meaning, namely, $10^5 = 100{,}000$.

Example 13–5

Find log 26.58 on your calculator.

Solution A common series of key strokes is ② ⑥ . ⑤ ⑧ (LOG), with the resulting displayed value of 1.424554977. Depending on the degree of accuracy required, this value can be rounded to 1.425.

Example 13–6

Use your calculator to find $\log^{-1} 1.425$.

Solution Key in ① . ④ ② ⑤ (INV) (LOG) and read the displayed value of 26.6072506. Note that this is fairly close to the number 26.58, whose common log was found in the previous example. The discrepancy is due to rounding.

Example 13–7

The power gain in decibels (dB)** is often denoted by G' and is given by the relationship $G' = 10 \log P_o/P_i$, where P_o is the output power of the device and P_i is its input power. Find the decibel power gain if the output power is 2.3 W and the input power is 35 mW.

$$G' = 10 \log \frac{2.3}{0.035} \approx 10 \log 65.714 \approx 18.177 \text{ dB}$$

The Natural Log

In the natural sciences and many engineering applications, the use of base e logarithms is much more common than the use of so-called common logs. Unfortunately, you will have to accept this fact on faith until you study calculus in advanced mathematics courses.

---

*Note that different calculators may use very different keystroke sequences in performing logarithmic and exponential operations. Read your calculator instruction manual carefully.
**We discuss decibels in a later section.

Exponents and Logarithms

Natural logs use the irrational number e, whose approximate value is 2.718, as a base. And because natural logs are so widely used, the notation $\ln x$ has become more popular than $\log_e x$, although both have precisely the same meaning; it is *not* incorrect to use the latter notation.

For example, to find $\ln 4.6$ on your calculator, key in [4] [.] [6] [ln], and read the displayed value of 1.526056304, which can be rounded to 1.526, depending on the degree of accuracy required. Conversely, to find e^3, for example, key in [3] [INV] [ln] and read 20.1 (approximately).

Example 13-8

Find $\ln 5.55$ on your calculator.

Solution Key in [5] [.] [5] [5] [ln] and read the displayed value of 1.713797928.

Example 13-9

Find N on your calculator if $\ln N = -0.2276$.

Solution Key in [.] [2] [2] [7] [6] [-/+] [INV] [ln] and read the displayed value of 0.796442773.

Change of Base

Although the scientific calculator has keys for performing operations with base 10 and base e, there may be times when we need to work with logs to other bases. For example, how can we calculate $\log_2 5$? Fortunately, there is an easy way to change the base from 2 (or any other number) to base 10, base e, or any other convenient base as necessary. The following derivation reveals the method of conversion.

Let $y = \log_b x$. Then, $b^y = x$, in exponential form. If we take the $\log_a$ of both sides of this equation, we get

$$y \log_a b = \log_a x$$

Substituting for y, we obtain

$$(\log_b x)(\log_a b) = \log_a x$$

Finally, we have Rule 13-4.

Rule 13-4 Change of Base

$$\log_b x = \frac{\log_a x}{\log_a b}$$

In the example posed earlier, let $a = 10$, $b = 2$, and $x = 5$. Then,

$$\log_2 5 = \frac{\log_{10} 5}{\log_{10} 2}$$

$$\approx \frac{0.69897}{0.30103} \approx 2.322$$

To verify this answer, we may use the power key [x^y] and enter [2] [x^y] [2] [.] [3] [2] [2] [=], reading the rounded displayed value 5.000.

Example 13-10

Find $\log_3 16$.

Solution From $\log_b x = (\log_a x)/(\log_a b)$, we write

$$\log_3 16 = \frac{\log_{10} 16}{\log_{10} 3}$$
$$= \frac{1.204119983}{0.477121254}$$
$$= 2.523719019$$

Check $\boxed{3}$ $\boxed{x^y}$ 2.523719019 $\boxed{=}$ 16.0000001

Note that we could have used the natural logarithm to get the same answer. From $\log_b x = (\log_a x)/(\log_a b)$ we write

$$\log_3 16 = \frac{\ln x}{\ln b}$$
$$= \frac{\ln 16}{\ln 3}$$
$$= \frac{2.772588722}{1.098612289}$$
$$= 2.523719013$$

Exercise Set 13-2

Questions

1. What are the two most widely used systems of logarithms, and what is the base of each?
2. What are four common notations for the inverse logarithm of a given number N to the base b?
3. What calculator key may be used to check the solution of $N = \log_b x$ if the base is neither 10 nor e?

Problems

Use the calculator to find each value.
1. $\log 82.3$
2. $\log \sqrt{12}$
3. $\log 3.71 \times 10^{-6}$
4. $\log 8.6^2$
5. $\log 24^{1/3}$
6. $\ln 21.0$
7. $\ln 0.003 \times 10^{-12}$
8. $\log_e \sqrt{e}$
9. $\log_e 15^2$
10. $\ln 3^{1/3}$

Use a calculator to find each value.
11. antilog 1.80071
12. $\log^{-1} 1.52284$
13. $10^{-0.6020}$
14. antilog -3

Exponents and Logarithms

15. $\log^{-1} 3\frac{1}{2}$
16. $e^{3.091}$
17. $e^{-6.9}$
18. $\frac{1}{2} e^{-2}$
19. $e^{\sqrt{6}}$
20. $e^8 \times e^{-2}$
21. An octave is defined as the interval between any two frequencies whose ratio is 2:1. Therefore, the interval, in octaves, between any two frequencies is given by:

$$\text{Octave interval} = \log_2\left(\frac{F}{f}\right)$$

where

$$F = \text{the higher frequency}$$
$$f = \text{the lower frequency}$$

   a. Write a new formula that gives the octave interval to base 10 rather than base 2.
   b. What is the octave interval if the lower frequency is 205 Hz and the upper frequency is 319 Hz?

22. For an inductor (L, in henrys) in series with a resistance (R, in ohms), the time (t, in seconds) for the instantaneous current (i, in amperes) to reach a certain value if the initial current is I (in amperes) is given by the expression

$$t = \left(\frac{-L}{R}\right) \ln\left(\frac{i}{I}\right)$$

   If $I = 265$ mA at $t = 0$, $L = 2$ H, $R = 10\ \Omega$, and $i = 120$ mA, what is the value of t?

23. The power gain of an antenna in decibels (dB) is given by $G_P = 10 \log(P_t/P_i)$, where P_t is the power in watts supplied to the antenna under test and P_i is the power supplied to a hypothetical isotropic antenna. What is the power gain if P_t is $\frac{1}{50}$ the isotropic antenna power?

24. The characteristic impedance (Z_0) of a coaxial transmission line is given by

$$Z_0 = \frac{138}{\sqrt{\epsilon_r}} \log \frac{D}{d} \text{ ohms}$$

   where ϵ_r is the dielectric constant of the insulating material, D is the diameter of the outer conductor, and d is the diameter of the inner conductor. If $\epsilon_r = 2.25$, $D = 0.255$ in., and $d = 0.024$ in., find the value of Z_0 in ohms.

25. The BASIC command log(X) returns the natural log of x. Using the change-of-base formula developed in this section, write a BASIC program that will compute the log L of any positive number N to any base b. Include a subprogram that will check the result according to $b^L = N$.

13–6 EXPONENTIAL EQUATIONS

An **exponential equation** is one in which the unknown appears as an exponent. In solving such equations, it is often an advantage to take logarithms of both sides and then proceed with the solution. For example, consider $3^x = 24$. Taking the common log of both sides gives $\log 3^x = \log 24$. Thus $x \log 3 = \log 24$, so $x = (\log 24)/(\log 3) \approx 2.8928$. Check this result using the power key $\boxed{x^y}$ on your calculator. Observe that $3^{2.8928} \approx 24$.

Note that we would have obtained the same answer if natural logs were taken rather than common logs. That is, taking the natural log of each side of the original equation gives

$$\ln 3^x = \ln 24$$
$$x \ln 3 = \ln 24$$
$$x = \frac{\ln 24}{\ln 3}$$
$$x \approx 2.8928$$

Example 13–11

Solve for x in the equation $5^{x-2} = 12$.

Solution Taking the log of each side gives $\log 5^{x-2} = \log 12$, or $(x - 2)\log 5 = \log 12$. Therefore, $x - 2 = (\log 12)/(\log 5)$, and $x = (\log 12)/(\log 5) + 2$. Finally, $x \approx 1.544 + 2 = 3.544$. Checking our answer, we get $5^{3.544-2} = 5^{1.544} \approx 12$.

Example 13–12

Solve for i in the equation $2(5^{i-1}) = 15^i$.

Solution Taking the log of each side and using properties of logarithms gives

$$\log 2 + (i - 1)\log 5 = i \log 15$$

From this,

$$\log 2 + i \log 5 - \log 5 = i \log 15$$
$$i \log 5 - i \log 15 = \log 5 - \log 2$$
$$i(\log 5 - \log 15) = \log 5 - \log 2$$
$$i = \frac{\log 5 - \log 2}{\log 5 - \log 15} \approx \frac{0.39794}{-0.47712} \approx -0.834$$

Check

$$2(5^{-0.834-1}) \stackrel{?}{=} 15^{-0.834}$$
$$2(0.05225) \stackrel{?}{=} 0.1045$$
$$0.1045 = 0.1045$$

13–7 LOGARITHMIC EQUATIONS

Students who have mastered simple algebraic equations of the type encountered in Chapter 4 should have little difficulty with **logarithmic equations,** provided one important fact is kept in mind:

Any expression following the word *log* cannot be operated on by algebraic means.

For example, if $0.5 = \log 2/x$, we *cannot* begin to solve this simply by multiplying both sides by x, obtaining $0.5x = \log 2$.

We may, however, begin by expressing our equation in exponential form as

$$10^{0.5} = \frac{2}{x}$$

Exponents and Logarithms

Solving for x,

$$x = \frac{2}{10^{0.5}}$$
$$x \approx 0.63246$$

Checking our answer in the original equation, we see that $\log(2/0.63246) \approx \log 3.16226 \approx 0.5$.

The following examples should make the general methods for solving logarithmic equations clear.

Example 13-13

Solve for N if $\log_3 N = 5$.

Solution Expressing the original equation in exponential form, we get

$$3^5 = N$$

from which $N = 243$.

Check

$$5 \stackrel{?}{=} \log_3 243$$

We may use the change-of-base formula as follows in order to verify our answer with the calculator:

$$\log_b x = \frac{\log_a x}{\log_a b}$$

$$\log_3 243 = \frac{\log_{10} 243}{\log_{10} 3}$$

$$\approx \frac{2.38561}{0.47712} \approx 5$$

Example 13-14

The *loudness* of a sound (L) in decibels (dB) is a psychological phenomenon related to the physical *intensity* (I) of the sound wave by the equation

$$L = 10 \log \frac{I}{k}$$

where k is the lowest intensity of a 1 kHz sound that can be heard by the average person, regarded to be 1×10^{-12} W/m². Solve the equation for I.

Solution We begin by dividing both sides by 10, obtaining

$$\frac{L}{10} = \log \frac{I}{k}$$

Writing our new equation in exponential form gives

$$10^{L/10} = \frac{I}{k}$$

from which

$$k 10^{L/10} = I$$

$$k \log^{-1} \frac{L}{10} = I$$

Example 13-15

The characteristic impedance (Z_0) of a parallel-wire transmission line with air dielectric is given by

$$Z_0 = 276 \log \frac{2D}{d} \text{ ohms}$$

where

D = center-to-center distance between conductors
d = diameter of the conductors

Solve the equation for d.

Solution We begin by dividing both sides by 276, obtaining

$$\frac{Z_0}{276} = \log \frac{2D}{d}$$

Expressing this in exponential form, we have

$$10^{Z_0/276} = \frac{2D}{d}$$

Therefore, $d = 2D/10^{Z_0/276}$, or $d = 2D/\log^{-1} Z_0/276$.
For example, if $Z_0 = 300 \ \Omega$ and $D = 0.30$ in., find the diameter of d.

$$d = \frac{2D}{10^{Z_0/276}} \approx \frac{0.6}{10^{1.08696}}$$

$$\approx \frac{0.6}{12.2169} \approx 0.049 \text{ in.}$$

Check Use the original formula:

$$276 \log \frac{2D}{d} = 276 \log \frac{0.6}{0.049}$$

$$\approx 276 \log 12.2449$$

$$\approx 276 \times 1.0880 \approx 300 \ \Omega$$

Example 13-16

The capacitor in a series RC circuit has an initial charge Q_0 coulombs. The charge Q on the capacitor at any time t after the circuit voltage is applied may be determined by solving the equation $\ln Q = (-t/RC) + \ln Q_0$.
Writing this equation in exponential form, we obtain

$$Q = e^{(-t/RC + \ln Q_0)}$$
$$= e^{-t/RC} \cdot e^{\ln Q_0}$$
$$= Q_0 e^{-t/RC}$$

Exercise Set 13-3

Problems

Solve each of the following exponential equations for the variable indicated. Check all answers.
1. $5^x = 4$
2. $3^b = 7$
3. $e^{2k} = 3.884$
4. $e^x = 2.222$
5. $12.2^{x+2} = 19^x$
6. $4^{x-1} = 12^x$
7. $5^{x-1} = 2$
8. $2^{a-2} = 4^2$
9. $12^x = 36$
10. $8^x = 102$
11. $5^{x+2} = e^{2x}$
12. $e^t = 2e^{3t-1}$
13. $e^{x+2} = 8.2$
14. $e^k = 9.02$
15. $10^{x+2} = 2^x$
16. $10^{x-1} = 15$
17. $e^{3x} = 88$
18. $e^{-2x} = 0.00721$
19. $10^e = e^{10x}$
20. $e^{10} = 10^{ex}$

Solve each of the following logarithmic equations for the variable indicated. Check all answers.

21. $3 \log_8 x = -2$
22. $2 \log_5 x = 5$
23. $\ln x + \ln 4 = 4$
24. $22 - \log x = \log 5$
25. $\frac{1}{3} \log(x + 5) + \log 3 = 1$
26. $\frac{1}{2} \log(a - 3) + \log 2 = \frac{1}{2}$
27. $5 \log_{32} x = -3$
28. $\frac{1}{2} \log e = x - 6$
29. $\ln 6 - \ln x = -3$
30. $\log 12 = \log x - \frac{2}{3}$
31. $\log_3 5x = 4$
32. $\log_2 \frac{x}{4} = 16$
33. $\log_2 I + \log_2 7 = \log_2 21$
34. $\log_3 t^2 - \log_3 t = 2$
35. $2 \log_2 5 - \log_2 Z = \log_2 40$
36. $\log_3 2 - \log_3 x = \frac{1}{2} \log_3 2$
37. $\log_7 Q + \log_7 (2Q - 4) = \log_7 5$
38. $\log_5 x + \log_5 (x + 1) = \log_5 12$
39. $\log_2 V + \log_2 (V + 2) = 3$
40. $\log x^{-1/2} + \log x = 1$

41. The universal time-constant curve in Figure 13–5(b) showing the number of time constants required for C to discharge through R may be obtained from the circuit in (a). The instantaneous voltage across the load Γ_R may be computed from

$$\Gamma_R = Ve^{-t/RC}$$

where V is the voltage at start of discharge ($t = 0$). If $V = 1$ V, $t = 220$ µs, and $\Gamma_R = 0.368$ V, find the value of C in microfarads.

Figure 13–5

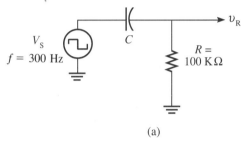

(a)

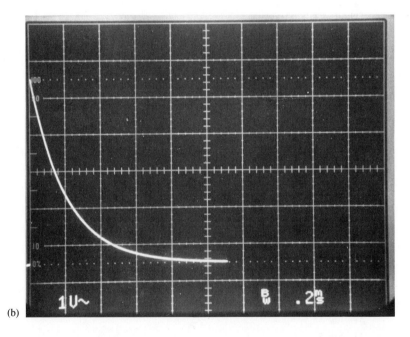

(b)

42. The law of exponential growth states that the amount of some quantity N increases in such a way that the rate of growth at each instant is proportional to the amount N present at that instant. This law is stated mathematically by

$$N = N_0 e^{kt} \quad (k > 0)$$

Suppose that the rate of world population growth is 1.89% per year and is expected to remain as such until the year 2000. Use the preceding formula to predict the population in the year 2000, given that the population in 1985 was 4.1 billion.

(**Hint:** $t = 2000 - 1985$ and $k = 0.0189$.)

43. A student without a calculator is stuck with the problem of determining which is larger: $\log_5 625$ or $\log_2 512$. Can you help this student solve the problem without a calculator?

Exponents and Logarithms

44. The characteristic impedance of coaxial cable is given by

$$Z_0 = \frac{138}{\sqrt{\epsilon_r}} \log \frac{D}{d} \text{ ohms}$$

where ϵ_r is the dielectric constant of the insulating material, D is the diameter of the outer conductor, and d is the diameter of the inner conductor. If $\epsilon_r = 2.25$, $D = .255$ in., and $Z_0 = 50\ \Omega$, find the diameter of the inner conductor.

13-8   DECIBELS*

Decibel power gain (G') is defined as a logarithmic unit expressing the ratio of one power to another and is given by

$$G \text{ (dB)} = G' = 10 \log \frac{P_o}{P_i}$$

where P_o is the output power and P_i is the input power. Historically, the common logarithm was chosen as the measurement scale, since the human ear perceives sound in approximately this same logarithmic manner. The coefficient 10 was chosen simply because the bel (the original unit) was too large and had to be divided up into 10 (deci-) parts. Hence, the unit decibel emerged as the unit for measuring power ratios.

Consider now that since power may be given by $P = V^2/R$, then the decibel power gain may be rewritten as

$$G' = 10 \log \frac{V_o^2/R_o}{V_i^2/R_i}$$

where R_i and R_o are the input and output resistances, respectively, of the device under test. This may also be written as

$$G' = 10 \log \frac{V_o^2}{V_i^2} \times \frac{R_i}{R_o}$$

Finally, from Rule 13–1 for the log of a product,

$$G' = 20 \log \frac{V_o}{V_i} + 10 \log \frac{R_i}{R_o}$$

Note that if $R_o = R_i$, then $10 \log 1 = 0$, and the preceding equation becomes

$$G' = A' = 20 \log \frac{V_o}{V_i}$$

This says that the decibel power gain (G') is numerically equal to the decibel voltage gain (A') *if and only if* the device input and output resistances are equal. This result is of great practical importance, since it implies that only a voltmeter is needed to compute decibel power gains if the resistances across which the voltage measurements were taken are identical.

---

*Adapted from "How to Speak dB-ese" in F. R. Monaco, *Introduction to Microwave Technology*, Columbus, Oh.: Merrill, 1989.

306 Advanced Topics in Electronics Mathematics

This last equation allows us to calibrate voltmeter scales for direct measurement in units of absolute power, either dBm or watts, if the readings are always taken across the *same* reference resistance as used to calibrate the meter scale. Therefore, if we want to make use of the voltage-measurement idea to determine absolute power by measurement, we must specify two things:

1. The value of P_i that is to be used as the comparison value.
2. The resistance value across which both V_o and V_i will be taken.

Let us see how this is done in practice.

The Simpson model 260 VOM in Figure 13–6 has a decibel scale that ranges from -20 dB to $+10$ dB and decibel measurements are taken with the selector switch on the 2.5 V AC scale. Moreover, a note in the lower-left corner of the instrument face specifies a 0 dB power level of 0.001 W across 600 Ω, thus fulfilling the essential requirements (1) and (2). But what does all this mean?

To begin with, note that from the formula for power, $P = V^2/R$, we may solve for V as

$$V = \sqrt{PR}$$

Therefore, from the specs given on the instrument face,

$$V = \sqrt{0.001 \times 600} = 0.775 \text{ V}$$

So, 0 dB will actually correspond to a little over ¾ V, which is the RMS voltage required to sustain a power of 1 mW across a 600 Ω load. Thus the Simpson 260 actually measures absolute power gains in dBm, where m stands for power above 1 mW, which is a common reference level.

On solving the logarithmic equation A (dB) $= 20 \log V_o/V_i$ for V_i, we see that a decibel power gain of $+10$ requires almost 2.5 V: $A' = 10$ dB $= 20 \log V_o/0.775$ V. Solving for V_o,

$$V_o = 0.775(\log^{-1} 0.5) \approx 2.45 \text{ V}$$

Figure 13–6

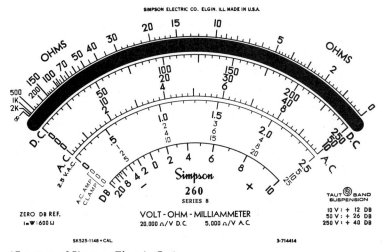

(Courtesy of Simpson Electric Co.)

Exponents and Logarithms

This is why the decibel scale may be read directly when the selector switch is on the 2.5 V AC range. Of course, we may make decibel measurements with the voltage-selector switch set on any other range, but we must add a correction factor, which is printed on the lower right of the meter scale.

These correction factors were derived as follows: Any voltage reading obtained with the selector on the 10 V range is $10/2.5 = 4$ times greater than the voltage reading on the 2.5 V range. Therefore, the decibel reading is greater by $20 \log 10/2.5 \approx 12$ dB. So, 12 dB must be added to any reading. In a similar way, any other correction factor may be found:

$$0.5 \text{ V AC range:} \quad 20 \log \frac{0.5}{2.5} \approx -14 \text{ dB}$$

$$50 \text{ V AC range:} \quad 20 \log \frac{50}{2.5} \approx +26 \text{ dB}$$

$$250 \text{ V AC range:} \quad 20 \log \frac{250}{2.5} = +40 \text{ dB}$$

Similar values may be found for any other AC voltage scale.

Decibel Voltage Gain (A')

By definition, the decibel is a power-ratio unit. However, it may be conveniently used as a voltage-ratio unit. The idea of using voltage measurements to indicate gain is a result of the ease and convenience of making simple voltage measurements instead of power measurements.

The formula used to compute decibel voltage gain is given by

$$A' = 20 \log \frac{V_o}{V_i}$$

which was derived from the earlier equation

$$G' = 10 \log \frac{V_o^2/R_o}{V_i^2/R_i}$$

used to define decibel power gain. The reason for using the coefficient of 20 is to ensure that $A' = G'$ whenever the circuit input and output impedances are matched, as mentioned earlier. When $R_i \neq R_o$, then the relationship $A' = G'$ no longer holds, and we must compute power and voltage gains separately using the appropriate formula for each:

$$G' = 10 \log \frac{P_o}{P_i} \quad \text{or} \quad A' = 20 \log \frac{V_o}{V_i}$$

Note that if the correction factor $10 \log R_i/R_o$ is applied to $20 \log V_o/V_i$, the result is decibel power gain (G'). For example, consider the situation shown in Figure 13–7. Here, we see that the input and output resistances are not equal, and so $A' \neq G'$. However, by applying the correction factor $10 \log R_i/R_o$ to A', we obtain G'—that is,

$$A' = G' = A' + 10 \log \frac{R_i}{R_o} = 15 \text{ dB} + (-4 \text{ dB}) = 11 \text{ dB}$$

Figure 13-7

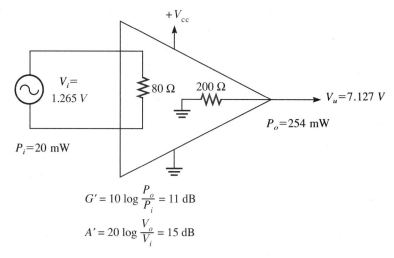

Which Decibel?

Many different power (voltage) levels and resistance values are used throughout the industry, and it is important for you to always know which one is being applied in your particular test situation. For example, we have just seen that 1 mW across 600 Ω is an important and common reference. However, in microwave work, 1 mW across 50 Ω is the standard. Whenever 1 mW is used as the reference, we use the *absolute* unit of power, the dBm. Similarly, dBmV indicates that 1 millivolt is the reference level, as in the case of some field-strength meter that also specifies 75 Ω as the reference impedance rather than 600 Ω. The unit of audio frequency signal level is the VU (volume unit), which has the same meaning as the dBm—i.e., 1 mW across 600 Ω. Other dB reference units are the dBV (1 V) and dBk (1 kilowatt (kW)). Anytime there is a unit following the dB abbreviation, we know we are dealing with *absolute,* not relative, gains.

Exercise Set 13-4

Problems

1. A technician wishes to determine the maximum power output at 1 kHz of an audio oscillator whose output impedance is 600 Ω. Using a Simpson 260, she measures +9 dB on the 10 V AC scale. What is the maximum power output of the oscillator in watts?
2. A technician measures 25 mV across the input of an audio amplifier whose input impedance is 2200 Ω. The audio output voltage is 0.62 V into an 8 Ω speaker.
   a. What is the dB power gain (G')?
   b. What erroneous decibel power gain (G') results using only voltage measurements?
   c. What needs to be done to get a true decibel power gain (G') using only voltage measurements?
3. An oscilloscope probe has a loss of 3 dB at 100 MHz. If the actual circuit voltage being measured is 12.5 V_P, what voltage will be displayed on the screen? (See Figure 13-8.)
4. A technician measures the stage gain of an amplifier as +9.5 dB (actually dBm) using a Simpson 260. The output impedance of the stage is 1850 Ω. What is the actual decibel power gain?

Exponents and Logarithms   309

5. The power output of a transmitter is 20 kW, but at the antenna, it is only 8 kW. What is the attenuation factor of an 18,000 ft transmission line connecting the transmitter with the antenna? (See Figure 13–9.)

Figure 13–8

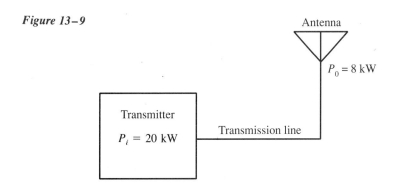

Figure 13–9

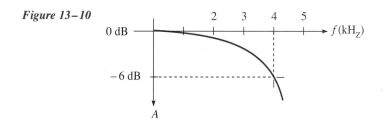

6. The response of a low-pass filter should be -6 dB at 4 kHz. If the output voltage is 0.25 V at 3 kHz and drops to 0.125 V at 4 kHz, is the filter working correctly? (See Figure 13–10.)

Figure 13–10

7. A certain stereo amplifier is guaranteed to have an output of 100 W per channel ± 1 dB. Is this necessarily a good purchase?
8. A multicavity klystron has a power gain of 87 dB. If this represents an output of 12 kW, how much input power is required?
9. The power gain of an amplifier is to be 16 dB when the output power is 3.6 W. If the input power is 5 mW, by how much must it be increased to obtain the desired gain?

Summary

Rules for Operating with Logarithms

Rule 13-1 Log of a Product
$$\log_b MN = \log_b M + \log_b N$$

Rule 13-2 Log of a Quotient
$$\log_b M/N = \log_b M - \log_b N$$

Rule 13-3 Log of a Power
$$\log_b M^q = q \times \log_b M$$

Corollary to Rule 13-3
$$\log_b b^p = p$$

Rule 13-4 Change of Base
$$\log_b x = \frac{\log_a x}{\log_a b}$$

Key Terms

exponential relationship
logarithmic relationship
common log
natural log
antilog
exponential equation
logarithmic equation
decibel

Upon completion of this chapter, you will have:

1. Learned how to specify electrical vectors as phasors.
2. Learned how to add plane vectors.

14 Introduction to Vectors

14-1 INTRODUCTION

If you want to express the fact that two automobiles *A* and *B* start at the same point, *A* goes north at 60 mi/h, and *B* goes east at 30 mi/h, what methods might you use? There are several possibilities. You might write,

A went north at 60 mi/h, and *B* went east at 30 mi/h.

Alternatively, you could write

A: 60, north
B: 30, east

Although both of these strategies provide all the necessary information, the first is wordy and the second is mathematically clumsy.

What is needed is a method of conveying all the information in a simple, unambiguous way. Consider the graphical representation in Figure 14-1.

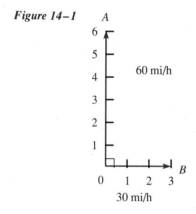

Figure 14-1

In this case, we have used the lengths of lines (e.g., 3 cm and 6 cm) to represent the magnitude (speed) of each automobile and have pointed each line in its proper compass direction. Note that Figure 14-1 provides information concerning magnitude (60 and 30) as well as direction (north and east). We say that any quantity having both magnitude as well as direction is a **vector** quantity. Conversely, a quantity having only magnitude is called a **scalar** quantity. All the quantities considered thus far in this book have been scalar quantities.

It is common practice to denote vectors by using the letters of the endpoints of the lines representing the quantities. In Figure 14-1, we might refer to the vectors as $\overrightarrow{OA}$ and $\overrightarrow{OB}$. The arrow (→) above each notation further emphasizes the fact that these are vector quantities.

In using lines to represent the magnitude of a vector, it is common to draw each vector to scale, using some convenient length. In Figure 14-1, we used 3 cm and 6 cm to represent 30 and 60 mi/h, respectively, but any other convenient units could have been used as well. A drawing of this type is called a **vector diagram**.

Introduction to Vectors

Example 14-1

Suppose that a line 2.2 in. in length represents a vector of 56 mi/h west. How long must a line representing 208 mi/h south be drawn?

Solution We can solve this problem by using a simple proportion.

$$\frac{2.2 \text{ in.}}{56 \text{ mi/h}} = \frac{x \text{ inches}}{208 \text{ mi/h}}$$

$$x = 208\left(\frac{2.2}{56}\right) \approx 8.17 \text{ in.}$$

14-2 VECTOR SUMMATION

Suppose two strings are attached to an object; also suppose you pull on one string with a force of 30 lb to the right while someone else pulls straight up on the other with a force of 60 lb. In what direction would the object move, and what would be the combined force acting along this direction? The situation has been drawn in Figure 14–2. Note that Figure 14–2 is identical to Figure 14–1 with the exception that the vectors $\overrightarrow{OA}$ and $\overrightarrow{OB}$ have been labeled in pounds rather than miles per hour.

With this picture before us, we can now answer the question originally posed. Intuitively, we see that the object will probably move a little to the right and upward. That's a good start in answering the direction part of the question, but what about the combined force on O? Can we simply add 60 lb and 30 lb to obtain 90 lb? If both forces were in the same direction, we could find the total force by simple addition (90 lb). Conversely, if the forces were exerted in opposite directions (180° apart), the total force would be their difference (30 lb). However, since these forces are at right angles to one another, simple arithmetic addition is not sufficient. Figure 14–3 shows the true state of affairs.

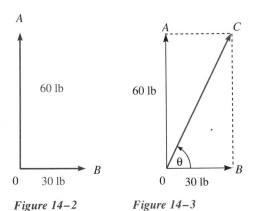

Figure 14–2   *Figure 14–3*

From this figure, we see that the **resultant,** or net force, is represented by the vector $\overrightarrow{OC}$ acting at some angle θ. Note, too, that the resultant is not the simple sum of the other two component vectors. Our problem, then, is one of determining the exact magnitude of the resultant and the direction (i.e., angle θ) taken by $\vec{OC}$.

It is apparent from Figure 14–3 that triangle OBC is a right triangle. As such, we can apply the Pythagorean theorem developed in Chapter 8 to determine the total length (i.e., resultant force) of $\vec{OC}$. Subsequently, we make use of the inverse trigonometric relationships given in Chapter 12 to determine θ. Here is the procedure we may follow:

$$\vec{OC} = \sqrt{\vec{OB}^2 + \vec{BC}^2} \qquad \text{where } \vec{BC} = \vec{OA}$$
$$\approx 67.08 \text{ lb}$$

$$\theta = \tan^{-1} \frac{\vec{BC}}{\vec{OB}}$$
$$= \tan^{-1} 2$$
$$\approx 63.435$$

Finally, we may write $\vec{OC} = 67.08 \angle 63.435°$ lb.

The foregoing procedure was simplified by the fact that the component forces were acting at right angles to each other. What can we do, however, in the case of those forces that do not act at 90° to one another? Consider, for example, the two forces shown in Figure 14–4. Here, it is clear that the resultant will act along some angle partway between the two component vectors and will be directed toward the right. The question once again before us is how to determine the magnitude and direction of the resultant vector.

In Figure 14–5, we have redrawn Figure 14–4 on a system of rectangular coordinates, with each component vector acting at its own angle with the x-axis. These vectors are specified as $\vec{OA} = 45 \angle 31°$ and $\vec{OB} = 105 \angle -24°$ lb. We can now determine the resultant in a manner similar to that used in connection with Figure 14–3. The general procedure is shown in Figure 14–6. Note that the resultant vector $\vec{OC}$ is the diagonal of parallelogram $OBCA$. This is similar to what we did in Figure 14–3. Here, of course, there are no 90° angles to simplify matters. However, as shown in Figure 14–7, we can use some simple plane geometry to create the right triangle ORC. The resultant $\vec{OC}$ is now seen to be given by $\sqrt{(OR)^2 + (RC)^2}$.

Our only problem is that we do not have lengths OR and RC directly. Note, however, that $OR = PB + BQ$ and $RC = OP - CQ$. Furthermore, as shown in Figure 14–8, the required lengths PB, BQ, OP, and CQ are merely the sides of the right triangles created by the component vectors $\vec{OA}$ and $\vec{OB}$. In other words, $\vec{OA}$ can be **resolved** into its **rectangular parts** BQ and CQ, and $\vec{OB}$ is similarly resolved into PB and OP.

We can now apply the trigonometric methods of Chapter 12 in order to resolve $\vec{OA}$ and $\vec{OB}$ into their rectangular component parts. For example, in right triangle BQC, we observe that $BQ = 45 \cos 31°$ and $CQ = 45 \sin 31°$. See Figure 12–13 to refresh your memory of these relationships. Similarly, $PB = 105 \cos 24°$ and $OP = 105 \sin 24°$. In particular, we have

$$BQ = 45 \cos 31° \approx 38.573$$
$$CQ = 45 \sin 31° \approx 23.177$$
$$PB = 105 \cos(-24°) \approx 95.922$$
$$OP = 105 \sin(-24°) \approx -42.707$$

Using this information, we may now write

$$OR = PB + BQ = 95.922 + 38.573 = 134.495$$
$$RC = OP + CQ = -42.707 + 23.177 = -19.53$$

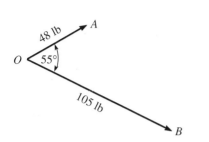

Figure 14–4

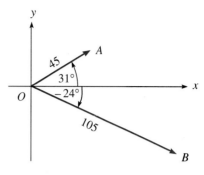

Figure 14–5

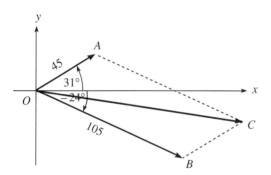

Figure 14–6

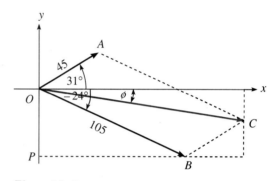

Figure 14–7

Figure 14–8

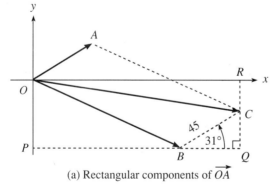

(a) Rectangular components of $\overrightarrow{OA}$

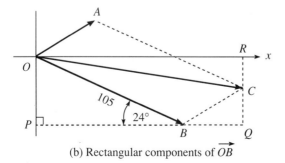

(b) Rectangular components of $\overrightarrow{OB}$

315

Finally, since $\vec{OC} = \sqrt{(OR)^2 + (RC)^2}$, we have
$$\vec{OC} = \sqrt{(134.495)^2 + (-19.54)^2}$$
$$\approx 135.907 \text{ lb}$$

Note that the negative angle ϕ shown in Figure 14–7 that $\vec{OC}$ makes with the positive x-axis is given by
$$-\phi = \tan^{-1}\frac{RC}{OR}$$
$$\approx \tan^{-1}(-0.1452)$$

Therefore, $\phi \approx -8.262°$.

We can now write the resultant vector as
$$\vec{OC} = 135.907 \angle -8.262° \text{ lb}$$

This procedure is summarized by Rule 14–1.

Rule 14–1

1. Draw the component vectors to scale on a rectangular coordinate system and include the resultant. Label each vector carefully, paying attention to the sign of angle θ.
2. Resolve each of the given vectors into its rectangular parts using the trigonometric relationships
$$x = A \cos \theta$$
$$y = A \sin \theta$$
   where
   A = magnitude of the given vector
   x = horizontal component
   y = vertical component
3. Use the component parts in step 2 to form the right triangle whose hypotenuse is the resultant.
4. Find the magnitude of the resultant using the Pythagorean theorem:
$$\text{Resultant} = \sqrt{H^2 + V^2}$$
   where H and V represent the lengths of the horizontal and vertical sides found in step 3.
5. Find the phase angle θ using the inverse tangent relationship
$$\theta = \tan^{-1}\frac{V}{H}$$
6. Check your solution against the drawing made in step 1.

Phasors

Figure 14–9 shows two 182 Hz sinusoidal voltages differing in phase by an angle of 78.5°. Since each of these voltages has a magnitude, they qualify as scalar quantities. But

Introduction to Vectors

do they represent vector quantities as well? If we stretch the definition of a vector somewhat, we can regard the phase angle as the difference in direction between each of these voltages. In this sense, then, voltages that differ in phase may also be regarded as vector quantities. However, in the field of electronics, it is preferred by many to refer to these "electrical vectors" as **phasors.** Therefore, from this point on, we will make reference to phasors when we discuss such electrical quantities that possess magnitude as well as phase characteristics. In Chapters 15 and 16, we will make more use of the notion of phasors as they apply to actual AC circuits. Moreover, we will develop a method of AC-circuit solution involving phasor algebra.

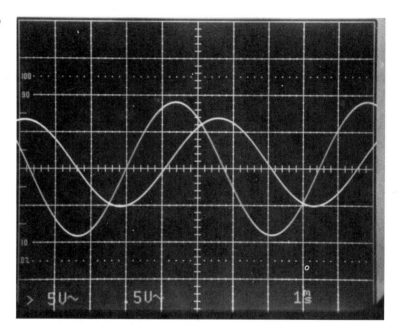

Figure 14–9

The following examples will make the ideas in this section clearer.

Example 14–2 Find the vector sum of $OA = 40 \angle 40°$ and $OB = 25 \angle 155°$. (See Figure 14–10.) Check your answer graphically.

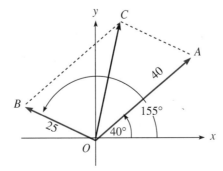

Figure 14–10

Solution We see that the horizontal components of the given vectors add to

$$40 \cos 40° + 25 \cos 155° \approx 7.984$$

The vertical components are

$$40 \sin 40° + 25 \sin 155° \approx 36.277$$

This information may now be combined in a drawing of the resultant vector $\overrightarrow{OC}$ (see Figure 14–11). We may now apply the Pythagorean theorem to find the magnitude of the resultant.

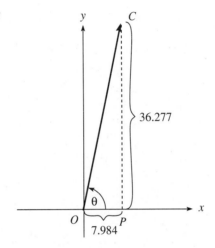

Figure 14–11

$$OC = \sqrt{(OP)^2 + (CP)^2} \approx 37.145$$

Finally, the angle θ is given by

$$\theta = \tan^{-1} \frac{CP}{OP} \approx 77.588°$$

The resultant is expressed as

$$\overrightarrow{OC} = 37.145 \angle 77.588°$$

Example 14–3 A motorboat is propelled at 28 mi/h due east across a river flowing south at a rate of 6.2 mi/h. How fast and in what direction will the boat travel?

Solution The situation is shown in Figure 14–12. Using the Pythagorean theorem, we see that the magnitude of the resultant is

$$OC = \sqrt{(OA)^2 + (OB)^2} \approx 28.678$$

The angle is given by

$$\theta = -\tan^{-1} OB/OA \approx -12.485°$$

Introduction to Vectors

Figure 14–12

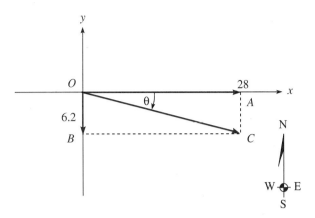

Example 14-4   Draw a vector diagram illustrating the amplitude and phase relationship of the two sine waves shown in Figure 14–13.

Figure 14–13

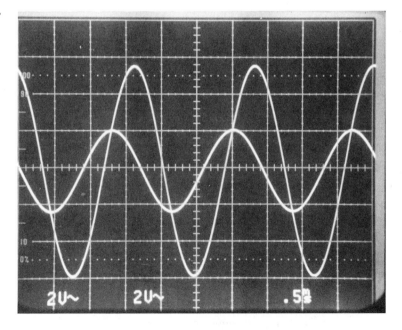

Solution   From the information shown on the display, the amplitude of the larger waveform is 5.6 V (2.8 divisions times 2 V/cm). That of the smaller waveform is 2 V (1.0 vertical division times 2 V/cm). Furthermore, we see that the smaller voltage leads the larger by 63.5°. The phase angle is determined by

320 Advanced Topics in Electronics Mathematics

solving the relationship

$$\frac{3.4 \text{ divisions}}{360°} = \frac{0.6 \text{ division}}{x \text{ degrees}}$$

from which $x \approx 63.5°$. The vector diagram is shown in Figure 14–14.

Figure 14–14

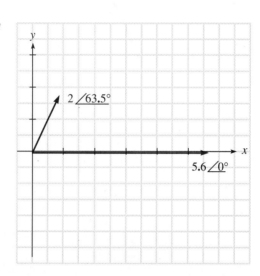

Exercise Set 14–1

Questions

1. What is the difference between a vector and a scalar quantity? Give an example of each.
2. What notation is commonly used to denote a vector?
3. What is meant by the resultant?
4. What is a phasor?
5. What is a vector diagram?

Problems

Find the sum of each of the following pairs of vectors. Draw each pair and the resultant to scale on a Cartesian coordinate system.

1. 48 ∠90°; 26 ∠63°
2. 163.2 ∠75°; 205.4 ∠−18°
3. 12 ∠30°; 18 ∠275°
4. 88 ∠22.6°; 107 ∠300°
5. 5 ∠330°; 2 ∠260°
6. 3.07 ∠10.91°; 2.19 ∠−11.2°
7. 3×10^{-8} ∠276°; 5×10^{-7} ∠8°
8. 3.07×10^{-6} ∠770°; 2×10^{-5} ∠−84°
9. 81 ∠23.4° ft; 1014 ∠−16.2° in.
10. 6.2 ∠−275° mi; 11 ∠38.2° km
11. The following BASIC program computes the vector sum of any two vectors whose resultant lies in quadrant I or IV.

Introduction to Vectors

```
10  PRINT"filename:    VECTOR-1"
20  PRINT
30  PRINT"Input the magnitude and angle of the first vector."
40  INPUT A,B
50  PRINT"Input the magnitude and angle of the second vector."
60  INPUT C,D
70  LET E=A*SIN(B*.017453292 )
80  LET F=A*COS(B*.017453292 )
90  LET G=C*SIN(D*.017453292 )
100 LET H=C*COS(D*.017453292 )
110 LET I=E+G
120 LET J=F+H
130 LET K=SQR(I^2+J^2)
140 LET M=ATN(I/J)*180/3.141592654
150 PRINT K;"is the magnitude, and";M;"is the angle of the resultant"
```

    a. Use this program to verify your solutions to Problems 1 through 10.
    b. Verify the results obtained in Figure 14–8.
12. Modify the BASIC program in Problem 11 so that it will print out the partial results of lines 70 through 120.
13. You are to fly to a city 1200 mi due north of your present location. Your true airspeed (speed of plane relative to the air) is 256 mi/h. If there is a continuous crosswind of 38 mi/h from the east, what must your actual heading be, and how long will the trip take?
14. a. Find the phase angle and magnitude of each voltage shown in Figure 14–15. Express the two voltages using phasor (vector) notation.
    b. Draw a phasor (vector) diagram showing the relationship between the two voltages.

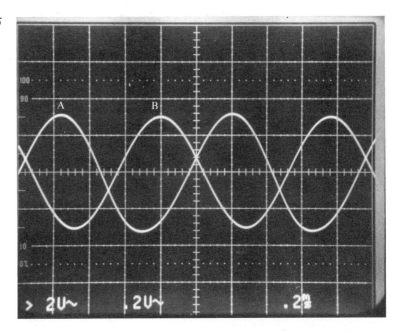

Figure 14–15

Key Terms

vector	resultant	rectangular parts
scalar	resolved	phasor
vector diagram		

Important Rules

Rule for Finding the Resultant
Rule 14–1

1. Draw the component vectors to scale on a rectangular coordinate system and include the resultant. Label each vector carefully, paying attention to the sign of angle θ.
2. Resolve each of the given vectors into its rectangular parts using the trigonometric relationships

$$x = A \cos \theta$$
$$y = A \sin \theta$$

where

A = magnitude of the given vector
x = horizontal component
y = vertical component

3. Use the component parts in step 2 to form the right triangle whose hypotenuse is the resultant.
4. Find the magnitude of the resultant using the Pythagorean theorem:

$$\text{Resultant} = \sqrt{H^2 + V^2}$$

where H and V represent the lengths of the horizontal and vertical sides found in step 3.

5. Find the phase angle θ using the inverse tangent relationship

$$\theta = \tan^{-1} \frac{V}{H}$$

6. Check your solution against the drawing made in step 1.

15
Series AC Circuits

16
Complex Algebra and Parallel AC Circuits

PART SEVEN
Advanced Applications of Mathematics to Electronics

Upon completion of this chapter, you will have learned how the fundamental concepts of algebra, logarithms, exponentials, and trigonometry are applied to the solution of series AC circuit problems.

15 Series AC Circuits

15-1 INTRODUCTION

In the previous chapter, it was observed that vectors (phasors) cannot be added *algebraically* unless they are either in the same or exactly opposite directions. For example, 10 $\angle 0°$ and 3 $\angle 0°$ may be summed algebraically to 13 $\angle 0°$. Similarly, 10 $\angle 180°$ and 3 $\angle 0°$ may be summed to 7 $\angle 180°$. On the other hand, if the phase angle is different than zero, we must resort to *vector summation* in order to determine the resultant or vector sum.

In electronics, there are two commonplace components that have the effect in an AC circuit of creating an electrical phase angle between the voltage and current operating in the device. These components, the inductor (L) and the capacitor (C), make it necessary to resort to phasor algebra in determining the various circuit currents and voltages. We cannot rely exclusively on ordinary algebra in the analysis of these types of circuits. Therefore, it is the purpose of this chapter to introduce the idea of AC phase relations in simple series circuits and to lay the groundwork for the discussion of phasor algebra that will be taken up in the next chapter.

15-2 THE RESISTIVE AC CIRCUIT

Alternating-current circuits obey the same laws and rules as direct-current circuits *provided* the voltage and current are always in phase. This situation ordinarily occurs only in circuits composed entirely of pure resistances. Figure 15–1 shows a simple AC resistive circuit. Note that there is no phase difference between V and I—that is, $\theta = 0°$. Part (c) of the figure shows the phasor diagram for the circuit.

Figure 15–1

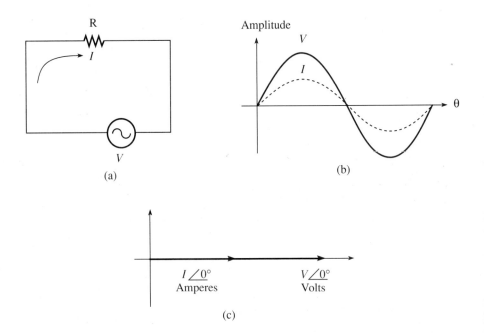

Series AC Circuits

In purely resistive AC circuits, the voltage and current will always be in phase. Therefore, Ohm's law and the DC circuit properties developed in Chapter 7 apply to AC resistive circuits as well.

To find the current (I) through the resistor in Figure 15–1, we can use Ohm's law and write $I = V/R$. Note that the arrow indicating current flow in Figure 15–1a has a direction that is entirely arbitrary. The arrow is merely a convenient way of showing the path taken by the current. As you know, the current under the influence of an alternating voltage reverses its direction each time the voltage reverses.

Suppose, for example, that the resistor has a value of 1000 Ω and is connected directly across the 120 V AC outlet in your home. The current through the resistor is simply $I = V/R = 120/1000 = 120$ mA. It may have occurred to you, however, to question this result, since the value of the alternating sine-wave voltage shown in Figure 15–1(b) is always varying. The voltage changes from zero to maximum and then back to zero again and reverses its polarity each alternation (half-cycle). To what specific value between these two extremes, then, does the 120 V refer? In order to answer this question, we must pause here for a moment to discuss the concept of effective value.

Effective AC Value

Suppose we were to place a 1000 Ω resistor across a 120 V DC source. The power dissipated as heat in the resistor is given by $P = V^2/R = 120^2/1000 = 14.4$ W. Now suppose we place the same 1000 Ω resistor across an AC voltage source whose peak value is 120 V. Will the same amount of power be dissipated in this situation? Since the AC voltage is varying continuously from zero to maximum and then back to zero twice each cycle, it stands to reason that an AC voltage with 120 V peak is not going to be as effective in heating the resistor as the same amount of DC voltage.

As stated before, the power dissipated by a resistor across a DC voltage source is given by $P = V^2/R$. Moreover, the instantaneous power of the same resistor across an AC voltage source is given by $p = v^2 R$. We know from Chapter 12 that the instantaneous value of a sine wave is given by $v = V_p \sin \theta$, where V_p is the peak value. If, for convenience of computation, we let $R = 1$ Ω and $V_p = 1$ V, then the instantaneous power may be written as $p = v^2 = \sin^2 \theta$. At this point, if we were to add up a great many of these instantaneous powers and divide by the number used to form the sum, we would have the average power that a 1 V peak AC voltage source could produce in a 1 Ω resistor.

Using a BASIC computer program, we have calculated the instantaneous power for every $(1/10)°$ for a total of 3601 data points, as shown in Table 15–1. As you can easily see, the average power is fairly close to 0.5 W. In fact, the more data points we use, the closer to the theoretical value of 0.5 we get. We conclude, therefore, that a 1 V peak AC sine-wave source will produce ½ W of average power in a 1 Ω resistor. It is apparent, though, that the formula $P = V^2/R$ does not give a correct value if we merely plug in the values of 1 V peak AC and 1 Ω of resistance. These figures give us 1 W rather than ½ W. We are forced to conclude that there must be some other value of voltage—call it the effective voltage, V_e—that is less than peak and that produces ½ W of power. In other words,

$$P = \frac{V_e^2}{R}$$

or, in our particular case where $R = 1$, $0.5 = V_e^2$, from which $V_e \approx 0.707$. Note that this same value of DC voltage—0.707 V—will produce the same ½ W of power in our 1 Ω

Table 15–1

```
RUN
Enter the number of degrees BEGINNING with? 0
Enter the number of degrees ENDING with? 360
Enter the degree INCREMENT? 0.1

DEGREES              SIN^2
 0                    0
  .1                  3.046171E-06
  .2                  1.218465E-05
  .3                  2.741532E-05
  .4                  4.8738E-05
  .5                  7.615244E-05
  .6                  1.096583E-04
  .7000001            1.492551E-04
  .8000001            1.949425E-04
      .                   .
      .                   .
      .                   .
358.3127              8.670379E-04
358.4127              7.673452E-04
358.5127              6.737348E-04
358.6127              5.861846E-04
358.7127              5.047438E-04
358.8127              4.293685E-04
358.9127              3.601011E-04
359.0127              2.969044E-04
359.1127              2.398138E-04
359.2127              1.887989E-04
359.3127              1.43888E-04
359.4127              1.050669E-04
359.5127              7.232806E-05
359.6127              4.568972E-05
359.7128              2.513766E-05
359.8128              1.068334E-05
359.9128              2.319049E-06

The average value of SIN^2 is .4999862
```

resistor. Another way of saying this is that 1 Volt peak AC (V_p) has the same effect as 0.707 Volt DC. Mathematically, we write this conclusion as

$$\frac{V^2}{R} = \frac{1}{2} \cdot \frac{V_p^2}{R}$$

or

$$V^2 = 0.5 V_p^2$$

Therefore,

$$V = V_e = 0.707 V_p$$

This last equation tells us that the *effective* AC voltage is actually 0.707 times the peak voltage.

Series AC Circuits

You may remember that there were three steps involved in obtaining this value of 0.707. We took the square root of the mean (average) squared values of many instantaneous powers. For this reason, the **effective value** of AC voltage (V_e) is frequently referred to as the **root mean square,** or **RMS,** value. In practice, the subscript is dropped, and we write the following.

Rule 15–1

$$V = 0.707 V_p$$

In Rule 15–1, the symbol V implies the RMS value.

As a final comment, we state that the effective AC current (I) can be obtained in the same manner and is computed using the same general formula.

Rule 15–2

$$I = 0.707 I_p$$

Whenever an AC voltage or current is given without qualification, it is the RMS value that is implied. Furthermore, unless otherwise stated specifically, AC voltage and current meters read RMS values.

Example 15–1   What is the peak voltage available at a 120 V AC electrical outlet?

Solution   From $V = 0.707 V_p$, we solve for V_p, which gives us $V_p = V/0.707 = 120/0.707 \approx 169.73$ V peak.

Example 15–2   What DC voltage is required to produce the same amount of power in a 5 kΩ resistor as a 6.3 V AC source?

Solution   Remember that the RMS AC voltage is the same as the DC voltage required to produce the same power in a given resistor. Therefore, 6.3 RMS V AC = 6.3 V DC.

Example 15–3   A power of 10.14 W is dissipated in a 600 Ω resistor across an AC (sinusoidal) voltage. What is the peak current in the circuit?

Solution   From $P = I^2 R$, we solve for the RMS current value, obtaining $I = \sqrt{P/R} = \sqrt{10.14/600} = 0.13$ A. Since this is the RMS value, we can solve $I = 0.707 I_p$ for I_p, obtaining $I_p = I/0.707 = 0.13/0.707 \approx 184$ mA.

Example 15–4

In the graph of $y = \sin\theta$, at what values of θ does the effective value of 0.707 occur during one period?

Solution From Figure 15–2, we see that $\theta = \sin^{-1} 0.707$. Since $0.707 = 1 \cdot 0.7071 = 1 - 0.7071$, $\theta = \sin^{-1} 0.707$ or $\theta = \sin^{-1}(-0.707)$. The values 0.707 and -0.707 occur at $45°$ and every $90°$ thereafter.

Example 15–5

What power is dissipated in a 1000 Ω resistor connected across a 120 V peak AC voltage source?

Solution

$$V = 0.707 V_p = 0.707 \times 120 = 84.84 \text{ V}$$

Therefore, $P = V^2/R = (84.84)^2/1000 \approx 7.2$ W.

Note that this is exactly one-half the power produced by a 120 V DC source across the same resistor.

Exercise Set 15–1

Problems

1. How much current is read on an AC ammeter if a 55 V peak AC source is connected across a 4700 Ω resistor?
2. A technician uses an oscilloscope to measure the AC voltage drop across a 2200 Ω resistive load. The measured value is 17.82 V peak-to-peak. What is the RMS value?
3. An AC voltage source has a peak value of 10 V. A DC voltage source also has this same value. If the DC voltage produces a temperature rise of 152°C in a certain resistor, what will the temperature rise be when the 10 Volt AC source is used across the same resistor?
4. A technician connects a cathode-ray oscilloscope (CRO) across an 1800 Ω resistor, as shown in Figure 15–3. A peak voltage of 40 is read from the scope. What is the RMS circuit current?
5. A 22 Ω resistor dissipates 50 W of power in an AC circuit. What is the value of the peak current through the resistor?
6. Write a BASIC computer program that determines the RMS current through any resistance connected across an AC voltage source whose peak value is known.

Inductors in Series and Parallel

The **self-inductances** of series-connected inductors add arithmetically (just like resistors in series) provided there is no magnetic coupling between them. We can write this result as follows.

Rule 15–3

$$L_t = L_1 + L_2 + \cdots + L_n$$

where L_t is the total inductance in henries (H).

Figure 15–4 shows the effect of cutting an inductor L_1 into two smaller inductors L_2 and L_3, separating them by some distance, and then reconnecting them in series. Surprisingly, the total self-inductance of L_2 and L_3 does not add up to L_1. In fact, it is actually less. The reason for this is apparent if we consider that the magnetic field about each turn of L_1 influences every other turn. If the inductor is physically separated into two

Figure 15–2

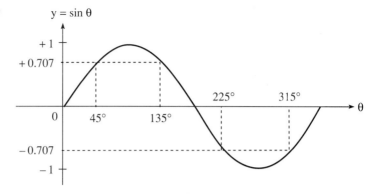

Figure 15–3

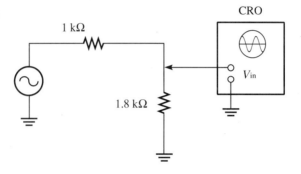

Figure 15–4

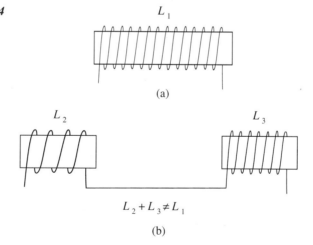

parts, it is clear that each turn on one coil no longer affects any of the turns on the other. In other words, there has ceased to be complete magnetic coupling between every turn of the original intact helix. In truth, then, we have created two entirely new inductors whose sum is $L_2 + L_3$, a different sum from the one when L_1 was a whole inductor unto itself.

In those cases where the inductors are close enough so that their magnetic fields interact, we say that there is **mutual inductance** (m) between them. In this situation, the total series inductance (L_t) is calculated as follows.

Rule 15–4

$$L_t = L_1 + L_2 \pm 2m$$

where the mutual inductance is given by $m = k\sqrt{L_1 \cdot L_2}$ and k is the **coefficient of coupling.**

The quantities m and k require some additional explanation. The coefficient of coupling (k) is simply the percent of the total magnetic flux (expressed as a decimal) linking the two inductors. For example, if 55% of the total flux links both inductors, the coefficient of coupling is $k = 0.55$. The value of m may be positive or negative depending on whether the fields aid ($+$) or oppose ($-$) one another. This in turn depends on how the two inductors are connected, the direction in which the turns are wound, and the physical orientation of the two inductors relative to one another.

For example, inductors L_2 and L_3 in Figure 15–4b are wound in the same direction and are said to be connected *series-aiding*. That is, the magnetic field of one aids the magnetic field of the other. If we assume for the moment that $k = 37\%$, $L_1 = 50$ mH, and $L_2 = 62$ mH, then the total inductance is

$$L_t = L_1 + L_2 + 2k\sqrt{L_1 \cdot L_2}$$
$$= 0.05 + 0.062 + 2 \cdot 0.37\sqrt{0.05 \cdot 0.062}$$
$$\approx 153 \text{ mH}$$

Note that the total series inductance is greater than the simple sum of 112 mH obtained when the inductors have no mutual inductance. Observe, too, that if the value of m were negative, the total inductance would be less than 112 mH.

It is possible to calculate the value of m as follows. When the inductors are connected series-aiding, their total inductance (L_a) is given by $L_a = L_1 + L_2 + 2m$. When connected series-opposing, their inductance (L_o) is $L_o = L_1 + L_2 - 2m$. The total inductance, then, is

$$L_a - L_o = (L_1 + L_2 + 2m) - (L_1 + L_2 - 2m) = 4m$$

Therefore,

$$m = \frac{L_a - L_o}{4}$$

Since the value of m is given by $k\sqrt{L_1 \cdot L_2}$, it is possible to calculate the coefficient of coupling as

$$k = \frac{m}{L_1 \cdot L_2}$$

The rate of growth of the total current into inductors connected in parallel is the sum of their individual current growths. Therefore, we can use the same reciprocal rule for

Series AC Circuits

their total inductance as was used for resistors in parallel. Again, assuming no coupling between the various inductors, we may write the following.

Rule 15–5

$$L_t = \frac{1}{1/L_1 + 1/L_2 + \cdots + 1/L_n}$$

For example, Figure 15–5 shows two inductors connected in parallel with $k = 0$.

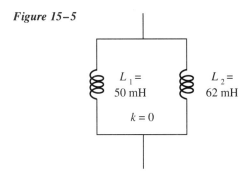

Figure 15–5

Using their individual self-inductances, we write

$$L_t = \frac{1}{1/0.05 + 1/0.062} \approx \frac{1}{20 + 16.13}$$
$$\approx 28 \text{ mH}$$

15–3 TRANSFORMERS

Figure 15–6 shows two ideal inductors with unity coupling. The dotted line represents the magnetic flux (ɸ) linking the two inductors. A sinusoidal voltage source (V_s) is connected across L_1, which is called the *primary* winding. As a result of the magnetic field established by the current flow in L_1, a voltage is also induced in L_2, which is called the *secondary* winding. A load resistor (*RL*) connected across L_2 will have a current through it as a result of the secondary voltage. The device just described is called a *transformer*.

Figure 15–6

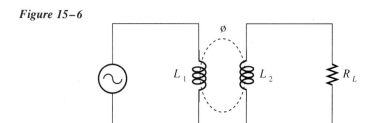

In the ideal transformer described, all the magnetic flux generated by a single turn of the primary cuts across each turn of the secondary. Therefore, the same voltage self-induced in any primary turn is also induced in any secondary turn. If the primary consists of just one turn and the secondary has two turns, the voltage induced in the secondary winding is twice that of the primary winding. In other words, the turns ratio is the same as the voltage ratio. This fact permits us to write the following proportion:

Rule 15-6

$$\frac{N_p}{N_s} = \frac{V_p}{V_s}$$

where

N_p = number of turns in primary
N_s = number of turns in secondary
V_p = voltage of the primary
V_s = voltage of the secondary

For example, a 200-turn secondary has 550 V across it when the voltage across the primary is 95 V. How many turns are there on the primary? On solving the above formula for N_p, we obtain

$$N_p = N_s \left(\frac{V_p}{V_s}\right)$$
$$= 200 \left(\frac{95}{550}\right)$$
$$\approx 34.5 \text{ turns}$$

This is an example of a *step-up transformer*, wherein the secondary voltage is higher than that applied to the primary. When the secondary voltage is less than the primary, the transformer is referred to as a *step-down transformer*.

In our ideal transformer, where $k = 1.0$ and there are no losses due to heating, the efficiency is 100%. Recall that efficiency (η) is defined as the ratio of the output of a device to its input expressed as a percent. It is not uncommon for actual transformers to have efficiencies exceeding 99%. Therefore, we know that the input power (P_i) and the output power (P_o) are equal for all practical purposes. Another way of writing this idea is with the equations $P_i = V_p \cdot I_p$ and $P_o = V_s \cdot I_s$. If we now equate these two expressions and rearrange the terms, we obtain another useful transformer relationship:

Rule 15-7

$$\frac{V_p}{V_s} = \frac{I_s}{I_p}$$

where

I_s = current in the secondary
I_p = current in the primary

For example, the transformer shown in Figure 15–7 has a step-down turns ratio of 3:1. The secondary voltage is 16 V. To find the current flowing in the primary winding, we use Ohm's law. The current in the secondary is given by Ohm's law as

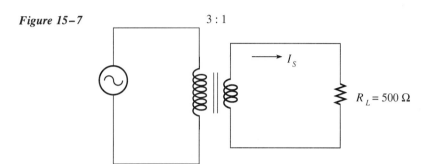

Figure 15–7

$$I_s = \frac{V_s}{RL} = \frac{16}{500} = 32 \text{ mA}$$

Since this is a step-down transformer with a 3:1 turns ratio, the primary voltage is

$$V_p = 3V_s = 3 \times 16 = 48 \text{ V}$$

Finally, we can solve the relationship $V_p/V_s = I_s/I_p$ for the primary current, obtaining

$$I_p = I_s\left(\frac{V_s}{V_p}\right) = 0.032\left(\frac{16}{48}\right)$$
$$\approx 10.7 \text{ mA}$$

We may derive still another useful transformer relationship that will allow us to match impedances. We divide both numerator and denominator of the left member of $V_p/V_s = I_s/I_p$ by the product $I_p \times I_s$ to get

$$\frac{\frac{V_p}{I_p I_s}}{\frac{V_s}{I_p I_s}} = \frac{I_s}{I_p}$$

Note, however, that the quotient V_p/I_p is the primary impedance Z_p and V_s/I_s is the secondary impedance Z_s. We can now write

$$\frac{\frac{Z_p}{I_s}}{\frac{Z_s}{I_p}} = \frac{I_s}{I_p}$$

which may be rearranged as

$$\frac{Z_p}{I_s} \cdot \frac{I_p}{Z_s} = \frac{I_s}{I_p}$$

or

$$\frac{Z_p}{Z_s} = \left(\frac{I_s}{I_p}\right)^2$$

By combining the two relationships derived earlier, we see that

$$\frac{N_p}{N_s} = \frac{V_p}{V_s} = \frac{I_s}{I_p}$$

which shows that

$$\left(\frac{N_p}{N_s}\right)^2 = \left(\frac{I_s}{I_p}\right)^2$$

Substituting this result into the value of Z_p/Z_s just given, we get the following relationship:

Rule 15–8

$$\frac{Z_p}{Z_s} = \left(\frac{N_p}{N_s}\right)^2$$

where

Z_p = impedance connected to primary
Z_s = impedance connected to secondary

Rule 15–8 states that the ratio of the impedances is the same as the square of the turns ratio. For example, consider the audio amplifier shown in Figure 15–8.

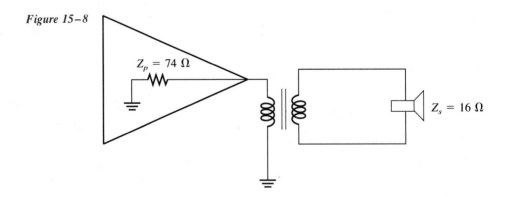

Figure 15–8

From $Z_p/Z_s = (N_p/N_s)^2$, we can solve for the turns ratio by taking the square root of each side, obtaining the turns ratio N:

$$N = \sqrt{\frac{74}{16}} \approx 2.15$$

Series AC Circuits

Therefore, for example, a transformer with 215 primary turns and 100 secondary turns will match the load impedance to the source, thereby transferring maximum power to the speaker.

15-4 INDUCTANCE IN AC CIRCUITS

An inductor has the property of causing the current through it to lag behind the applied AC sine-wave voltage by 90°. This effect is shown in Figure 15–9a. Since time is measured along the *x*-axis beginning at zero, it is clear that the voltage waveform (V) has already completed 90° when the current waveform (I) is just starting at 0°. Therefore, the current lags behind the voltage by 90°.

Figure 15–9

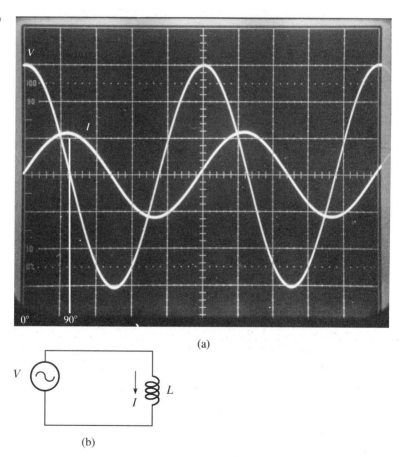

The cause of this current lag is the counterelectromotive force (Cemf) induced in the inductor windings themselves each time the magnetic field surrounding the coil changes in response to the varying sinusoidal voltage. The Cemf thus produced tends to send the current back in the opposite direction. This has the net effect of delaying the start of the current by 90°.

Figure 15–10 makes use of vector representation to show the 90° lag in the current. If we conceive of the two phasors V and I rotating counterclockwise about the origin O, with V always separated from I by a right angle, it is apparent that V will always be ahead of I by 90°.

It may have occurred to you from all that has just been said that any property strong enough to delay the current must also be strong enough to oppose its flow. This is true, and the opposition provided to the flow of sinusoidal current is called **inductive reactance,** symbolized as X_L. The magnitude of the inductive reactance is measured in ohms and depends on how fast the magnetic field is changing and also on the amount of inductance (L) measured in units of henries (H). The inductance itself depends on the physical dimensions of the helix (length, diameter, etc.). And the rate of change of the magnetic field depends on the frequency (f) of the AC sine-wave voltage across the coil. We can put all these facts together mathematically to come up with a formula for X_L:

> **Rule 15–9**
>
> $$X_L = 2\pi f L \quad \text{ohms}$$

It is evident from this formula that the magnitude of the inductive reactance varies directly with f and L. Increasing (decreasing) either quantity increases (decreases) X_L. Moreover, unlike resistance, X_L is frequency dependent. That is, a 1 kΩ resistor is always a 1 kΩ resistor no matter what the frequency is. However, if a certain inductor has an inductive reactance of 1 kΩ, changing the frequency—even slightly—will result in a new value of X_L. Figure 15–11 shows the linear relationship between X_L and the frequency (f) for a given value of L in henries.

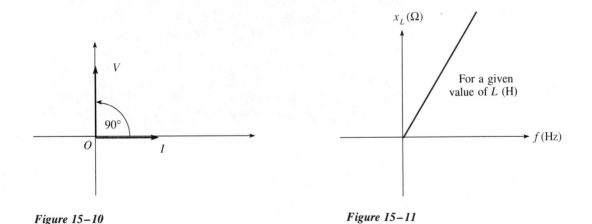

Figure 15–10 *Figure 15–11*

Example 15–6

What is the inductive reactance of a 50 mH inductor at (a) 30 MHz? (b) 1.6 kHz? (c) 256 MHz?

Solution
a. $X_L = 2\pi f L = 2\pi(30 \times 10^6)(0.05) \approx 9.425 \ \mu\Omega$
b. $X_L = 2\pi f L = 2\pi(1.6 \times 10^3)(0.05) \approx 503 \ \Omega$
c. $X_L = 2\pi f L = 2\pi(256 \times 10^6)(0.05) \approx 80.42 \ \mu\Omega$

Series AC Circuits

Example 15-7

An inductor is to have an inductive reactance of 2.5 kΩ at a frequency of 182.6 kHz. What must be the value of the inductance?

Solution Solving $X_L = 2\pi f L$ for L, we get

$$L = \frac{X_L}{2\pi f} = \frac{2500}{2\pi(182.6 \times 10^3)} \approx 2.2 \text{ mH}$$

Exercise Set 15-2

Problems

1. A 58 μH inductor is connected in series with another inductor. The coefficient of coupling is $k = 0.62$. If the mutual inductance (m) is 63.4×10^{-6}, what is the value of the other inductor?
2. When two inductors are connected series-aiding, their total inductance is 8.5 H. When connected series-opposing, the total inductance is 6.05 H. What is the mutual inductance of the combination?
3. In Problem 2, the coefficient of coupling is 14.3%. If one of the inductors has a value of 5.5 H, what is the value of the other?
4. Using the reciprocal formula for inductors in parallel, derive the formula for two inductors in parallel as given by

$$L_t = \frac{L_1 \cdot L_2}{L_1 + L_2}$$

5. Solve the formula derived in Problem 4 for (a) L_2 in terms of L_1 and L_t and (b) L_1 in terms of L_2 and L_t.
6. The primary of a transformer operating from 117 V has 1003 turns. If the secondary voltage is 12.6 V, how many turns are on the secondary?
7. The primary of the circuit shown in Figure 15-12 is to be protected by a fuse rated 10% higher than the nominal line current. What ampere rating should the fuse have?

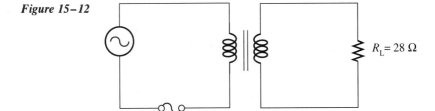

Figure 15-12

Fuse

8. A 50 Ω source impedance is to be matched to a 3.5 Ω load impedance using a transformer having a 378 turn primary. How many turns must the secondary have?
9. An ideal 497 mH inductor has an inductive reactance of 1250 Ω. At what frequency is the inductor being used?
10. Write a BASIC computer program for each of the following.
    a. The total inductance of two inductors when the coefficient of coupling is known.
    b. The coefficient of coupling between two inductors when the mutual inductance is known.
    c. In the relationship

    $$\frac{N_p}{N_s} = \frac{V_p}{V_s} = \frac{I_s}{I_p}$$

    find the sixth value when five values are given.
    d. The turns ratio required to match source and load impedances.
    e. The inductive reactance at any values of f and L.

340 Advanced Applications of Mathematics to Electronics

15–5 CAPACITORS IN SERIES AND PARALLEL

The electrical charge stored in a capacitor is given by

$$Q = CV$$

where Q is the charge in coulombs,* C is the capacitance in farads, and V is the applied voltage. When capacitors are connected in series, each one is charged by electrostatic induction, since there is no actual electron flow through the dielectric material separating the plates. Therefore, the charge is *always the same for any capacitor* in the series string.

From $Q = CV$, we may express the sum of the voltages across any individual series capacitor as

$$V = \frac{Q}{C}$$

and the total applied voltage may be expressed as

$$V_t = \frac{Q}{C_t}$$

where C_t is the total capacitance of the series combination. It is evident from Kirchhoff's voltage law that the total voltage (V_t) must equal the sum of the individual voltages across each capacitor. This fact allows us to write

$$\frac{Q}{C_t} = \frac{Q}{C_1} + \frac{Q}{C_2} + \cdots + \frac{Q}{C_n}$$

If we now divide through by Q, we get

$$1/C_t = 1/C_1 + 1/C_2 + \cdots + 1/C_n$$

On solving this equation for C_t, we obtain the following.

Rule 15–10

$$C_t = \frac{1}{1/C_1 + 1/C_2 + \cdots + 1/C_n}$$

This leads us to conclude that capacitors in series add in the same way as resistors in parallel—that is, according to the reciprocal of the individual values.

Example 15–8 What is the total capacitance of the circuit shown in Figure 15–13?

$$C_t = \frac{1}{\dfrac{1}{33 \times 10^{-12}} + \dfrac{1}{0.001 \times 10^{-6}} + \dfrac{1}{0.0000022}}$$

$$\approx \frac{1}{30.3 \times 10^9 + 1 \times 10^9 + 454.5 \times 10^3}$$

$$\approx 31.95 \times 10^{-12} = 31.95 \text{ pF}$$

---
*A coulomb is the charge carried by 6.24×10^{18} electrons.

Series AC Circuits

Figure 15-13

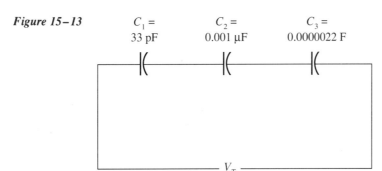

In the case of capacitors in parallel, the voltage across each capacitor is the same. However, the charge stored in each may be different, and it depends on the value of the capacitor as given by the formula $Q = CV$. The total charge is the simple sum of all the charges stored in each capacitor. This fact allows us to write

$$C_t \cdot V_t = C_1 \cdot V_t + C_2 \cdot V_t + \cdots + C_n \cdot V_t$$

On dividing through by V_t, we obtain the following.

Rule 15-11

$$C_t = C_1 + C_2 + \cdots + C_n$$

In other words, capacitors in parallel add just like resistors in series.

For example, what is the total capacitance on the circuit shown in Figure 15-14?

Figure 15-14

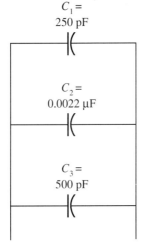

$$C_t = 250 \times 10^{-12} + 0.0022 \times 10^{-6} + 500 \times 10^{-12}$$
$$= 2.95 \times 10^{-9} = 2950 \text{ pF}$$

15-6 CAPACITANCE IN AC CIRCUITS

A capacitor has the property of causing the current to lead the applied voltage by 90°. This effect is shown in Figure 15–15. The cause of this phenomenon in an AC circuit may be understood by considering what happens during the first quarter cycle of the applied voltage. Since there is no charge on the capacitor at the instant the voltage is applied, the capacitor appears as a short circuit, and the current flow is maximum. As more and more electrons flow onto one plate of the capacitor (and leave the other), a counterelectromotive force builds up that opposes the flow of any additional current, since like charges repel. Therefore, just as the current is going to zero, the voltage across the plates of the capacitor is at a maximum. From this point on, the current leads the applied voltage by 90°. Figure 15–16 shows the phasor representation.

The counterelectromotive force across the capacitor produces an opposition to the flow of alternating current. This opposition is called **capacitive reactance** (X_C), and its value in ohms depends on the value of the capacitor as well as the frequency of the applied AC. Mathematically, we can state these facts using the following formula.

Rule 15–12

$$X_C = \frac{1}{2\pi f C}$$

Since the capacitive and frequency terms are in the denominator, it is apparent that an increase in either will cause a decrease in the value of the capacitive reactance. The inverse relationship between frequency and capacitive reactance is shown in Figure 15–17 for a fixed value of C.

We notice two things that distinguish this curve from that of inductive reactance shown in Figure 15–11. First, the capacitive reactance does not vary directly with the frequency, as is the case with inductive reactance. The relationship between X_C and f is an inverse one; as f goes up, X_C goes down and vice versa. Second, we see that capacitive reactance is a nonlinear relationship, not a straight line as seen with inductive reactance. The reason for this can be seen if we rewrite the equation for capacitive reactance as the proportion $X_C \propto 1/f$. Table 15–2 shows the values of $1/f$ as f gets progressively smaller and smaller.

Table 15–2

f	$1/f$
7	0.143
6	0.17
5	0.2
4	0.25
3	0.33
2	0.5
1	1
½	2
⅓	3
¼	4
⅕	5
⅙	6
⅐	7

Figure 15–15

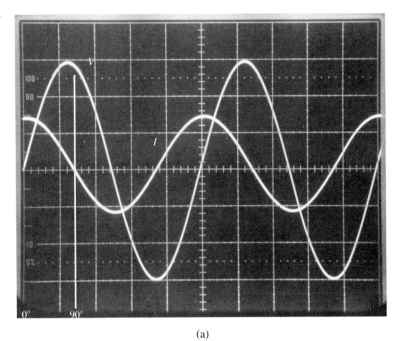

(a)

(b)

Figure 15–16

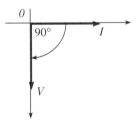

Figure 15–17

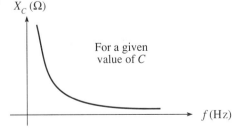

Figure 15–18 shows the nonlinear curve obtained by plotting the values in Table 15–2. It is apparent that this is quite different from the straight line obtained for inductive reactance.

Exercise Set 15–3

Problems

1. A technician requires 11 pF of capacitance in a circuit. She already has a 33 pF capacitor and a 22 pF capacitor. What value of capacitor must be added in series in order to obtain the required value?
2. Using the reciprocal formula for capacitors in series, derive the formula for two capacitors in series, as given by

$$C_t = \frac{C_1 \cdot C_2}{C_1 + C_2}$$

3. Use the formula derived in Problem 2 twice in order to verify the answer to Problem 1.
4. What is the capacitive reactance of a 250 pF capacitor at 30 MHz?
5. It is desired to double X_C at a fixed frequency by changing the value of the capacitor. What must the value of the new capacitor be in terms of the original value?
6. Two ideal 5 μF capacitors in series are placed across a 10 V DC source and fully charged as shown in Figure 15–19a. The capacitors are then removed from the source, placed in parallel, and connected across a third uncharged 5 μF capacitor, as shown in (b). What voltage will be measured across the parallel combination?
7. Write a BASIC computer program that will calculate X_C for any given values of f and C.

15–7 RC TIME CONSTANT

Consider the circuit shown in Figure 15–20. When S_1 is first closed, the current through the circuit is at a maximum because the capacitor appears as a short circuit. Therefore, the voltage across the resistor (V_r) is the same as V_s, and the voltage across the capacitor (V_c) is zero. These conditions exist for only an instant, however, since V_c increases as the capacitor charges. But as C charges, the current gets smaller and smaller. Consequently, the voltage drop across R also becomes less and less. After a short time (called the *transient response time*), the current falls to zero, V_r becomes zero, and $V_c = V_s$. The circuit is then said to have reached its *steady state*. A voltmeter across R reads zero; a voltmeter across C reads V_s; and an ammeter anywhere in the series circuit reads zero. The voltage transient conditions are summarized in Figure 15–21.

You might now ask yourself how much time is required for the steady state to be reached. The answer is, for all practical purposes, five *time constants*. But what exactly is a time constant? To answer this question, we ask you to accept the fact that as a capacitor is charged or discharged through a resistor, the current and voltages change at an exponential rate. This is a fact of nature, not of humans. And it always happens in the same way according to specific mathematical laws. In fact, the voltage drop across the resistor (V_r) when C is discharging is always given by the exponential equation $V_r = Ve^{-t/RC}$, where e is the natural log base, t is in seconds, R is in ohms, and C is in farads.

In Chapter 13 (logarithms), you were introduced to the idea of exponential change of voltage across a resistor in the RC circuit of Problem 41. Here we reproduce the same circuit as Figure 15–22. Note that the circuit is driven by a square-wave voltage that produces the same effect as the switch used in connection with Figure 15–20. Observe

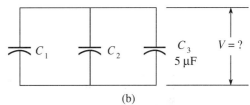

(a)

(b)

Figure 15–18

Figure 15–19

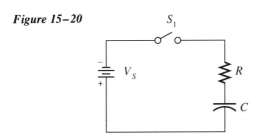

Figure 15–20

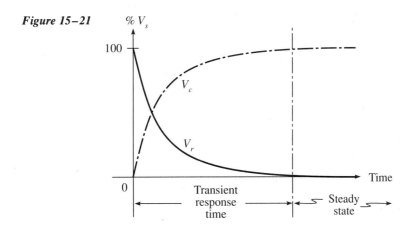

Figure 15–21

that the horizontal sweep of the scope used to take the photo is set at 0.2 ms/div (lower right corner of photo). It is clear, then, that the steady state is reached in 1 ms (5 div × 0.2 ms/div = 0.001 s).

Figure 15–22

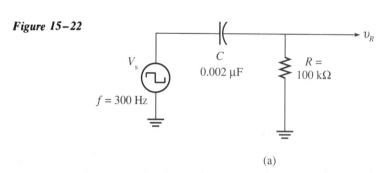

(a)

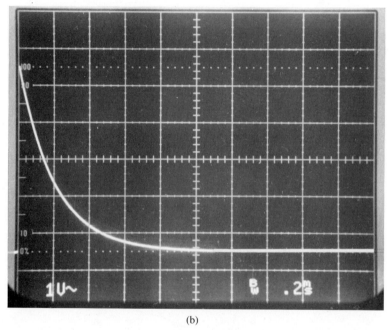

(b)

Consider now what happens if we multiply the resistance in ohms by the capacitance in farads of any series RC circuit. Since $Q = CV$, then $C = Q/V$. Furthermore, $Q = (I)(t)$, and $V = (I)(R)$. Therefore, $C = Q/V = It/IR = t/R$, and $RC = R(t/R) = t$, where t is in seconds.

Rule 15–13

$$t = RC$$

For example, in Figure 15–22, $C = 0.002$ μF and $R = 100$ kΩ. Thus, $t = R \times C = 200$ μs and $5RC = 1$ ms. This is the same result obtained experimentally using the circuit in Figure 15–22. In electronics, the product of R and C is called the **time constant**; it is defined as the time required for the circuit values to change by a certain amount. Five time

Series AC Circuits

constants are always required for a circuit to reach its *steady state* (for all practical purposes), no matter what the values of R and C. More specifically, one time constant is the time required for the voltage across C to charge to 63% of V_s or fall to 37% of V_s. In Figure 15–22, we see that after one scale division (one time constant), the voltage appears to be about 1.8 V, since the scope's vertical sensitivity is set at 1 V/div (lower left corner of photo). In applying the formula $V_r = Ve^{-t/RC}$ where $t = 0.2$ ms, $RC = 200 \times 10^{-6}$, and $V = 5$ V (5 scale divisions), we see that the voltage has indeed fallen to 1.839 V. Note, too, that 1.839/5 corresponds to 37%, as stated. This percentage can also be read directly from the graticule's left vertical scale in Figure 15–22.

Figure 15–23 shows the universal time constant graphs. We can use these graphs to determine the voltage across C or R at any time. For example, suppose the value of C in Figure 15–22 is 0.05 μF, $R = 30$ kΩ, and $V = 3$ V. What is the voltage across the capacitor on charge after two time constants? From curve (a), we see that the voltage has attained 86% of its full charge. Therefore, 86% × 3 V = 2.58 V. Note that it was not necessary to compute the actual time constant of 1.5 ms.

Figure 15–23

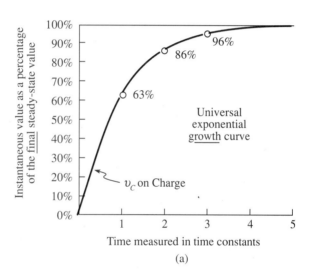

(a)

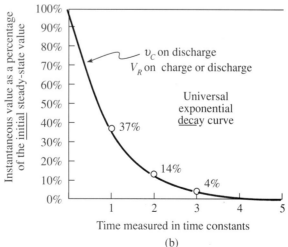

(b)

348 Advanced Applications of Mathematics to Electronics

We can use curve (b) to determine the voltage across R on discharge. For example, after two time constants, the voltage across R is about 14% of V_s, or $0.14 \times 3 = 0.42$ V. This result is consistent with that obtained from the formula $V_r = Ve^{-t/RC}$, where $t = 0.003$ s (two time constants), $R = 30$ kΩ, $C = 0.05$ μF, and $V_s = 3$ V. Applying the discharge formula, we obtain

$$V_r = (3)e^{-0.0030/0.0015}$$
$$= 3e^{-2}$$
$$\approx 0.41 \text{ V}$$

Exercise Set 15–4

Problems

1. A time constant of 4.7 μs is required using a 0.001 μF capacitor. What value of series resistor is required?
2. From Figure 15–23a, determine the voltage across the capacitor 1.5 time constants after a 12 V source has been applied.
3. What is the voltage across a resistor one time constant after a 470 μF capacitor charged to 122 V begins to discharge?
4. What value of resistance must be added to a 5.1 kΩ resistor in series with a 220 pF capacitor in order to get a time constant of 1.342 μs?
5. Write a BASIC computer program for calculating the voltage drop across R for given values of V, t, R, and C as C discharges through R.

15–8 IMPEDANCE OF SERIES CIRCUITS

Consider the circuit shown in Figure 15–24. If we apply Ohm's law to finding the voltage drops across R and L, we obtain $V_R = (I)(R) = 27.3$ V and $V_L = (I)(X_L) = 12.41$ V. Adding these two voltages gives almost 40 V. How is it possible to have the voltages sum to 40 V when only 30 V are applied to the circuit? Is Kirchhoff's voltage law in error? The answer to this seemingly contradictory result is that the voltages (phasors) across R and L are not in phase. And, as we learned in Chapter 14, phasors differing by other than 0° or 180° cannot be added algebraically. They must be treated according to the rules of vector (phasor) addition.

Earlier in this chapter, we showed the phasor relationship between the voltage and current of a purely resistive circuit (Figure 15–1c). Later, the voltage and current relationships for an inductive circuit were shown (Figure 15–10). For comparison, we have duplicated these phasor diagrams here as Figure 15–25. Note that in both parts (a) and (b) of Figure 15–21, the current phasor (I) has been drawn along the positive abscissa. Further, observe that V_R is in phase with I and V_L leads I by 90°, as was shown earlier in the chapter. Part (c) of the figure depicts the phasor summation of the two component voltages V_R and V_L by the simple application of the Pythagorean theorem. Applying this same theorem to Figure 15–24, we obtain

$$V = \sqrt{V_R^2 + V_L^2} = \sqrt{(27.3)^2 + (12.41)^2} \approx 30 \text{ V}$$

Hence, we see that there is no real contradiction between our vector result and Kirchhoff's law as applied earlier to DC circuits.

On looking at Figure 15–25c, it may have occurred to you that since the voltage drops are proportional to the ohmic oppositions of R and X_L, we might just as easily have labeled the vectors as R and X_L rather than V_R and V_L. In so doing, we can now apply the

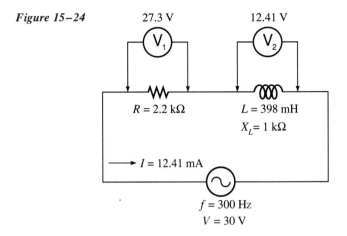

Figure 15-24

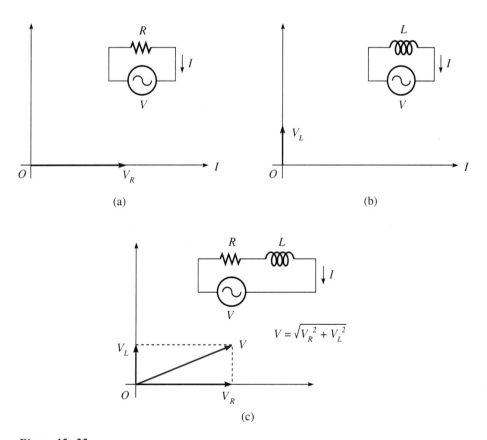

Figure 15-25

Pythagorean theorem to R and X_L just as we did previously with V_R and V_L. This time, however, our result will be the total circuit opposition to AC current. This opposition is called **impedance** (symbol Z) and is measured in ohms. The general method is shown in Figure 15–26. Applying the Pythagorean formula to the circuit in Figure 15–24, we obtain

$$Z = \sqrt{R^2 + X_L^2} = \sqrt{(2200)^2 + (1000)^2} \approx 2417 \; \Omega$$

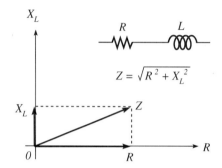

Figure 15–26

In a circuit containing capacitance rather than inductance, the voltage lags the current by 90°, and so the phasor diagram is as shown in Figure 15–27a. In part (b) of the figure, we have once again relabeled the axes in terms of ohmic opposition. It should be apparent, then, that the formula for the circuit impedance is given by $Z = \sqrt{R^2 + X_C^2}$ ohms.

Many circuits contain capacitance and inductance as well as resistance. Such a circuit is shown in Figure 15–28. According to the component and frequency values, $X_C = 3 \; k\Omega$ and $X_L = 2.5 \; k\Omega$. The phasor diagram is shown in Figure 15–29a. We observe that since X_C and X_L are 180° out of phase, they may be subtracted algebraically, giving $3 \; k\Omega - 2.5 \; k\Omega = 500 \; \Omega$. Since X_C dominates in this circuit, the net reactance is capacitive, as shown in Figure 15–29b. This gives the following rule.

Rule 15–14

$$Z = \sqrt{R^2 + (X_C - X_L)^2}$$

Consequently, in this case the total circuit impedance is

$$Z = \sqrt{R + (X_C - X_L)^2} = \sqrt{4700^2 + 500^2} \approx 4726.5 \; \Omega$$

Since the difference between the two reactances is squared, the order in which the subtraction is made makes no difference. The result will always be a positive number.

The total current in a series AC circuit can be found by the usual application of Ohm's law, replacing R by Z. For example, in Figure 15–24, the total circuit current is given by $I = V/Z = 30/2417 \approx 12.41$ mA. Other forms of Ohm's law may be similarly adapted, as in $V = (I)(Z)$ and $Z = V/I$. The case of power in an AC circuit is taken up in the next section.

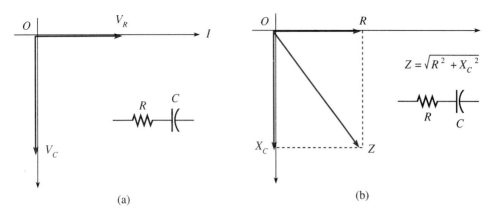

Figure 15–27

Figure 15–28

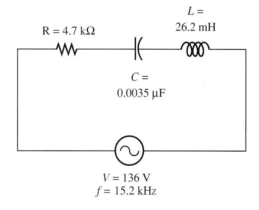

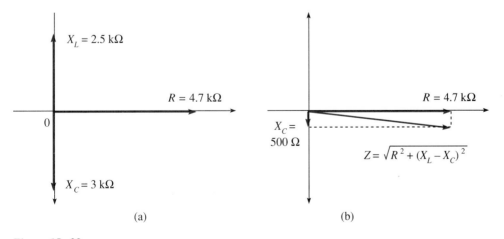

Figure 15–29

Phase Angle

The angle between the phasors representing the source voltage (V_s) and total circuit current (I_t) is called the **phase angle** (symbolized by the Greek letter theta, θ). In a purely resistive circuit the value of θ is 0°. In a purely reactive circuit, $\theta = \pm 90°$. For any other combination of circuit elements, $-90° < \theta < +90°$. For example, in Figure 15–30, the phase angle is shown to be +56°. From our previous study of trigonometry, we know that θ is given by the following.

Rule 15–15

$$\theta = \tan^{-1}\frac{V_L}{V_R} \quad \text{or} \quad \theta = \tan^{-1}\frac{X_L}{R}$$

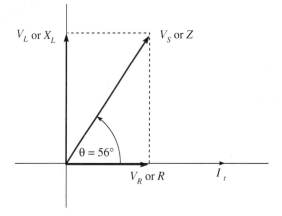

Figure 15–30

For example, if $X_L = 1631 \ \Omega$ and $R = 1.1 \ k\Omega$, then

$$\theta = \tan^{-1}\frac{1631}{1100}$$

and $\theta \approx +56°$.

Example 15–9

Consider the circuit shown in Figure 15–31a. The technician takes the photo shown in part (b) of the figure and determines that the phase angle is −25.5°. What method was used?

Solution Beginning with the photo in part (b), we set up the proportion

$$\frac{6.7 \text{ divisions}}{360°} = \frac{1.2 \text{ divisions}}{\phi}$$

Solving for ϕ,

$$\phi = 1.2 \times \frac{360}{6.7} \approx 64.5°$$

From Figure 15–32, we see that the angle ϕ is actually the angle between V_c and V_s. However, θ is $-90° + \phi = -25.5°$, which is the phase angle between I_t (or V_R) and V_s.

Figure 15-31

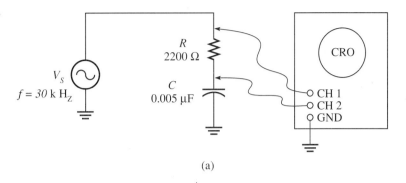

(a)

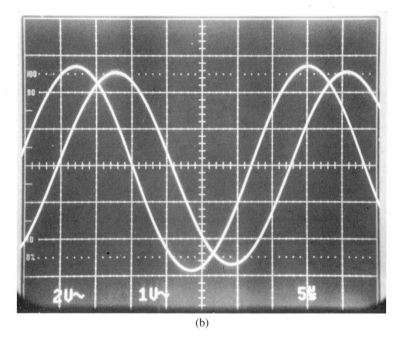

(b)

Figure 15-32

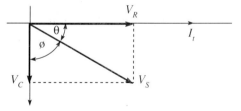

354 Advanced Applications of Mathematics to Electronics

We can verify this result as follows.

$$X_C = 1061 \; \Omega$$
$$R = 2.2 \; k\Omega$$

Therefore,

$$-\theta = \tan^{-1}\frac{1061}{2200}$$

from which $\theta \approx -25.8°$. This represents an error of less than 1.2%.

Exercise Set 15-5

Problems

1. Find the total impedance of the circuit shown in Figure 15–33.

Figure 15–33

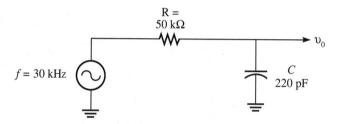

2. The 59 μH inductor shown in Figure 15–34 has an internal resistance of 3 kΩ. What is the circuit current?

Figure 15–34

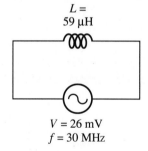

3. In Problem 2, what percent of error is made in the calculation of the total current if the internal resistance is neglected?

Hint Percent of error (%e) is calculated as

$$\%e = \frac{|N - n|}{N} \times 100$$

where

N = the true or assumed-true value
n = the comparison quantity

4. In Problem 1, what new value of C is required to increase the circuit impedance by 12%?

5. In Problem 1, what value of inductance must be inserted in series with C in order to offset its effect exactly?

6. What is the source frequency in the circuit shown in Figure 15–35?

Series AC Circuits

Figure 15-35

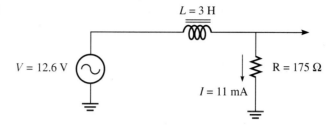

7. What is the phase angle in Problem 6?
8. What is the phase angle specified under the conditions of Problem 5?
9. Write a BASIC program that will compute the impedance and phase angle given the frequency, resistance, inductance, and capacitance.
10. Write a BASIC program for computing the value of C in Problem 4 required for either an increase or decrease in Z.

15-9 SERIES RESONANCE

You may recall from an earlier section that the curve of inductive reactance is linear, whereas that of capacitive reactance is nonlinear. These two curves have been superimposed on one another and redrawn in Figure 15-36. It is clear that at the point of intersection, $X_C = X_L$. From what has been said earlier, it has probably occurred to you that this is the point where the inductive and capacitive reactances cancel each other out, leaving only resistance. That is, at this point we have a condition where

$$2\pi f L = \frac{1}{2\pi f C}$$

On solving this equation for frequency (f), we obtain the following.

Rule 15-16

$$f = \frac{1}{2\pi \sqrt{LC}}$$

This frequency is called the **resonant frequency** for given values of L and C.

Figure 15-36

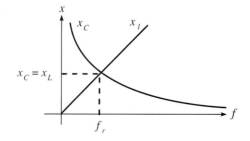

Although Figure 15–36 shows only two curves, you should realize that there are an infinite number of each, since there are theoretically an infinite number of possible L and C values. Moreover, each combination exhibits one and only one resonant frequency value. For example, consider the circuit shown in Figure 15–37. At a frequency of 100 kHz, $X_C = 1592\ \Omega$ and $X_L = 850\ \Omega$. Under these conditions, the impedance is $768.5\ \Omega$, and the line current is 26 mA. However, if we change the frequency to 137.5 kHz, then X_L and X_C are equal and opposite. The equivalent circuit is a 200 Ω resistor across the voltage source, as shown in Figure 15–38.

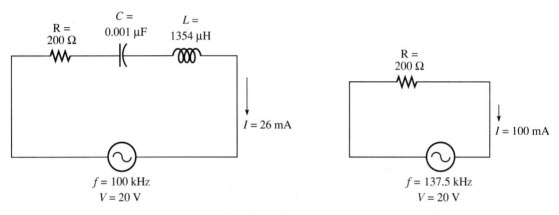

Figure 15–37

Figure 15–38

At the resonant frequency of 137.5 kHz, the line current increases to 100 mA, and $Z = R = 200\ \Omega$. It remains as an exercise for you to verify the circuit conditions shown in Figure 15–37 for both frequencies.

Exercise Set 15–6

Problems

1. What is the resonant frequency of the circuit shown in Figure 15–28?
2. In order for maximum power to be transferred from source to load, the internal resistance (R_i) of the source must equal the resistance of the load (R_L), and the reactive effects must cancel. In the circuit in Figure 15–39, the source impedance is 352 kΩ. What are the values of R_L and L for maximum power transfer?

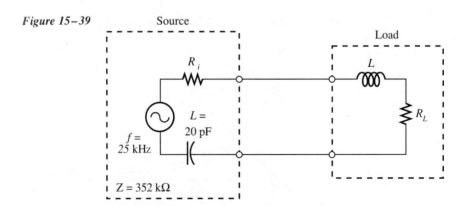

Figure 15–39

Series AC Circuits

3. The resonant frequency of a circuit is 343.82 kHz. The value of the capacitor is 3900 pF. What is the value of the inductor?
4. A 50 mH RF choke coil and a 0.0022 µF capacitor are connected in series. The internal resistance of the coil is 68 Ω. If the source voltage is 6.3 V, what is the current at the resonant frequency?
5. In the circuit shown in Figure 15–40, what value of C is required in order to produce resonance?

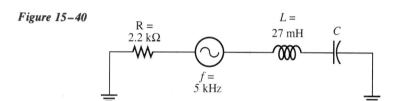

Figure 15–40

6. Write a BASIC program that will compute the resonant frequency f_r given the values of L and C. Extend this program to find any variable in the resonant-frequency formula given the other two.

15–10   POWER IN AC CIRCUITS

In your study of direct current circuits, you learned that the power dissipated could be obtained by any of the following three formulas:

$$P = VI$$
$$P = \frac{V^2}{R}$$
$$P = I^2 R$$

Each of these power formulas has been applied to the various circuit elements of Figure 15–41, and the results are tabulated in Table 15–3.

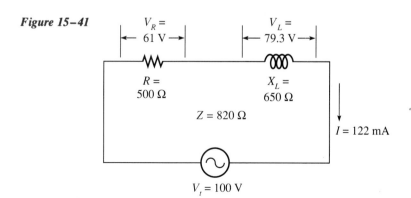

Figure 15–41

Table 15–3

	Applied to Entire Circuit	Applied to Resistor Only	Applied to Inductor Only
$P = VI$	$P = V_t I$ $= 100 \times 0.122$ $= 12.2$	$P = V_R I$ $= 61 \times 0.122$ $\approx 7.44 \text{ W}$	$P = V_L I$ $= 79.3 \times 0.122$ ≈ 9.67
$P = \dfrac{V^2}{R}$	$P = \dfrac{V_t^2}{Z}$ $= \dfrac{100^2}{820}$ ≈ 12.2	$P = \dfrac{V_R^2}{R}$ $= \dfrac{61^2}{500}$ $\approx 7.44 \text{ W}$	$P = \dfrac{V_L^2}{X_L}$ $= \dfrac{79.3^2}{650}$ ≈ 9.67
$P = I^2 R$	$P = I^2 Z$ $= (0.122)^2 820$ ≈ 12.2	$P = I^2 R$ $= (0.122)^2 500$ $\approx 7.44 \text{ W}$	$P = I^2 X_L$ $= (0.122)^2 650$ ≈ 9.67

Since there are three different values of power shown in the table (12.2, 7.44, and 9.67), it is reasonable to ask which is correct. Recall that in a resistor, the current and voltage are always in phase. Therefore, the product of source voltage V_t and line current I always results in the true average power. In the case of reactance, however, the voltage and current do not rise and fall together; hence their RMS values are reached at different times. Consequently, their product does not equal the true average power even though the product of their instantaneous values does represent the power at some particular moment.* Therefore, the only column representing true average power is the second, where the power is given as 7.44 W. Note that the values shown in the first and third columns are not specified in watts, since this unit is reserved for true power only. The figure in the first column (12.2) is called **apparent power** and is measured in units of volt-amperes (VA). The value in the third column (9.67) is the **reactive power** and has units of VAR (volt-ampere reactive). Neither of these last two values represents true power, but they are related to it as shown in Figure 15–42. These relationships are summarized in Rule 15–17.

Rule 15–17

Apparent power: $\quad P = V_t I; \quad P = \dfrac{V_t^2}{Z}; \quad P = I^2 Z$

Reactive power: $\quad P = V_X I; \quad P = \dfrac{V_X^2}{X}; \quad P = I^2 X$

True power: $\quad\quad\, P = V_R I; \quad P = \dfrac{V_R^2}{R}; \quad P = I^2 R$

---

*For a full discussion of instantaneous power relationships, refer to R. Bartkowiak, *Electric Circuit Analysis*, New York: Harper & Row, 1985, Chapter 12.

Figure 15-42

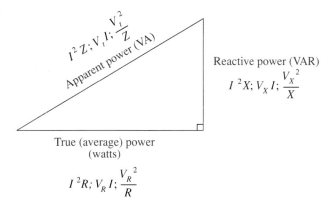

Note that the three types of power are related by the Pythagorean theorem—that is

$$\text{Apparent power} = \sqrt{(\text{true power})^2 + (\text{reactive power})^2}$$

From Table 15–3, we see that

$$12.2 \approx \sqrt{(7.44)^2 + (9.67)^2} \text{ VA}$$

or

$$7.44 \approx \sqrt{(12.2)^2 - (9.67)^2} \text{ W}$$

In Figure 15–42, the angle θ is the phase angle between the applied voltage and total line current, as stated in a previous section. Therefore, if we have the apparent power in volt-amperes, we know from our study of trigonometric relationships that the true power is given by the following rule.

Rule 15–18

$$\text{True power} = P = VI \cos \theta$$

The value of cos θ is referred to as the **power factor** and can take on any value between 0 and 1.

Rule 15–19

$$PF = \cos \theta$$

Note that if θ = 0°, then the true power and apparent power are equal. However, if θ = 90°, then the true power is zero; the circuit is purely reactive. In Figure 15–41, we see that $\theta = \tan^{-1} X_L/R$. Therefore, θ ≈ 52.43°. The power factor, then, is cos 52.43° ≈ 0.61, and the true power is $VI \cos \theta = 100 \times 0.122 \times 0.61 \approx 7.44$ W.

The power factor of a consumer's load is extremely important to utility companies supplying electrical power. Consider two factories supplied with single-phase, 220 V service. Company 1 uses 500 A and has a power factor of 1.0 (θ = 0°). The power consumed is 500 × 220 = 110 kVA = 110 kilowatts (kW). If the utility company charges 6.8¢ per kilowatt-hour (kWh), it receives $7.48 per hour for power. On the other hand, company 2 draws the same current at the same voltage for the same 110 kVA but has a power factor of 0.55, for instance ($\theta \approx 56.6°$). Hence, the actual power consumed is only 500 × 220 × 0.55 = 60.5 kW, and the power company receives only $4.11 for exactly the same services provided. After all, it is the kVA load (not the actual power) that determines wire size, transformer capacity, and similar demand factors. It is natural to expect, then, that utility companies add a monthly penalty charge for low-power-factor energy consumption in order to offset operating costs. It is good business sense to maintain a power factor as close to unity as possible. In practice, the power factor falls short of the ideal 1.00, but it is seldom less than about 50% (0.50).

In order to bring a low power factor up to an acceptable level, a power-factor correction is made by adding a reactance of the type opposite the load reactance across the supply line. For example, consider an inductive load with a 0.61 power factor. As shown in Figure 15–43, a capacitor connected across the line will serve to cancel all or part of the reactive power without changing any of the original load demand characteristics (i.e., current or voltage). In general, it is expensive in terms of initial costs to bring the power factor up to exactly 1.00. Therefore, based on economic considerations, the power factor might be increased to 0.95, for example. Consider the circuit shown in Figure 15–44 depicting a load dissipating 100 W of power, with a power factor of 0.61. What value of capacitor must be connected across the load in order to increase the power factor to 0.95? Figure 15–45 shows the conditions before and after. We see that

$$VAR_1 = P \tan 52.41° \approx 130$$

and

$$VAR_2 = P \tan 18.19° \approx 32.9$$

Therefore,

$$VAR_1 - VAR_2 = 130 - 32.9 = 97.1$$

From $P = V^2/X_C$,

$$X_C = \frac{V^2}{P} = \frac{(120)^2}{97.1} \approx 148.3 \ \Omega$$

From $X_C = 1/2\pi fC$,

$$C = \frac{1}{2\pi f X_C} \approx 18 \ \mu F$$

Exercise Set 15–7

Problems

1. In Figure 15–46, determine the true, apparent, and reactive powers. Verify your answers using the power triangle shown in Figure 15–33.
2. In Problem 1, what is the power factor?
3. What power is dissipated in the circuit in Figure 15–47?

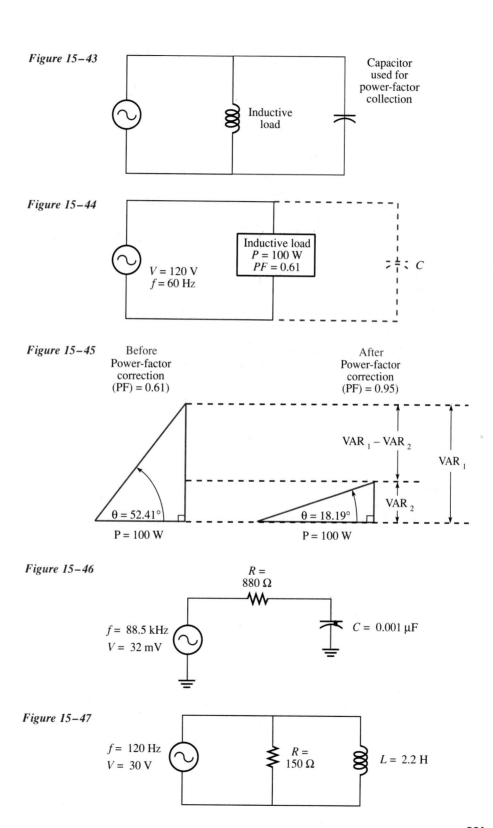

4. a. What is the true power dissipated in the circuit shown in Figure 15–48?
   b. What is the power factor?
   c. What power is dissipated when the circuit is brought to resonance?
   d. What is the power factor at resonance?

Figure 15–48

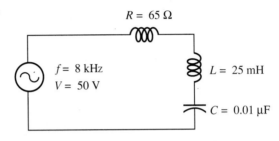

5. A circuit consists of an inductive load across a voltage source of 67.5 V, 400 Hz. The power factor of the circuit is 0.52. What value of capacitor is required to raise the power factor to 0.89?
6. Write a BASIC program that will compute the true, reactive, and apparent powers and phase angle for a series AC circuit.
7. Write a BASIC program that will calculate the value of capacitance required to raise the power factor by a specified amount.

Key Terms

effective value
root mean square (RMS)
self-inductance
mutual inductance
coefficient of coupling

inductive reactance
capacitive reactance
time constant
impedance
phase angle

resonant frequency
apparent power
reactive power
power factor

Important AC Formulas

Effective (RMS) Value of Voltage and Current
Rule 15–1

$$V = 0.707 V_p$$

Rule 15–2

$$I = 0.707 I_p$$

Inductors in Series ($k = 0$)
Rule 15–3

$$L_t = L_1 + L_2 + \cdots + L_n$$

Inductors in Series ($k \neq 0$)
Rule 15–4

$$L_t = L_1 + L_2 \pm 2m$$

where

$$m = k\sqrt{L_1 \cdot L_2}$$

Series AC Circuits

Inductors in Parallel
Rule 15–5
$$L_t = \frac{1}{1/L_1 + 1/L_2 + \cdots + 1/L_n}$$

Transformer Relationships
Rule 15–6
$$\frac{N_p}{N_s} = \frac{V_p}{V_s}$$

where

N_p = number of turns in primary
N_s = number of turns in secondary
V_p = voltage of the primary
V_s = voltage of the secondary

Rule 15–7
$$\frac{V_p}{V_s} = \frac{I_s}{I_p}$$

where

I_s = current in the secondary
I_p = current in the primary

Rule 15–8
$$\frac{Z_p}{Z_s} = \left(\frac{N_p}{N_s}\right)^2$$

where

Z_p = impedance connected to primary
Z_s = impedance connected to secondary

Inductive Reactance
Rule 15–9
$$X_L = 2\pi f L \quad \text{ohms}$$

Capacitors in Series
Rule 15–10
$$C_t = \frac{1}{1/C_1 + 1/C_2 + \cdots + 1/C_n}$$

Capacitors in Parallel
Rule 15–11
$$C_t = C_1 + C_2 + \cdots + C_n$$

Capacitive Reactance
Rule 15–12
$$X_C = \frac{1}{2\pi f C} \quad \text{ohms}$$

RC Time Constant
Rule 15–13
$$t = RC$$

Impedance of Series AC Circuit
Rule 15–14
$$Z = \sqrt{R^2 + (X_C - X_L)^2}$$

Phase Angle
Rule 15–15
$$\theta = \tan^{-1}\frac{V_L}{V_R} \quad \text{or} \quad \theta = \tan^{-1}\frac{X_L}{R}$$

Resonant Frequency
Rule 15–16
$$f = \frac{1}{2\pi\sqrt{LC}}$$

Power Relationships in AC Circuit
Rule 15–17

Apparent power: $\quad P = V_t I; \quad P = \dfrac{V_t^2}{Z}; \quad P = I^2 Z$

Reactive power: $\quad P = V_X I; \quad P = \dfrac{V_X^2}{X}; \quad P = I^2 X$

True power: $\quad P = V_R I; \quad P = \dfrac{V_R^2}{R}; \quad P = I^2 R$

Power Factor
Rule 15–18
$$\text{True power} = P = VI \cos\theta$$

Rule 15–19
$$PF = \cos\theta$$

Upon completion of this chapter, you will have:
1. Learned the structure and characteristics of complex numbers and how to add, subtract, multiply, and divide them.
2. Learned how to convert complex numbers from polar form to rectangular form and vice versa.
3. Learned how complex numbers are used to specify resistance, reactance, and impedance in AC circuits.
4. Learned how to use complex algebra to solve parallel and series-parallel AC circuits.

16 Complex Algebra and Parallel AC Circuits

16-1 INTRODUCTION

Just when you were beginning to think you had mastered enough algebra to survive, along comes something called *complex* algebra to make life miserable again. Well, happily, complex doesn't mean complicated. In fact, it's fairly simple once you get the hang of it. But why the need for complex algebra in the first place? If we look around at the real world, we see that a large part of it utilizes AC devices connected in parallel or series-parallel combinations. For example, virtually everyone has AC supplied to their homes, and house wiring is essentially parallel. Moreover, devices that are plugged in to circuits represent various arrangements of series-parallel loads. These devices include refrigerators, microwave ovens, television receivers, personal computers, air conditioners, washing machines, and so forth.

Unfortunately, however, you cannot completely analyze the circuits in these devices using the simple methods developed so far. For example, consider the circuit shown in Figure 16-1. Can you determine the total impedance between points A and B? Are you able to calculate the total line current? Can you determine the phase angle? As you feverishly search your brain for a simple formula such as $Z = \sqrt{R^2 + X^2}$, which we developed in Chapter 15, let me assure you there is, alas, no such simple equation into which you can just start plugging numbers. Of course, the formula just shown will, as you know, give the individual branch impedances, but then how would you combine the branches? Simple addition doesn't work. What now? Well, first things first. Incidentally, the total impedance of this circuit turns out to be about 478 Ω. Later, we'll see exactly how this result was obtained, but first we must talk about other kinds of numbers.

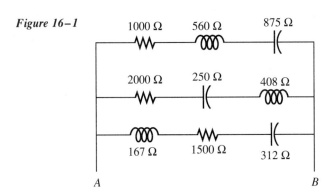

Figure 16-1

16-2 IMAGINARY NUMBERS

In order to talk about complex algebra, we must first talk about complex numbers. But in order to do that, we need to discuss two other kinds of numbers, *real* and *imaginary*, and how they fit together to make up a complex number.

Luckily, you already know what real numbers are. Recall that in Chapter 2 we defined a real number as any number you can find on the number line. Remember the number line? It is shown again in Figure 16-2. Try to think of a number you can't find on the number line. The truth is, almost every number you can think of (and even a few you can't) appears somewhere as a point on the number line. In fact, there are an infinite

Complex Algebra and Parallel AC Circuits

number of such points, hence, numbers. Would you be surprised, though, to learn that there are also an infinite number of numbers you *can't* find on this line? They're called **imaginary numbers,** and they fit on another line. Before we show that line, let's talk about what an imaginary number is.

Figure 16-2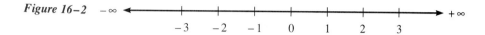

Suppose you want to find $\sqrt{16}$. You might write $\sqrt{16} = \pm 4$. But suppose you want to find $\sqrt{-16}$. What would your answer be? Obviously, the answer cannot be either -4 or $+4$, since the square of either number yields $+16$, not -16. Moreover, the answer cannot be *both* $+4$ and -4 even though their product is -16 because the square root of a number is a single number that when multiplied by itself gives the original number.

On the surface, it would seem as though we have reached a dead end. Happily, however, mathematics is an invention, not a discovery, and so we are free to do whatever is required to solve a problem, provided it is consistent with everything that has gone before. That being the case, mathematicians were quick to invent a new kind of number that would provide an answer to the preceding question. The new type of number is called an *imaginary* number, and the logic behind its creation goes something like this. If the numeral $+4$ shown in Figure 16-3a were multiplied by -1, it would end up 180° on the other side of 0 as a -4, as shown in Figure 16-3b, since, by the rules of multiplication developed in Chapter 3, $(+4)(-1) = -4$.

Figure 16-3

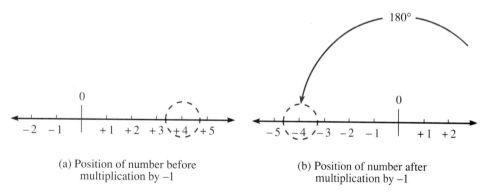

(a) Position of number before multiplication by –1

(b) Position of number after multiplication by –1

Note that the situation shown in Figure 16-3 attaches a geometrical interpretation to the operation of multiplying a real number by -1, namely, rotation of the real number through 180°. It is probably apparent to you that in the process of rotating through 180°, the real number $+4$ must have passed through 90°. Since this is so, the following question might arise: By what number would we need to multiply $+4$ (or any other real number) in order to have it rotate just to the halfway point of 90°? Well, whatever this number is, let's call it N. Multiplication by N twice will obviously get you to 180°. This is simply another way of saying $N \times N = -1$ because each N corresponds to just one 90° rotation.

Therefore, as shown in Figure 16–4, multiplication of $+4$ by N will rotate the number to a new axis located 90° from the real-number line. This new axis is called an *axis for imaginary numbers*, since there is obviously no real number (N) that, when squared, yields -1. That is, there is no *real* solution to the problem:

$$\text{If } N^2 = -1, \quad N = ?$$

Consequently, any number affected by N must also be imaginary, and we plot these new numbers on their own axis.

In order to be consistent with the real-number system, N must have some value such that $N^2 = -1$. After a few minutes of head scratching, you are forced to conclude that the only possible value of N is

$$N = \sqrt{-1}$$

This answer probably occurs to you because you know that

$$N \times N = \sqrt{-1} \times \sqrt{-1} = -1$$

That is,

$$N^2 = -1$$

In mathematics, this imaginary quantity (N) is symbolized by the lowercase letter i (for imaginary) and is often called the *imaginary operator* because of the manner in which it operates on other numbers. In electronics, however, since the letter i stands for the instantaneous value of an alternating current, we use the lowercase letter j instead. Therefore, the **operator j,** as it is called, is given as

$$j = \sqrt{-1}$$

where

$$j^2 = -1$$

In the previous example (Figure 16–4), our new number would be written as $+j4$ or simply $j4$, with the plus sign understood. See Figure 16–5. Note that the operator j always *precedes* the numeral, indicating that the j *operates* on the number, as defined.

If multiplying by j rotates the real number 90° and multiplying twice by j rotates the number by 180°, it should be obvious that multiplying three times by j causes a number to be rotated through 270° to the negative part of the imaginary axis. Finally, multiplying four times by j brings us back to where we started—that is, to the positive real axis. We express these facts mathematically as follows:

$$j = \sqrt{-1}$$
$$j^2 = -1$$
$$j^3 = j^2 j = -j$$
$$j^4 = j^2 j^2 = +1$$

The geometrical interpretation of these facts is shown in Figure 16–6, where we have used the j operator on four real numbers: 3, 4, 5, and 6.

In Figure 16–6, note that rotation of $+5$ counterclockwise by j^3 through 270° to $-j5$ is the same as rotating $+5$ clockwise through 90°. We conclude, therefore, that operation by $-j$ rotates a real number in a *clockwise* direction, and operation by $+j$ rotates the number in a *counterclockwise* direction.

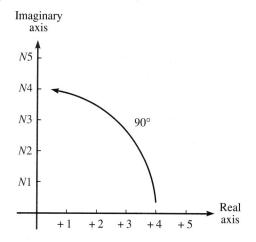

Figure 16–4

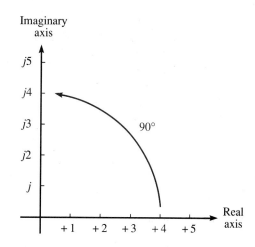

Figure 16–5

Figure 16–6

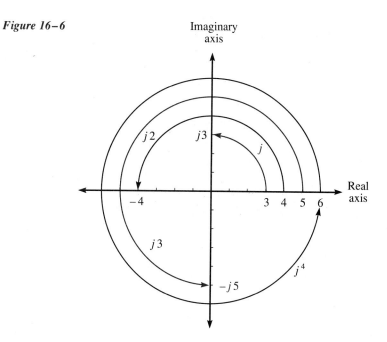

Advanced Applications of Mathematics to Electronics

The significance of the foregoing discussion is that we now have a new set of Cartesian coordinates wherein real numbers are plotted along the horizontal axis (sometimes called the real axis, or *x*-axis), and imaginary numbers are plotted along the vertical axis (sometimes called the imaginary axis, or *y*-axis). This new system of rectangular coordinates is shown in Figure 16–7.

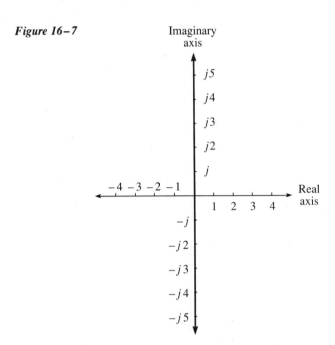

Figure 16–7

If we return to the question posed earlier in this section—that is, what is $\sqrt{-16}$—we now see that the problem may be reduced to

$$\sqrt{-16} = \sqrt{-1} \times \sqrt{16} = j4$$

where the answer, $j4$, is imaginary and is shown in Figure 16–5.

Example 16–1

$\sqrt{-49} = j7$

Example 16–2

$-\sqrt{-16I^2} = -j\sqrt{16I^2} = -j4I$

Exercise Set 16–1

Questions

1. Define the term *imaginary operator* in your own words.
2. The answer to the problem $\sqrt[3]{-27}$ is a *real* number, not an imaginary number. Explain why this is the case for any *odd* root. (*Note:* $\sqrt[3]{-27} = -3$)
3. Why are all *even* roots of negative numbers imaginary numbers?

Complex Algebra and Parallel AC Circuits

4. Explain why the two axes (real and imaginary) are 90° apart rather than some other angle.
5. Figure 16–6 showed that $j^4 = +1$. How would you use this fact to show that $j^{1001} = j^{992}j^9 = +j$?

Problems

1. Using the rectangular coordinate system, show graphically that the real number -3 operated on by j^7 becomes the imaginary number $j3$.
2. Using the rectangular coordinate system, show graphically that the imaginary number $-j4$ becomes the real number -4 when operated on by $-j^5$.
3. Determine the answer to Problem 1 mathematically using the j operator.
4. Determine the answer to Problem 2 mathematically.
5. Use the principle of rationalizing the denominator to show that $1/j = -j$.

Determine the answer to each of the following problems.

6. $\sqrt{-81}$
7. $-\sqrt{-R^2}$
8. $-\sqrt{-5}$
9. $\sqrt{\dfrac{144}{-\pi}}$
10. $3\sqrt{-18}$

16–3 COMPLEX NUMBERS

Now that we have an understanding of what an imaginary number is and why imaginary numbers require a separate axis, we are ready to delve into the concept of complex numbers. By definition, a **complex number** is a number having a real part and an imaginary part. The first question that may occur to you is, On which axis do we plot these numbers? The answer is, Neither.

Look at the point shown as P in Figure 16–8. We note that the point P is three real units to the right of 0 and five imaginary units above it. Note, too, that the position occupied by P is unique. That is, no other number can fit there. Therefore, P is uniquely determined by the coordinates $+3$ and $+j5$. This point P represents a complex number, and its magnitude is $3 + j5$.

Figure 16–8

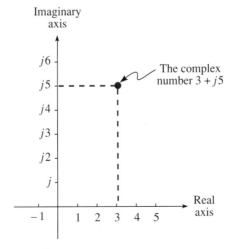

The observant student might inquire, Where did the counterclockwise rotation of 90° caused by j come into play? One way to visualize P as having arrived at the position shown is to consider two real vectors (one 3 units long, and the other 5 units long) placed nose-to-tail, as shown in Figure 16–9. If the vector whose length is 5 units is now affected one time by j, this vector will be rotated 90° counterclockwise, and its head will come to rest at the point P. This is shown in Figure 16–10, where the axes have been labeled with the more customary letters x and y.

Figure 16–9

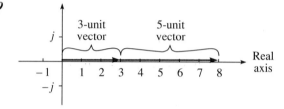

Figure 16–10

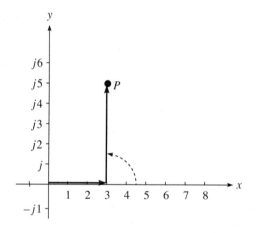

In Figure 16–11, we have reflected the point P in the y-axis, the origin, and the x-axis, producing the points Q, R, and S, respectively. Note that the coordinates of these points indicate the magnitude of the complex numbers and are logically consistent with the definition of the j operator. It is left as an exercise for the student to show, in a manner similar to that used with Figures 16–9 and 16–10, that points Q, R, and S do, in fact, represent the complex numbers indicated.

Before leaving this section, it is important to observe that a pure imaginary number of the form $\pm jb$ may also be considered as a complex number whose real part is 0. Moreover, a single real number may be considered to be a complex number whose imaginary part is 0.

Exercise Set 16–2

Questions

1. How would you define a complex number in your own words?
2. The complex number $a + jb$ does not represent an arithmetic sum. Explain what this means.
3. Explain why the polarities (signs) of each part of a complex number determine the quadrant in which it lies.

Figure 16–11

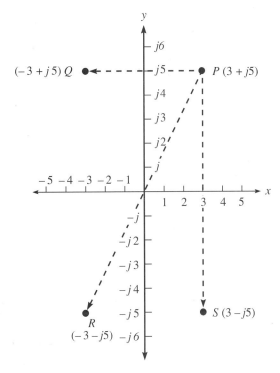

Problems

1. Plot the following complex numbers as points on an xy-coordinate axis system. Clearly label each point.
  a. $P(-6 - j6)$ **b.** $Q(-\pi - j\sqrt{2})$ **c.** $R(1 + j)$
  d. $S(-j12)$ **e.** $T(-\frac{1}{2})$ **f.** $U(5 - j\sqrt{3})$

2. If a straight line is drawn between the origin and the point P in Figure 16–8, find the angle between the line and the x-axis.
3. Find the length of the straight line in Problem 2.
4. Find the length of the straight line drawn between the origin and each point listed in Problem 1.
5. Find the angle between the straight line joining each point in Problem 1 with the origin and the x-axis.

16–4 COMPLEX NUMBERS APPLIED

We have finally arrived at that point in our discussion where we might ask ourselves the all-important question, So what? Admittedly, it may have been difficult to see where all this complex number business is heading. Now, however, we will apply everything that we have been learning.

Recall from our study of series AC circuits in Chapter 15 that you were told to plot resistance along the x-axis and reactance along the y-axis. You were then told to complete the rectangle and find the impedance as the length of the diagonal using the Pythagorean identity

$$Z = \sqrt{R^2 + X^2}$$

What you were actually doing was treating resistance as a *real* quantity, reactance as an *imaginary* number, and impedance as a *complex* number. For example, consider the situation in Figure 16–8 representing the complex number $3 + j5$. We may consider this to represent $3\,\Omega$ of resistance and $5\,\Omega$ of inductive reactance connected in series, as shown in Figure 16–12. The length of the diagonal represents the impedance found by

$$Z = \sqrt{R^2 + X_L^2} = \sqrt{3^2 + 5^2} \approx 5.83\,\Omega$$

Moreover, the phase angle θ is given as

$$\theta = \tan^{-1}\frac{5}{3}$$
$$\approx 59°$$

Figure 16–12

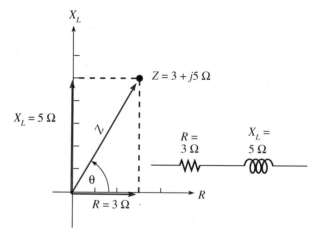

It may have occurred to you, then, that we actually have two methods of representing the situation of a series AC circuit consisting of resistance and inductive reactance:

1. Impedance $= \sqrt{R^2 + X_L^2}$ ohms
2. Impedance $= R + jX_L$ ohms

The first method might be called the *geometrical* representation, whereas the second might be referred to as the *complex* representation. For example, in Figure 16–12, we may represent the circuit in either of these equivalent ways:

1. $Z = \sqrt{3^2 + 5^2} \approx 5.83\,\Omega$
2. $Z = 3 + j5\,\Omega$

In solving AC circuit problems, it is a much simpler process, as will be shown, to specify all impedances using the complex-number form. The reason for this is quite simple. Impedances expressed in complex-number form may be operated upon (i.e., added, subtracted, multiplied, and divided) as ordinary binomials by simply applying the algebraic rules we learned earlier. Such is not the case, however, when the impedances are expressed in the geometric (Pythagorean) manner.

We now show how complex numbers may be manipulated by algebraic means.

16–5 ADDITION AND SUBTRACTION OF COMPLEX NUMBERS

The rule for adding or subtracting complex numbers is simple and probably somewhat obvious.

> **Rule 16–1 Adding or Subtracting Complex Numbers**
> Add (subtract) the real parts and the imaginary parts separately.

Example 16–3 $(3 + j5) + (-2 + j12) = 1 + j17$

Example 16–4 $-(4 - j2) - (5 + j) = (-4 + j2) - (5 + j) = -9 + j$

Example 16–5 $(1500 - j400) + (450 + j185) = 1950 - j215$

From what was said in Chapter 15 concerning the graphical display of reactances and here concerning the graphical representation of complex numbers, the following should now be evident:

1. *Inductive reactance* is always represented as *positive:* that is, $X_L = +jx$.
2. *Capacitive reactance* is always represented as *negative:* that is, $X_C = -jx$.

In Example 16–5, if the two complex numbers represented two impedances in series, then the total series impedance consists of 1950 Ω of resistance and 215 Ω of capacitive reactance. This is shown in Figure 16–13.

Figure 16–13

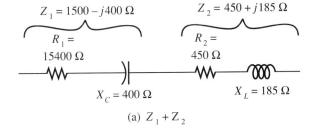

(a) $Z_1 + Z_2$

(b) Equivalent series circuit

16-6 MULTIPLICATION AND DIVISION OF COMPLEX NUMBERS

The multiplication of complex numbers is accomplished by treating them as ordinary binomials.

> **Rule 16-2 Multiplying Complex Numbers**
> Treat the complex numbers as binomials, remembering $j^2 = -1$.

Example 16-6

$$(2 + j4)(-3 + j2) = -6 + j4 - j12 + j^2 8$$
$$= -6 - j8 - 8 \quad \text{Since } j^2 = -1$$
$$= -14 - j8$$

Example 16-7

Find the product of the two impedances $Z_1 = 1.5k - j200 \; \Omega$ and $Z_2 = 1k + j50 \; \Omega$.

Solution

$$(1500 - j200)(1000 + j50) = 1.5 \times 10^6 + j75 \times 10^3 - j200 \times 10^3 - j^2 1 \times 10^4$$
$$= 1.5 \times 10^6 - j125 \times 10^3 + 1 \times 10^4 \quad \text{Since } j^2 = -1$$
$$= 1.51 \times 10^6 - j125 \times 10^3$$

The equivalent circuit consists of a 1.51 MΩ resistance in series with 125 kΩ of capacitive reactance.

Division of complex numbers can be accomplished by treating them as ordinary binomials and rationalizing the denominator using the conjugate of the denominator. The *conjugate* of the complex number $a + jb$ is $a - jb$. The concept of rationalizing the denominator was presented in Chapter 6.

> **Rule 16-3 Dividing Complex Numbers**
> Treat them as ordinary binomials and rationalize the denominator using the conjugate of the denominator.

Example 16-8

$$\frac{2 - j4}{3 + j2} = \frac{2 - j4}{3 + j2} \times \frac{3 - j2}{3 - j2} \quad 3 - j2 \text{ is the conjugate of the denominator.}$$
$$= \frac{6 - j4 - j12 + j^2 8}{9 - j^2 4}$$
$$= \frac{6 - j16 - 8}{9 + 4} \quad \text{Remember, } j^2 = -1.$$
$$= \frac{-2 - j16}{13}$$
$$= -\frac{2}{13} - j\frac{16}{13}$$

Complex Algebra and Parallel AC Circuits

Example 16-9   Find the admittance of $Z = 200 - j42\ \Omega$.

Solution   Recall that admittance Y is the reciprocal of impedance; that is,

$$Y = \frac{1}{Z}$$

$$\frac{1}{200 - j42} = \frac{1}{200 - j42} \times \frac{200 + j42}{200 + j42} = \frac{200 + j42}{200 - j^2 42}$$

$$\approx 4.79 \times 10^{-3} + j1 \times 10^{-3}\ \text{S}$$

Exercise Set 16-3

Questions

1. What is the advantage of treating impedances as complex numbers?
2. In your own words, state the rules for adding or subtracting numbers in complex form.
3. In your own words, state the rules for multiplying complex numbers.
4. In your own words, state the rules for dividing numbers in complex form.
5. State the mathematical relationship between impedance (Z) and admittance (Y).
6. What is the sign of the j operator for inductive and capacitance reactances expressed in complex form?
7. Why are resistances treated as real numbers and reactances as imaginary numbers?
8. In what way is the geometrical (Pythagorean) representation of an impedance the same as the complex representation? How do they differ?

Problems

1. Find the equivalent impedance circuit of the following series-connected impedances:

$$Z_1 = 1.1k - j200\ \Omega$$
$$Z_2 = 5k + 0.9k\ \Omega$$
$$Z_3 = 500 - j855\ \Omega$$

Draw the schematic diagram of the equivalent series impedance, and express its value in complex form.

2. What is the admittance of the equivalent series circuit in Problem 1?

Perform the indicated operation in each of the following problems.

3. $(5 - j7) - (0.44 + j6)$
4. $-(-j11) + (8.5 + j)$
5. $(\frac{1}{3} - j\frac{3}{5}) + (2 + j0.6)$
6. $(100 - j67) - (450 + j22)$
7. $(23 - j40)(5 - j20)$
8. $(-j2)(10 - j2)$
9. $-(-3 - j4)(-5 + j6)$
10. $(0.25 + j0.7)(0.38 - j1.007)$
11. $\dfrac{-j}{2 + j4}$
12. $\dfrac{1}{200 - j65}$
13. $\dfrac{3 - j7}{0.5 + j2}$
14. $\dfrac{1000}{25 - j50}$
15. $(4 - j8) + \dfrac{2 + j5}{4 - j4}$

16-7 THE POLAR FORM OF A COMPLEX NUMBER

As you have no doubt observed in completing the last set of problems, the division of one complex number by another can be quite tedious using the algebraic method. Fortunately, complex numbers can be expressed in another form that makes the division a simple matter using a hand-held digital calculator. Before we explain this procedure, however, we will take time out to explain the new form of the complex number and its relationship to the familiar form with which we have been dealing up to this point.

Figure 16–12 is repeated here as Figure 16–14. Note, as before, the total impedance of the circuit can be expressed in either the geometric form or the complex form. The complex form $R + jX$ is often referred to as the **rectangular form** of a complex number because its pictorial representation resembles a rectangle, as seen in Figure 16–14. Notice, however, that the impedance of the series circuit is also completely specified in terms of the resultant phasor shown as Z in the figure and the phase angle θ. In other words, the total impedance of 5.83 Ω and the phase angle of 59° also determine the circuit

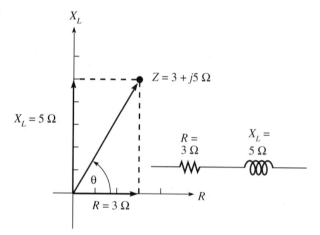

Figure 16–14

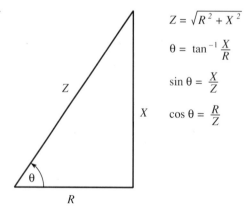

Figure 16–15

$$Z = \sqrt{R^2 + X^2}$$

$$\theta = \tan^{-1}\frac{X}{R}$$

$$\sin\theta = \frac{X}{Z}$$

$$\cos\theta = \frac{R}{Z}$$

Complex Algebra and Parallel AC Circuits

impedance. Therefore, we may write the impedance as 5.83 ∠59° Ω. This form of the impedance is called the **polar form** because it uses polar coordinates—that is, a radius vector (phasor) and an angle. In other words, the impedance may be written as $Z \angle \theta$ Ω. We will see that multiplication and especially division are handled much easier when the numbers are expressed in polar form. First, though, we show the simple mathematical relationship between the two forms. Refer to Figure 16–15.

From the right triangle shown in Figure 16–15, we make the following observations:

Rule 16–4

1. Given a complex number in rectangular form, $R + jX$, we can convert to the polar form $Z \angle \theta$ using the following Pythagorean and trigonometric relationships:

$$Z = \sqrt{R^2 + X^2} \quad \text{and} \quad \theta = \tan^{-1}\frac{X}{R}$$

2. Given a number in polar form, $Z \angle \theta$, we can convert to rectangular form using the trigonometric relationships

$$R = Z \cos \theta \quad \text{and} \quad X = Z \sin \theta$$

These formulas follow directly from the basic relationships shown in Figure 16–15 for the right triangle.

Example 16–10

Convert the rectangular form of an impedance, $Z = 100 - j35$, to polar form.

Solution From Rule 16–1, $Z = \sqrt{100^2 + 35^2} \approx 106$ and $\theta = \tan^{-1} X/R = \tan^{-1} 35/100 \approx 19.29°$. Therefore, $Z = 106 \angle 19.29°$ Ω.

Example 16–11

Convert the polar form of an impedance, $Z = 208 \angle -44°$ Ω, to rectangular form.

Solution $R = Z \cos \theta = 208 \cos -44° \approx 149.6$ Ω and $X = Z \sin \theta = 208 \sin -44° \approx -144.5$ Ω. Therefore, $Z = 149.6 - j144.5$ Ω.

Exercise Set 16–4

Questions

1. Explain what is meant by the rectangular form of a complex number.
2. Explain what is meant by the polar form of a complex number.
3. Write out all the right-triangle relationships required to convert between the three forms of a series impedance.

Problems

Using a coordinate axis system, plot each of the complex numbers.
1. $100 \angle -55°$
2. $55 - j20$
3. $67 \angle 30°$
4. $120 + j90$
5. In Problems 1 through 4, convert each complex number to the other form using Rule 16–1.

16-8 CALCULATOR CONVERSIONS

In an earlier section, we showed that both multiplication and division of complex numbers in rectangular form are possible using simple, though often tedious, algebraic methods. Now that we have developed another form of the complex number—namely, the polar form—we can now state the simple rules for multiplying or dividing these numbers.

> **Rule 16-5**
> 1. To multiply two numbers in polar form, simply multiply the magnitudes and add the angles:
> $$M \angle A° \times N \angle B° = MN \angle A + B°$$
> 2. To divide two numbers in polar form, simply divide the magnitudes and subtract the angles:
> $$\frac{M \angle A°}{N \angle B°} = \frac{M}{N} \angle A - B°$$

Example 16-12

$26 \angle 80° \times 3 \angle -20° = 78 \angle 60°$

Example 16-13

$$\frac{-41 \angle -63°}{16 \angle 22°} \approx -2.56 \angle -85°$$

Note, however, that numbers in polar form cannot be added or subtracted. In order to perform these operations, we must convert numbers in polar form into their rectangular equivalents. Since such back-and-forth conversions are common in the solution of practical AC parallel circuits, we now spend some time developing a degree of facility with these conversions on the digital calculator.

While no particular scientific calculator is recommended, many use the following keystrokes in making rectangular-to-polar conversions. As mentioned in Chapter 1, you must carefully read the instruction booklet that comes with your particular calculator.

Example 16-14

Convert $12 - j42$ to polar form.

Solution The following keystrokes and displays are typical:

Keystrokes	Display
[1] [2]	12
[INV]	12
[R→P]	12
[4] [2] [−]	−42
[=]	43.68065933
[X↔Y]	−74.0546041

Therefore, $12 - j42 \approx 43.7 \angle -74°$.

Complex Algebra and Parallel AC Circuits   381

Example 16-15 Convert $-56 \angle 27°$ into rectangular form.

Solution

Keystrokes	Display
5 6 −	−56
INV	−56
P→R	−56
2 7	27
=	−49.89636536
X↔Y	−25.42346799

Therefore, $-56 \angle 27° \approx -49.9 - j25.4$.

Exercise Set 16-5

Questions

1. In your own words, state the rules for multiplying complex numbers in polar form.
2. In your own words, state the rules for dividing complex numbers in polar form.
3. Although complex numbers in rectangular form may be either multiplied or divided, explain what the principal advantage is in first converting such numbers to polar form.

Problems

Use your calculator to convert each of the following complex numbers into the other form.

1. $-\pi - j88$
2. $1 \angle 90°$
3. $254 + j275$
4. $0.06 \angle -30°$
5. $-18.65 + j104.59$
6. $\sqrt{2} \angle -45°$
7. $12408.66 + j88.0982$
8. $-\frac{1}{3} - j\frac{5}{8}$
9. $30 \angle -75°$
10. $-28 \angle -200°$
11. $9 - j9$
12. $0.88 - j0.505$

16-9 COMPLEX PARALLEL IMPEDANCES

In Chapter 7, we learned how to determine the total resistance of two or more resistors in parallel by adding their individual conductances and then taking the reciprocal to find the total resistance. Recall that the process used gave us the general formula

$$R_t = \frac{1}{1/R_1 + 1/R_2 + \cdots + 1/R_n}$$

In parallel AC circuits, the same process holds true, but we use admittance Y instead of conductance. In other words,

$$Z_t = \frac{1}{1/Z_1 + 1/Z_2 + \cdots + 1/Z_n}$$

where $Z_1, Z_2, \ldots$ are complex numbers. Recall that we had mentioned earlier that admittance and impedance are reciprocally related as follows:

$$Y = \frac{1}{Z} \quad \text{and} \quad Z = \frac{1}{Y}$$

Therefore, in the solution of parallel AC circuits, the first thing we must do is to find the reciprocals of all the parallel branch impedances. That is, we must have these impedances changed into polar form so that we can divide 1 by each one. Now, perhaps, you can appreciate the reason why we spent so much time learning how to change from rectangular to polar form using the calculator. Subsequent steps in the solution involve further conversions to rectangular form and then back to polar form again. Since this gets a little complicated using just words, we will first work through one example completely; then an algorithm will be outlined for the general solution of any parallel AC circuit problem.

Example 16-16

Consider the circuit shown at the beginning of this chapter. See Figure 16-16.

Figure 16-16

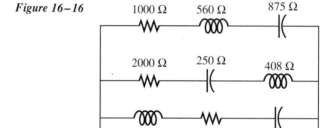

The first thing we must do is list the branch impedances as complex numbers in rectangular form.

$$Z_1 = 1000 + j560 - j875 \ \Omega$$
$$Z_2 = 2000 + j408 - j250 \ \Omega$$
$$Z_3 = 1500 + j167 - j312 \ \Omega$$

The next step is to convert each impedance into its polar form:

$$Z_1 = 1048 \ \angle -17.48° \ \Omega$$
$$Z_2 = 2006 \ \angle 4.52° \ \Omega$$
$$Z_3 = 1507 \ \angle -5.52° \ \Omega$$

The reciprocal of each of these polar-form impedances can now be taken to obtain the admittance in polar form:

$$Y_1 = \frac{1 \ \angle 0°}{1048 \ \angle -17.48°} \approx 954.2 \times 10^{-6} \ \angle 17.48° \ \text{S}$$

$$Y_2 = \frac{1 \ \angle 0°}{2006.23 \ \angle 4.52°} \approx 498.4 \times 10^{-6} \ \angle -4.52° \ \text{S}$$

$$Y_3 = \frac{1 \ \angle 0°}{1507 \ \angle -5.52°} \approx 663.6 \times 10^{-6} \ \angle 5.52° \ \text{S}$$

Complex Algebra and Parallel AC Circuits

In the next step, we need to add these individual branch admittances. However, numbers in polar form cannot simply be added, so we convert them to rectangular form in order to do the addition.

$$Y_1 = 910.1 \times 10^{-6} + j286.6 \times 10^{-6} \text{ S}$$
$$Y_2 = 496.8 \times 10^{-6} - j39.3 \times 10^{-6} \text{ S}$$
$$Y_3 = 660.5 \times 10^{-6} + j63.8 \times 10^{-6} \text{ S}$$

Therefore,

$$Y_t = Y_1 + Y_2 + Y_3 \approx 2.067 \times 10^{-3} + j311.1 \times 10^{-6} \text{ S}$$

We may now convert this total admittance to polar form in order to find the total impedance given by $Z_t = 1/Y_t$.

$$Y_t = 2.09 \times 10^{-3} \angle 8.56° \text{ S}$$

The impedance, then, is

$$Z_t = \frac{1}{Y_t} \approx 478.5 \angle -8.56° \text{ }\Omega$$

Finally, on changing this to rectangular form, the equivalent series circuit is revealed:

$$Z_t = 473 - j71.2 \text{ }\Omega$$

The equivalent series circuit is shown in Figure 16–17.

Figure 16–17

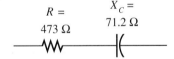

$R = 473 \text{ }\Omega$ $X_C = 71.2 \text{ }\Omega$

The following steps are listed as an algorithm for the general solution of such parallel circuits.

Rule 16–6

1. Convert each branch impedance to polar form: $Z \angle \theta°$.
2. Find the admittance (Y) of each branch by taking reciprocals: $Y = (1 \angle 0°)/(Z \angle \theta)$.
3. Convert each polar admittance into rectangular form: $Y \angle \theta = G \pm jB$.*
4. Find the total admittance Y_t by adding the individual admittances in step 3.
5. Convert the total admittance to polar form: $Y_t \angle \theta°$.
6. Take the reciprocal of the total admittance in order to find the total impedance: $Z_t = (1 \angle 0°)/(Y_t \angle \theta°)$.
7. Change Z_t to rectangular form to reveal the equivalent series circuit: $Z_t = R \pm jX \text{ }\Omega$.

---

*The real part of admittance is conductance (G) and the imaginary part is called susceptance (B). Compare this with impedance, where the real part is resistance (R) and the imaginary part is called reactance (X).

Exercise Set 16-6

Questions

1. Using your own words, write the steps taken in the solution of the general parallel AC circuit.
2. Define the following terms:
   a. Admittance
   b. Susceptance
3. For your particular calculator, write out the steps for converting from rectangular to polar and vice versa. Use the format shown below:

$R \to P$		$P \to R$	
KEYSTROKE	DISPLAY	KEYSTROKE	DISPLAY

Problems

1. The following two series impedances are connected in parallel: $Z_1 = 1000 - j159 \, \Omega$ and $Z_2 = 42 + j63 \, \Omega$. Find each of the following.
   a. The equivalent series circuit
   b. The total impedance
   c. The phase angle
   d. The total line current if the applied voltage is 12 V sinusoidal
   e. The actual power dissipated by the circuit under the same conditions in part (d)
2. If $Z_1 = 86 - j80 \, \Omega$ and $Z_2 = 92 - j104 + j77 \, \Omega$ are connected in parallel, find each of the following.
   a. The equivalent series circuit
   b. The total impedance
   c. The phase angle
   d. The total line current if the applied voltage is 120 V, 60 Hz
   e. The actual power dissipated by the circuit under the same conditions in part (d)
3. Find the power dissipated in the circuit shown in Figure 16–18.

Complex Algebra and Parallel AC Circuits

Figure 16–18

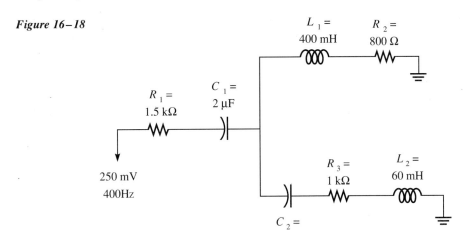

4. If the impedances $Z_1 = 148 \angle 5.1° \, \Omega$ and $Z_2 = 48.7 \angle 42.8° \, \Omega$ are connected in parallel, find the total impedance of the circuit.
5. For the circuit shown in Figure 16–19, derive a general formula for the solution for such series-parallel circuits.

   Hint: From Chapter 7, recall the formula for two resistances in parallel.

Figure 16–19

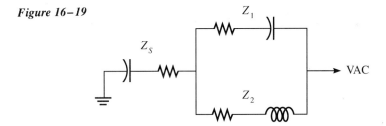

6. In the circuit shown in Figure 16–20, draw the equivalent circuit found in the box labeled $Z_1 = ?$.

Figure 16–20

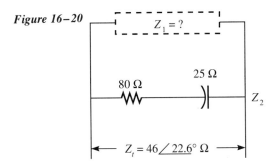

7. The following impedances are connected in parallel: $Z_1 = 112 \angle 20°$ Ω and $Z_2 = 77.7 - j42$ Ω. Find the equivalent series circuit.
8. If 32 V sinusoidal are applied to the input terminals of the circuit in Problem 6, find the total line current and the total power dissipated in the circuit.
9. If an impedance of $55.8 \angle -26.4°$ Ω is connected in series with the circuit shown in Problem 6, find the total impedance of the series-parallel combination.
10. A certain series circuit has an admittance of

$$Y = 0.00032 + j0.00024 \text{ S}$$

If the frequency of the applied sinusoidal voltage is 300 MHz, find the value of the inductance required to "tune out" (cancel) the susceptance.

11. The following BASIC computer program may be used to convert an impedance in rectangular form to polar form.

```
10 PRINT"Enter the value of the resistance R."
20 INPUT R
30 PRINT"Enter the value of the reactance X, paying"
40 PRINT"careful attention to the sign (+ or -)."
50 INPUT X
60 LET Z=SQR(R^2+X^2)
70 LET A=ATN(X/R)*57.2957795l
80 PRINT"The value of the impedance and angle are";Z;A
90 END
```

Write a BASIC program for finding the impedance of a two-branch parallel AC circuit having R and X in both branches.

Use the BASIC program developed in Problem 11 to calculate the following parallel AC circuits.
12. $Z_1 = 120 + j847$ Ω and $Z_2 = 86 - j80$ Ω
13. $Z_1 = 82 - j150$ Ω and $Z_2 = 250 + j420$ Ω
14. $Z_1 = 38 + j71$ Ω and $Z_2 = 104 - j66$ Ω
15. Use the BASIC program developed in Problem 11 to verify your answer to Problems 6 and 7.
16. Write a BASIC program to solve the circuit in Figure 16–1.

Key Terms

imaginary number
operator j

complex number
rectangular form

polar form

Important Formulas and Procedures

Operating with Complex Numbers

Rule 16–1 Adding or Subtracting Complex Numbers
Add (subtract) the real parts and the imaginary parts separately.

Rule 16–2 Multiplying Complex Numbers
Treat the complex numbers as binomials, remembering $j^2 = -1$.

Rule 16–3 Dividing Complex Numbers
Treat the complex numbers as binomials and rationalize the denominator using the conjugate of the denominator.

Complex Algebra and Parallel AC Circuits

Relationship Between Complex Numbers in Rectangular and Polar Form

Rule 16–4

1. Given a number in rectangular form, $R + jX$, we can convert to the polar form $Z \angle \theta$ using the following Pythagorean and trigonometric relationships:

$$Z = \sqrt{R^2 + X^2} \quad \text{and} \quad \theta = \tan^{-1}\frac{X}{R}$$

2. Given a number in polar form, $Z \angle \theta$, we can convert it to rectangular form using the trigonometric relationships

$$X = Z \sin \theta \quad \text{and} \quad R = Z \cos \theta$$

Figure 16–21

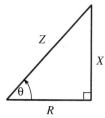

Multiplying and Dividing Numbers in Complex Form

Rule 16–5

1. To multiply two numbers in polar form, simply multiply the magnitudes and add the angles; for example,

$$M \angle A° \times N \angle B° = MN \angle A + B°$$

2. To divide two numbers in polar form, simply divide the magnitudes and subtract the angles:

$$\frac{M \angle A°}{N \angle B°} = \frac{M}{N} \angle A - B°$$

The General Solution of a Parallel AC Circuit

Rule 16–6

1. Convert each branch impedance to polar form: $Z \angle \theta°$.
2. Find the admittance (Y) of each branch by taking reciprocals: $Y = (1 \angle 0°)/(Z \angle \theta°)$.
3. Convert each polar admittance into rectangular form: $Y \angle \theta = G + jB$.
4. Find the total admittance Y_t by adding the individual admittances in step 3.
5. Convert the total admittance to polar form: $Y_t \angle \theta°$.
6. Take the reciprocal of the total admittance in order to find the total impedance: $Z_t = (1 \angle 0°)/(Y_t \angle \theta°)$.
7. Change Z_t to rectangular form to reveal the equivalent series circuit: $Z_t = R \pm jX\ \Omega$.

APPENDIXES

A
Conversion Factors

B
Resistor Color Codes

C
A Proof of the Theorem of Pythagoras

D
Powers of Ten

E
Mathematical Symbols

F
American Wire Gauge Conductor Sizes

G
The Greek Alphabet

H
Review of Boolean Algebra and Karnaugh Maps

I
Rules Review

J
Answers to Selected Problems

A Conversion Factors

Mass	
1 gram	10^{-3} kilogram
	6.854×10^{-5} slug
1 kilogram	1000 grams
	0.06854 slug
1 slug	14.59 kilograms
	14,590 grams
1 metric ton	1000 kilograms

Velocity	
1 foot/minute	0.3048 meter/minute
	0.011 364 mile/hour
1 foot/second	1097 kilometers/hour
	18.29 meters/minute
	0.6818 mile/hour
1 kilometer/hour	3281 feet/hour
	54.68 feet/minute
	0.6214 mile/hour
1 kilometer/minute	3281 feet/minute
	37.28 miles/hour
1 knot	6076 feet/hour
	101.3 feet/minute
	1.852 kilometers/hour
	30.87 meters/minute
	1.151 miles/hour
1 meter/hour	3.281 feet/hour
1 mile/hour	1.467 feet/second
	1.609 kilometers/hour

Area	
1 acre	4047 square meters
	43 560 square feet
1 are	0.024 71 acre
	1 square dekameter
	100 square meters
1 hectare	2.471 acres
	100 ares
	10 000 square meters
1 square foot	144 square inches
	0.092 90 square meter
1 square inch	6.452 square centimeters
1 square kilometer	247.1 acres
1 square meter	10.76 square feet
1 square mile	640 acres
	2.788×10^7 square feet
	2.590 square kilometers

Volume	
1 board-foot	144 cubic inches
1 bushel (U.S.)	1.244 cubic feet
	35.24 liters
1 cord	128 cubic feet
	3.625 cubic meters
1 cubic foot	7.481 gallons (U.S. liquid)
	28.32 liters
1 cubic inch	0.01639 liter
	16.39 milliliters
1 cubic meter	35.31 cubic feet
	10^6 cubic centimeter
1 cubic millimeter	6.102×10^{-5} cubic inch
1 cubic yard	27 cubic feet
	0.7646 cubic meter
1 gallon (imperial)	277.4 cubic inches
	4.546 liters
1 gallon (U.S. liquid)	231 cubic inches
	3.785 liters
1 kiloliter	35.31 cubic feet
	1.000 cubic meter
	1.308 cubic yards
	220 imperial gallons
1 liter	10^3 cubic centimeters
	10^6 cubic millimeters
	10^{-3} cubic meter
	61.02 cubic inches

Angles	
1 degree	60 minutes
	0.01745 radian
	3600 seconds
	2.778×10^{-3} revolution
1 minute of arc	0.01667 degree
	2.909×10^{-4} radian
	60 seconds
1 radian	0.1592 revolution
	57.296 degrees
	3438 minutes
1 second of arc	2.778×10^{-4} degree
	0.01667 minute

Length	
1 angstrom	1×10^{-10} meter
	1×10^{-4} micrometer (micron)
1 centimeter	10^{-2} meter
	0.3937 inch
1 foot	12 inches
	0.3048 meter
1 inch	25.4 millimeters
	2.54 centimeters
1 kilometer	3281 feet
	0.5400 nautical mile
	0.6214 statute mile
	1094 yards
1 light-year	9.461×10^{12} kilometers
	5.879×10^{12} statute miles
1 meter	10^{10} angstroms
	3.281 feet
	39.37 inches
	1.094 yards
1 micron	10^4 angstroms
	10^{-4} centimeter
	10^{-6} meter
1 nautical mile (International)	8.439 cables
	6076 feet
	1852 meters
	1.151 statute miles
1 statute mile	5280 feet
	8 furlongs
	1.609 kilometers
	0.8690 nautical mile
1 yard	3 feet
	0.9144 meter

Pressure

1 atmosphere	1.013 bars
	14.70 pounds/square inch
	760 torrs
	101 kilopascals
1 bar	10^6 baryes
	14.50 pounds-force/square inch
1 barye	10^{-6} bar
1 inch of mercury	0.033 86 bar
	70.73 pounds/square foot
1 pascal	1 newton/square meter
1 pound/square inch	0.068 03 atmosphere

Power

1 British thermal unit/hour	0.2929 watt
1 Btu/pound	2.324 joules/gram
1 Btu-second	1.414 horsepower
	1.054 kilowatts
	1054 watts
1 horsepower	42.44 Btu/minute
	550 footpounds/second
	746 watts
1 kilowatt	3414 Btu/hour
	737.6 footpounds/second
	1.341 horsepower
	10^3 joules/second
	999.8 internal watt
1 watt	44.25 footpounds/minute
	1 joule/second

Force

1 dyne	10^{-5} newton
1 newton	10^5 dynes
	0.2248 pound
	3.597 ounces
1 pound	4.448 newtons
	16 ounces
1 ton	2000 pounds

Energy	
1 British thermal unit	1054 joules
	1054 wattseconds
1 foot-pound	1.356 joules
	1.356 newtonmeters
1 joule	0.7376 foot-pound
	1 wattsecond
1 kilowatthour	3410 British thermal units
	1.341 horsepowerhours
1 newtonmeter	0.7376 footpounds
1 watthour	3.414 British thermal units
	2655 footpounds
	3600 joules

B Resistor Color Codes

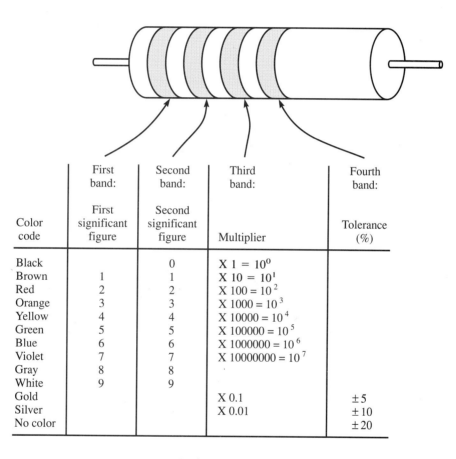

Color code	First band: First significant figure	Second band: Second significant figure	Third band: Multiplier	Fourth band: Tolerance (%)
Black		0	$\times 1 = 10^0$	
Brown	1	1	$\times 10 = 10^1$	
Red	2	2	$\times 100 = 10^2$	
Orange	3	3	$\times 1000 = 10^3$	
Yellow	4	4	$\times 10000 = 10^4$	
Green	5	5	$\times 100000 = 10^5$	
Blue	6	6	$\times 1000000 = 10^6$	
Violet	7	7	$\times 10000000 = 10^7$	
Gray	8	8		
White	9	9		
Gold			$\times 0.1$	± 5
Silver			$\times 0.01$	± 10
No color				± 20

C A Proof of the Theorem of Pythagoras

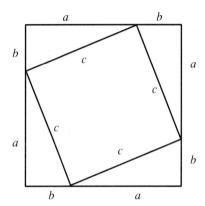

1. The area of the large square is $(a + b)^2$.
2. The area of the small square is c^2.
3. The difference between the areas of these two squares is the area of the four little triangles,

$$4 \times \frac{1}{2}ab, \text{ or } 2ab$$

4. A mathematical expression relating these facts is given by

$$(a + b)^2 - c^2 = 4\left(\frac{1}{2}ab\right)$$
$$a^2 + 2ab + b^2 - c^2 = 2ab$$
$$a^2 + 2ab - 2ab + b^2 = c^2$$
$$\text{Therefore, } a^2 + b^2 = c^2$$

D Powers of Ten

$0.000\,000\,000\,000\,000\,001 = 10^{-18}$ = ten to the negative *eighteenth* power = atto   a
$0.000\,000\,000\,000\,001 = 10^{-15}$ = ten to the negative *fifteenth* power = femto f
$0.000\,000\,000\,001 = 10^{-12}$ = ten to the negative *twelfth* power = pico   p
$0.000\,000\,001 = 10^{-9}$ = ten to the negative *ninth* power = nano   n
$0.000\,001 = 10^{-6}$ = ten to the negative *sixth* power = micro μ
$0.001 = 10^{-3}$ = ten to the negative *third* power = milli   m
$1 = 10^{0}$ = ten to the *zero* power = unit
$1\,000 = 10^{3}$ = ten to the *third* power = kilo   k
$1\,000\,000 = 10^{6}$ = ten to the *sixth* power = mega M
$1\,000\,000\,000 = 10^{9}$ = ten to the *ninth* power = giga   G
$1\,000\,000\,000\,000 = 10^{12}$ = ten to the *twelfth* power = tera   T
$1\,000\,000\,000\,000\,000 = 10^{15}$ = ten to the *fifteenth* power = peta   P
$1\,000\,000\,000\,000\,000\,000 = 10^{18}$ = ten to the *eighteenth* power = exa   E

E Mathematical Symbols

	× or ·	multiplied by
	÷ or /	divided by
	+	positive, plus, add, OR
	−	negative, minus, subtract
	±	positive or negative, plus or minus
	∓	negative or positive, minus or plus
= or ∷		equals
	≡	identity
≈ or ≅		is approximately equal to
	≠	does not equal
	>	is greater than
	≫	is much greater than
	<	is less than
	≪	is much less than
	≧	greater than or equal to
	≦	less than or equal to
	∴	therefore
	∠	angle
	⊥	perpendicular to
	∥	parallel to
	$|n|$	absolute value of n
	△	increment of
	%	percent
	∝	is proportional to

F American Wire Gauge Conductor Sizes

AWG Number	Diameter (mils)*	Ohms per 1000 ft†	AWG Number	Diameter (mils)*	Ohms per 1000 ft†
0000	460.0	0.04901	19	35.89	8.051
000	409.6	0.06180	20	31.96	10.15
00	364.8	0.07793	21	28.46	12.80
0	324.9	0.09827	22	25.35	16.14
1	289.3	0.1239	23	22.57	20.36
2	257.6	0.1563	24	20.10	25.67
3	229.4	0.1970	25	17.90	32.37
4	204.3	0.2485	26	15.94	40.81
5	181.9	0.3133	27	14.20	51.47
6	162.0	0.3951	28	12.64	64.90
7	144.3	0.4982	29	11.26	81.83
8	128.5	0.6282	30	10.03	103.2
9	114.4	0.7921	31	8.928	130.1
10	101.9	0.9989	32	7.950	164.1
11	90.74	1.260	33	7.080	206.9
12	80.81	1.588	34	6.305	260.9
13	71.96	2.003	35	5.615	329.0
14	64.08	2.525	36	5.000	414.8
15	57.07	3.184	37	4.453	523.1
16	50.82	4.016	38	3.965	659.6
17	45.26	5.064	39	3.531	831.8
18	40.30	6.385	40	3.145	1049.0

*1 mil—0.001 inch.
†Copper at 20°C.

G The Greek Alphabet

Name	Capital	Lower-case	Commonly Used to Designate
Alpha	A	α	Angles, area, coefficients
Beta	B	β	Angles, flux density, coefficients
Gamma	Γ	γ	Conductivity
Delta	Δ	δ	Variation, density
Epsilon	E	ε	Base of natural logarithms
Zeta	Z	ζ	Uppercase: impedance
Eta	H	η	Lowercase: efficiency
Theta	Θ	θ	Angles
Iota	I	ι	
Kappa	K	κ	Dielectric constant, susceptibility
Lambda	Λ	λ	Lowercase: wavelength
Mu	M	μ	Lowercase: micro, amplification factor, permeability
Nu	N	ν	Reluctivity
Xi	Ξ	ξ	
Omicron	O	o	
Pi	Π	π	Lowercase: ratio of circumference to diameter ≈ 3.1416
Rho	P	ρ	Lowercase: resistivity
Sigma	Σ	σ	Capital: summation
Tau	T	τ	Time constant
Upsilon	Υ	υ	
Phi	Φ	φ	Magnetic flux, phase angle
Chi	X	χ	
Psi	Ψ	ψ	Dielectric flux, phase difference
Omega	Ω	ω	Capital: ohms; lowercase: angular velocity

H Review of Boolean Algebra and Karnaugh Maps

The following information is provided for the purpose of reviewing theorems of Boolean algebra and simplifying Boolean expressions using Karnaugh (pronounced "kar-nō") maps. Since either of these topics could easily fill a complete book, only the more salient aspects are covered here. Other resources* are available for the interested reader.

There are over a dozen versions of Karnaugh map (K-map) configurations in existence. And although individual differences do exist in terms of format, all K-maps of a given expression produce identical results. The version we illustrate here was chosen because it embodies most of the major principles of all map strategies.

Since the purpose of Karnaugh maps is to simplify Boolean expressions, it is best that we begin our brief discussion with a review of Boolean algebra.

BASIC LOGICAL OPERATIONS

In the following definitions of the three fundamental logical operations, 1 and 0 are the two binary states true-false, on-off, yes-no, and the like.

1. The basic logical AND operation between two Boolean variables A and B is shown in the following truth table.

Input		Output
A	B	$A \cdot B$
0	0	0
0	1	0
1	0	0
1	1	1

---

*For example, see L. Nashelsky, *Introduction to Digital Technology*, 3d ed. New York: John Wiley, 1983.
W. Streib, *Digital Circuits*. S. Holland, Ill: Goodheart-Willcox, 1989.

2. The basic logical OR operation between two Boolean variables A and B is shown in the following truth table.

Input		Output
A	B	$A + B$
0	0	0
0	1	1
1	0	1
1	1	1

3. The logical INVERSE operation changes the state of the Boolean variable A to $\overline{A}$ (read as "A-not").

Input	Output
A	$\overline{A}$
0	1
1	0

ELECTRONIC LOGIC GATES

Each logical operation is performed with electronic devices called *gates*. Gates are usually ICs that may employ active as well as passive devices and components. Gates made from discrete components are practically non-existent. The symbols for the three logical gates are shown in Figure H–1.

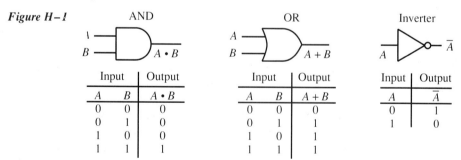

Figure H–1

Although only two inputs are shown for the AND and OR gates, there is no theoretical limit. However, only one input is allowed for the INVERTER. Each gate has only *one* output.

THEOREMS OF BOOLEAN ALGEBRA

The operations between Boolean variables can often be simplified by applying a few important theorems and postulates. Table H–1 presents these theorems; Table H–2 presents postulates.

Review of Boolean Algebra and Karnaugh Maps

Table H–1

Theorem
Commutative law $A + B = B + A$ $A \cdot B = B \cdot A$
Associative law $(A + B) + C = A + (B + C)$ $(A \cdot B) \cdot C = A \cdot (B \cdot C)$
Distributive law $A \cdot (B + C) = A \cdot B + A \cdot C$ $A + (B \cdot C) = (A + B) \cdot (A + C)$
Identity law $A + A = A$ $A \cdot A = A$
Negation law $\overline{(\overline{A})} = \overline{A}$ $\overline{(\overline{A})} = A$
Redundance law $A + A \cdot B = A$ $A \cdot (A + B) = A$ $0 + A = A$ $1 \cdot A = A$ $1 + A = 1$ $0 \cdot A = 0$ $\overline{A} + A = 1$ $\overline{A} \cdot A = 0$ $A + \overline{A} \cdot B = A + B$ $A \cdot (\overline{A} + B) = A \cdot B$
De Morgan's theorem $\overline{(A + B)} = \overline{A} \cdot \overline{B}$ $\overline{(A \cdot B)} = \overline{A} + \overline{B}$

Table H–2

Postulate
$0 \cdot 0 = 0$ $1 + 1 = 1$ $0 + 0 = 0$ $1 \cdot 1 = 1$ $1 \cdot 0 = 0 \cdot 1 = 0$ $1 + 0 = 0 + 1 = 1$

KARNAUGH MAPPING

A Karnaugh map is a mechanical technique used to simplify Boolean expressions. For purposes of brevity, only two- and three-variable K-maps are discussed.

Basically, the format used for the map follows from the fact that a Boolean expression in n variables creates a matrix of 2^n squares. For examples, a two-variable expression creates a map of $2^2 = 4$ squares, three variables produce $2^3 = 8$ squares, and so forth.

Each square represents a unique AND combination of the variables. Only AND operations can be represented *within* a given square. Two squares are required for ORing the variables. For example, Figure H–2 shows a map for a two-variable expression.

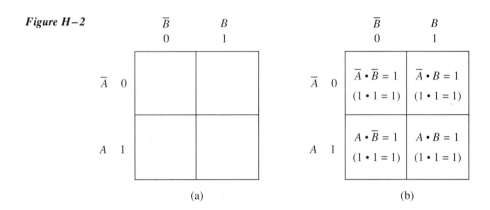

Figure H–2

Part (b) of the figure shows the ANDed operations of each combination of variables. Note that each combination indicates the logical state wherein the given Boolean expression equals 1. That is, the logical product of the two variables is always $1 \cdot 1 = 1$, as shown. Since this is always the case, a 1 placed in any individual square represents the logical product of the variables in question. For example, Figure H–3a represents the logical expression

$$\overline{A} \cdot \overline{B} + A \cdot B$$

Whenever squares appear adjacent (either horizontally or vertically), however, the resulting Boolean expression can be simplified by removing any complementary variables. Whatever is left is the simplified expression. For example, Figure H–3b shows the Boolean expression

$$A \cdot \overline{B} + A \cdot B$$

If we remove B and its complement, the remaining expression becomes

$$A + A = 1 = A$$

That is,

$$A \cdot \overline{B} + A \cdot B = A$$

As another example, Figure H–4 shows three adjacent squares representing the expression

$$\overline{A} \cdot \overline{B} + A \cdot \overline{B} + A \cdot B$$

By striking out all the complements, we are left with the simplified expression

$$\overline{B} + A$$

In a three-variable Karnaugh map, there are $2^3 = 8$ squares, as shown in Figure H–5. Figure H–5b shows the ANDed expressions. The logical state combinations of 0 and 1 must be as shown in order to simplify an expression by grouping of adjacent squares.

Review of Boolean Algebra and Karnaugh Maps

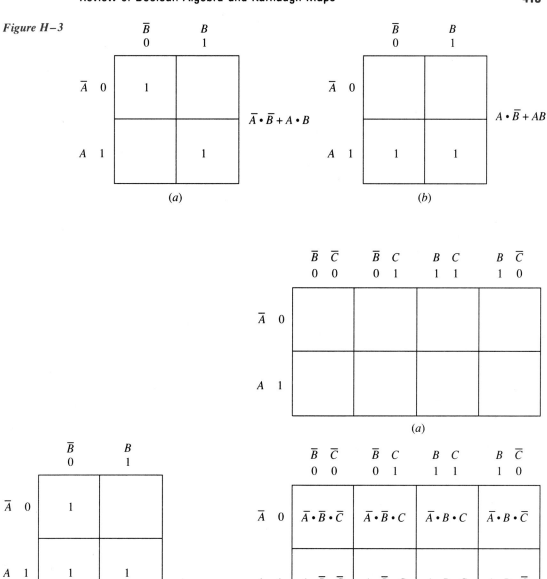

Figure H–3

$\overline{A} \cdot \overline{B} + A \cdot \overline{B} + AB$

Figure H–4

Figure H–5

Any group of four adjacent squares simplifies to a single-variable term. Any group of two adjacent squares simplifies to a two-variable term. A single square represents a three-variable term. See the examples in Figure H–6.

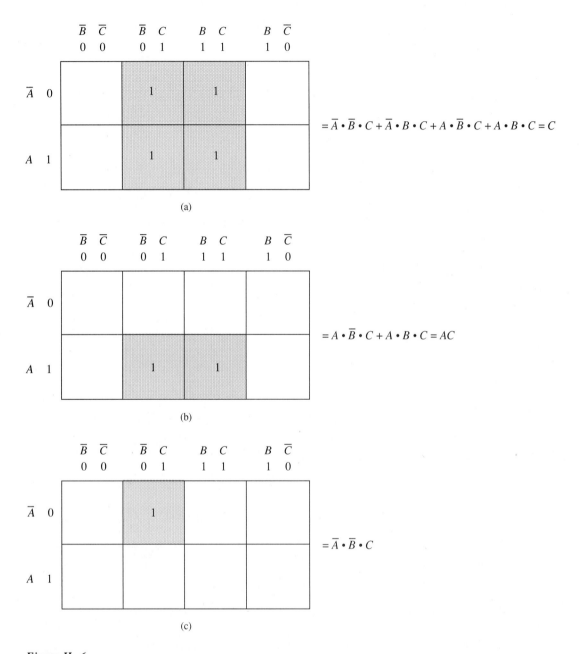

Figure H–6

Figure H–7 illustrates how end squares are regarded as adjacent. Note that the matrix can be rolled into a cylinder such that the outermost squares are, in fact, adjacent.

Review of Boolean Algebra and Karnaugh Maps

Figure H–7

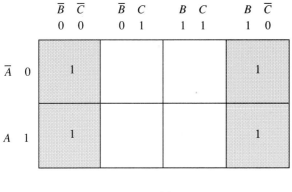

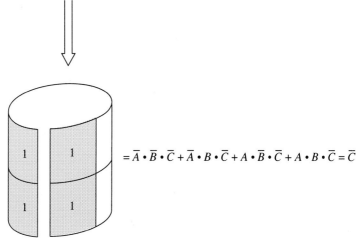

Figure H–8 shows an example of a simplified three-variable Boolean expression.

Figure H–8

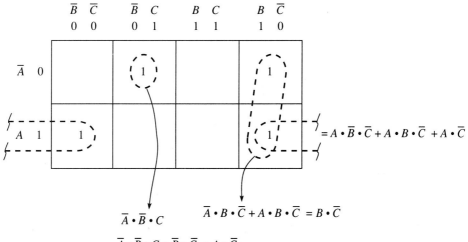

Rules Review

CHAPTER 2

Rules of Algebraic Addition

Rule 2–1 To add two or more numbers algebraically having the same sign, find the sum of their absolute values, and affix the common sign.

Rule 2–2 To add two numbers differing in sign, find the difference between their absolute values, and affix the sign of the number having the larger absolute value.

Rule 2–3 To add two or more numbers differing in sign, first find the sum of all the positive numbers using Rule 2–1, then find the sum of all the negative numbers by applying Rule 2–1 again. Finally, obtain the sum of the two resulting numbers using Rule 2–2.

Rule for Algebraic Subtraction

Rule 2–4 To subtract one signed number (the subtrahend) from another (the minuend), change the sign of the subtrahend, and add algebraically.

Rules for Working with Symbols of Grouping

Rule 2–5 Whenever symbols of grouping occur one within the other (i.e. are nested), we remove the innermost symbols first.

Rule 2–6 To remove symbols of grouping preceded by a plus sign, simply remove the symbols without making any other changes.

Rule 2–7 To remove symbols of grouping preceded by a minus sign, change the sign of each term within the symbols.

Rule 2–8 To enclose terms within symbols of grouping preceded by a plus (+) sign, simply enclose the terms without making any other changes.

Rule 2–9 To enclose terms within symbols of grouping preceded by a minus (−) sign, change the sign of each term to be included within the group.

CHAPTER 3

The Sign of a Product

Rule 3–1 The product of two like-signed numbers is positive.

Rule 3–2 The product of two unlike-signed numbers is negative.

Rule 3–3 Rules 3–1 and 3–2 are to be applied in succession when there are more than two factors.

Corollary to Rule 3–3 The product of an *even* number of negative factors is positive. The product of an *odd* number of negative factors is negative.

Laws of Exponents

Rule 3–4 Exponential Law of Multiplication
To find the product of two or more exponential terms having the same base, add their exponents.
$$b^p b^q = b^{p+q}$$

Rule 3–5 Law for the Power of a Power
$$(b^p)^q = b^{pq}$$

Rule 3–6 Law for the Power of a Product
$$(bc)^p = b^p c^p$$

Rule for Multiplying Two or More Monomials

Rule 3–7
1. Multiply the numerical coefficients together, paying careful attention to the sign of the result.
2. Multiply the literal factors using the rule for exponents.
3. Write the results of steps 1 and 2 as a single product.

Rule for Multiplying a Multinomial by a Monomial

Rule 3–8 To multiply a multinomial by a monomial, multiply each term of the multinomial by the monomial, according to Rule 3–7, and write each partial product as a term of the new multinomial.

Rule for Multiplying Multinomials

Rule 3–9 To multiply multinomials, multiply every term of one by every term of the other, and write each resulting product as a term of the new multinomial.

Rule 3–10 Law of Exponential Division
To divide exponential terms having the same base, subtract the exponent in the denominator from the exponent in the numerator.
$$\frac{b^p}{b^q} = b^{p-q} \quad (b \neq 0)$$

Rule 3–11 Law of Negative Exponents
$$b^{-p} = \frac{1}{b^p} \quad (b \neq 0)$$

Rule 3–12 The Law of Zero Exponents
$$b^0 = 1 \quad (b \neq 0)$$

The Sign of a Quotient

Rule 3–13 The quotient of like-signed terms is positive.
Rule 3–14 The quotient of unlike-signed terms is negative.

Rule for the Division of Monomials

Rule 3–15 To divide one monomial by another:
1. Divide the numerical coefficients, paying careful attention to the sign of the result.
2. Divide the literal factors using the rule for the division of exponents.
3. A single quotient is formed as the product of steps 1 and 2.

Rule for Dividing a Multinomial by a Monomial

Rule 3–16 Divide each term of the multinomial by the monomial according to Rule 3–7, and write each resulting partial quotient as a term of the new multinomial.

Rules Review

Rule for Dividing Multinomials

Rule 3–17
1. Arrange the terms of both multinomials in either an increasing or decreasing pattern of some common literal factor.
2. Divide the first term of the dividend by the first term of the divisor. This is the first term of the quotient.
3. Multiply the first term of the quotient by the entire divisor, and subtract this new expression from the appropriate terms of the dividend.
4. The result of step 3 creates a new dividend, which is then divided by the divisor, and steps 2 and 3 are repeated until either (a) there is no longer a remainder or (b) the remainder can no longer be divided by the divisor.

CHAPTER 4

Axioms

Axiom 4–1 If the same number is added to equal numbers, the sums are still equal.
Axiom 4–2 If the same number is subtracted from equal numbers, the remainders are still equal.
Axiom 4–3 If equal numbers are multiplied by the same number, the products are equal.
Axiom 4–4 If equal numbers are divided by the same nonzero number, the quotients are still equal.
Axiom 4–5 If two numbers are equal to the same (or equal) numbers, then they are equal to each other.
Axiom 4–6 If two numbers are equal, they both may be raised to the same power without altering the equality.
Axiom 4–7 If two numbers are equal, like roots may be extracted from both members without destroying the equality.

Rules for Solving Equations

Rule 4–1 Any term may be *transposed* (moved) from one member to the other simply by changing its sign.
Rule 4–2 Any term appearing on *both* sides of an equation and preceded by the *same* sign may simply be dropped from the equation.
Rule 4–3 The signs of all the terms of an equation may be changed without destroying the equality.

Rules for Solving Inequalities

Rule 4–4 All axioms that apply to equalities also apply to inequalities, except Axiom 4–4.
Rule 4–5 Whenever both members of an inequality are multiplied or divided by a negative number, *the sense* of the inequality must be reversed.

Guidelines for Solving Word Problems

1. Whenever possible, *draw a picture* of the details of the problem, labeling what is known and what is to be found.
2. *Translate* each significant word or phrase into a sign, symbol, or literal number.
3. *Solve* for the required quantity.
4. *Check* your answer against the conditions given in the problem.

CHAPTER 5

Law of Fractional Exponents

Rule 5–1

$$b^{p/q} = \sqrt[q]{b^p} \qquad q \neq 0$$

Appendix I

Removal of Common Monomial Factor
Rule 5–2
1. Determine the GCF of all terms by inspection.
2. Divide the given multinomial by the GCF.
3. Indicate the product of the quotient and the GCF as the required factors of the given expression.

Special Products and Their Factors
Square of the sum: $(a + b)^2 = a^2 + 2ab + b^2$
Square of the difference: $(a - b)^2 = a^2 - 2ab + b^2$
Product of the sum and difference: $(a + b)(a - b) = a^2 - b^2$

Factoring Trinomials of the Form $x^2 + (a + b)x + ab$

1. The *first* term of the trinomial is the product of the *first* terms of the binomials.

$$(x + a)(x + b) = x^2 + (ax + bx) + ab$$

2. The *last* term of the trinomial is the product of the *last* terms of the binomials.

$$(x + a)(x + b) = x^2 + (ax + bx) + ab$$

3. The *middle* term of the trinomial is the *sum* of the products of the *innermost* and *outermost* terms of the binomials.

$$(x + a)(x + b) = x^2 + (ax + bx) + ab$$

Factoring Trinomials of the Form $acx^2 + (ad + bc)x + bd$

1. The *first* term of the trinomial is the product of the *first* terms of the binomials.

$$(ax + b)(cx + d) = acx^2 + (ad + bc)x + bd$$

2. The *last* term of the trinomial is the product of the *last* terms of the binomials.

$$(ax + b)(cx + d) = acx^2 + (ad + bc)x + bd$$

3. The *middle* numerical coefficient of the trinomial is the *sum* of the products of the *innermost* and *outermost* numerical coefficients of the binomials.

$$(ax + b)(cx + d) = acx^2 + (ad + bc)x + bd$$

Rules Review

CHAPTER 6

Important Properties of Fractions

Rule 6–1 Any two of the three signs of a fraction can be changed without changing the value of the fraction.
Rule 6–2 The same *nonzero* number can be used either to multiply or divide *both terms* of a fraction without changing its value.
Rule 6–3 Factors common to both numerator and denominator can be canceled.
Rule 6–4 Fractions can be reduced to lowest terms by factoring out and canceling all factors common to both numerator and denominator.

Rule for Adding or Subtracting Fractions

Rule 6–5
1. Find the LCD.
2. Divide the LCD by the denominator of the first fraction, then multiply the quotient by the numerator of that same fraction.
3. Write the product formed in step 2 as the first term of the new numerator.
4. Repeat steps 2 and 3 for the other fractions involved.
5. Find the algebraic sum or difference of the products formed in steps 3 and 4, and write the result over the LCD.

Rule for Multiplying Fractions

Rule 6–6
1. Find the product of all the numerators.
2. Find the product of all the denominators.
3. Form the final answer by writing the result of step 1 over the result of step 2.

Rule for Dividing Fractions

Rule 6–7
1. Invert the fraction used as the divisor.
2. Multiply according to the rules for fractions.

CHAPTER 7

Total Resistance of a Series Circuit

Rule 7–1

$$R_t = \sum_{k=1}^{n} R_k = R_1 + R_2 + \cdots + R_n$$

The Voltage Divider Formula

Rule 7–2

$$V_m = V_s \frac{R_m}{\Sigma R}$$

Total Power in a Series or a Parallel Circuit

Rule 7–3

$$P_t = \sum_{k=1}^{n} P_k = P_1 + P_2 + \cdots + P_n$$

Appendix I

Total Resistance of a Parallel Circuit

Rule 7-4

$$R_t = \frac{1}{\frac{1}{R_1} + \frac{1}{R_2} + \cdots + \frac{1}{R_n}}$$

The Current Divider Formula

Rule 7-5

$$I_x = I_t \frac{G_x}{G_t}$$

Rule for Analysis of Units

Rule 7-6 Arrange each pertinent factor as a fraction in such a way that unwanted terms cancel, leaving the desired units intact.

CHAPTER 8

The Pythagorean Theorem

Rule 8-1 In any right triangle (one that contains a 90° angle), the lengths of the three sides are related by the equation

$$c^2 = a^2 + b^2$$

Rule 8-2

$$c = \sqrt{a^2 + b^2}$$

Distance Between Two Points

Rule 8-3

$$D = \sqrt{(x_1 - x_2)^2 + (y_1 - y_2)^2}$$

The 45-45 Triangle

Rule 8-4
1. Both acute angles (those less than 90°) are the same and equal 45°.
2. Both the legs (sides) are equal.
3. The length of the hypotenuse is always $\sqrt{2} \approx 1.414$ times longer than either leg.

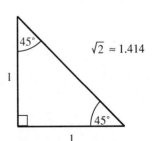

Rules Review

The 30-60 Triangle
Rule 8–5
1. The side opposite the 30° angle is half the hypotenuse.
2. The side opposite the 60° angle is $\sqrt{3} \approx 1.732$ times longer than the side opposite the 30° angle.

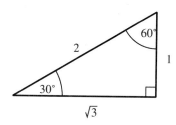

Slope of a Line
Rule 8–6

$$\text{Slope} = m = \frac{\text{rise}}{\text{run}} = \frac{(y_1 - y_2)}{(x_1 - x_2)}$$

Point-Slope Form
Rule 8–7

$$y = m(x - x_1) + y_1$$

Slope-Intercept Form
Rule 8–8

$$y = mx + b$$

CHAPTER 9

Solution by Addition and Subtraction
Rule 9–1
1. Eliminate one of the variables in both equations by adding or subtracting the two given equations or multiples of these equations.
2. Solve the resulting linear equation for the value of its variable.
3. Substitute this value back into *either* of the original equations, and solve for the remaining variable.

Three Cases of the Simultaneous Solution
Rule 9–2
1. One unique solution (Lines cross.)
2. No solution (Lines are parallel.)
3. Infinite number of solutions (Lines coincide.)

Appendix I

Solution by Substitution
Rule 9–3
1. Solve either equation for one of the variables in terms of the other.
2. Substitute the result from step 1 into the other original equation.
3. Solve the equation from step 2.
4. Substitute the solution in step 3 into either original equation and solve for the other variable.

Solution by Comparison
Rule 9–4
1. Solve both equations for the same variable.
2. Equate the results from step 1.
3. Solve for the variable from step 2.
4. Substitute the solution in step 3 into either original equation and solve for the remaining variable.

Solution with Determinants

To solve a system of two equations in two unknowns:

$$a_1 x + b_1 y = c_1$$
$$a_2 x + b_2 y = c_2$$

Rule 9–5

$$x = \frac{\begin{vmatrix} c_1 & b_1 \\ c_2 & b_2 \end{vmatrix}}{\begin{vmatrix} a_1 & b_1 \\ a_2 & b_2 \end{vmatrix}} = \frac{b_2 c_1 - b_1 c_2}{a_1 b_2 - a_2 b_1}$$

$$y = \frac{\begin{vmatrix} a_1 & c_1 \\ a_2 & c_2 \end{vmatrix}}{\begin{vmatrix} a_1 & b_1 \\ a_2 & b_2 \end{vmatrix}} = \frac{a_1 c_2 - a_2 c_1}{a_1 b_2 - a_2 b_1}$$

To solve a system of three equations in three unknowns:

$$a_1 x + b_1 y + c_1 z = d_1$$
$$a_2 x + b_2 y + c_2 z = d_2$$
$$a_3 x + b_3 y + c_3 z = d_3$$

Rule 9–6

$$x = \frac{\begin{vmatrix} d_1 & b_1 & c_1 \\ d_2 & b_2 & c_2 \\ d_3 & b_3 & c_3 \end{vmatrix}}{\begin{vmatrix} a_1 & b_1 & c_1 \\ a_2 & b_2 & c_2 \\ a_3 & b_3 & c_3 \end{vmatrix}} = \frac{b_2 c_3 d_1 + b_1 c_2 d_3 + b_3 c_1 d_2 - b_2 c_1 d_3 - b_3 c_2 d_1 - b_1 c_3 d_2}{a_1 b_2 c_3 + a_3 b_1 c_2 + a_2 b_3 c_1 - a_3 b_2 c_1 - a_1 b_3 c_2 - a_2 b_1 c_3}$$

$$y = \frac{\begin{vmatrix} a_1 & d_1 & c_1 \\ a_2 & d_2 & c_2 \\ a_3 & d_3 & c_3 \end{vmatrix}}{\begin{vmatrix} a_1 & b_1 & c_1 \\ a_2 & b_2 & c_2 \\ a_3 & b_3 & c_3 \end{vmatrix}} = \frac{a_1 c_3 d_2 + a_3 c_2 d_1 + a_2 c_1 d_3 - a_3 c_1 d_2 - a_1 c_2 d_3 - a_2 c_3 d_1}{a_1 b_2 c_3 + a_3 b_1 c_2 + a_2 b_3 c_1 - a_3 b_2 c_1 - a_1 b_3 c_2 - a_2 b_1 c_3}$$

Rules Review

$$z = \frac{\begin{vmatrix} a_1 & b_1 & d_1 \\ a_2 & b_2 & d_2 \\ a_3 & b_3 & d_3 \end{vmatrix}}{\begin{vmatrix} a_1 & b_1 & c_1 \\ a_2 & b_2 & c_2 \\ a_3 & b_3 & c_3 \end{vmatrix}} = \frac{a_1b_2d_3 + a_3b_1d_2 + a_2b_3d_1 - a_3b_2d_1 - a_1b_3d_2 - a_2b_1d_3}{a_1b_2c_3 + a_3b_1c_2 + a_2b_3c_1 - a_3b_2c_1 - a_1b_3c_2 - a_2b_1c_3}$$

CHAPTER 10

The General Quadratic Relationship
Rule 10–1

$$y = ax^2 + bx + c \quad (a \neq 0)$$

Rule 10–2

The y-intercept of the equation $y = ax^2 + bx + c$ is the point $(0, c)$.

The Axis of Symmetry
Rule 10–3

$$x = \frac{-b}{2a}$$

Key Points Used in Graphing Quadratic Relationships
Rule 10–4
1. y-intercept: $y = c$.
2. x-intercepts: Set $y = 0$ and solve the resulting equation for x.
3. axis of symmetry: $x = -b/2a$.
4. vertex: $x = -b/2a$. The y value is the corresponding value obtained for this x.

The Quadratic Formula
Rule 10–5

$$x = \frac{-b \pm \sqrt{b^2 - 4ac}}{2a}$$

The Discriminant
Rule 10–6
1. If $b^2 - 4ac = 0$, roots are real and equal.
2. If $b^2 - 4ac > 0$, roots are real and unequal.
3. If $b^2 - 4ac < 0$, the roots are imaginary roots. (The graph does not cross the x-axis.)

CHAPTER 12

Radian Measure

$$180° = \pi \text{ rad}$$
$$1 \text{ rad} \approx 57.29577951°$$
$$1° \approx 0.017453292 \text{ rad}$$

Appendix I

Trigonometric Relationships in a Right Triangle

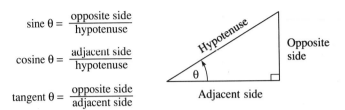

$$\text{sine } \theta = \frac{\text{opposite side}}{\text{hypotenuse}}$$

$$\text{cosine } \theta = \frac{\text{adjacent side}}{\text{hypotenuse}}$$

$$\text{tangent } \theta = \frac{\text{opposite side}}{\text{adjacent side}}$$

Graphs of the Trigonometric Relationships

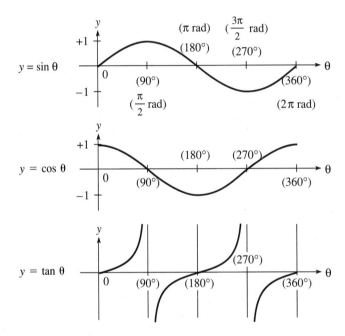

Properties of $y = A \sin(Bx - C)$ and $y = A \cos(Bx - C)$

$$\text{Amplitude} = A$$
$$\text{Period} = \frac{2\pi}{B} \text{ radians}$$
$$\text{Phase shift} = \frac{-C}{B} \text{ radians}$$

CHAPTER 13

Rules for Operating with Logarithms

Rule 13–1 Log of a Product

$$\log_b MN = \log_b M + \log_b N$$

Rule 13–2 Log of a Quotient

$$\log_b(M/N) = \log_b M - \log_b N$$

Rule 13–3 Log of a Power

$$\log_b M^q = q \times \log_b M$$

Corollary to Rule 13–3

$$\log_b b^p = p$$

Rule 13–4 Change of Base

$$\log_b x = \frac{\log_a x}{\log_a b}$$

CHAPTER 14

Rules for Finding the Resultant

1. Draw the component vectors to scale on a rectangular coordinate system and include the resultant. Label each vector carefully, paying attention to the sign of angle θ.
2. Resolve each of the given vectors into their rectangular parts using the trigonometric relationships

$$x = A \cos \theta$$
$$y = A \sin \theta$$

where

A = magnitude of the given vector
x = horizontal component
y = vertical component

3. Use the component parts in step 2 to form the right triangle whose hypotenuse is the resultant.
4. Find the magnitude of the resultant using the Pythagorean theorem

$$\text{Resultant} = \sqrt{H^2 + V^2}$$

where H and V represent the lengths of the horizontal and vertical sides found in step 3.
5. Find the phase angle θ using the inverse tangent relationship

$$\theta = \tan^{-1}\frac{V}{H}$$

6. Check your solution against the drawing made in step 1.

CHAPTER 15

Effective (RMS) Value of Voltage and Current

Rule 15–1

$$V = 0.707 V_p$$

Rule 15–2

$$I = 0.707 I_p$$

Appendix I

Inductors in Series ($k = 0$)
Rule 15–3
$$L_t = L_1 + L_2 + \cdots + L_n$$

Inductors in Series ($k \neq 0$)
Rule 15–4
$$L_t = L_1 + L_2 \pm 2m$$

where
$$m = k\sqrt{L_1 \cdot L_2}$$

Inductors in Parallel
Rule 15–5
$$L_t = \frac{1}{1/L_1 + 1/L_2 + \cdots + 1/L_n}$$

Transformer Relationships
Rule 15–6
$$\frac{N_p}{N_s} = \frac{V_p}{V_s}$$

where

N_p = number of turns in primary
N_s = number of turns in secondary
V_p = voltage of the primary
V_s = voltage of the secondary

Rule 15–7
$$\frac{V_p}{V_s} = \frac{I_s}{I_p}$$

where

I_s = current in the secondary
I_p = current in the primary

Rule 15–8
$$\frac{Z_p}{Z_s} = \left(\frac{N_p}{N_s}\right)^2$$

where

Z_p = impedance connected to primary
Z_s = impedance connected to secondary

Inductive Reactance
Rule 15–9
$$X_L = 2\pi f L \quad \text{ohms}$$

Rules Review

Capacitors in Series
Rule 15–10

$$C_t = \frac{1}{1/C_1 + 1/C_2 + \cdots + 1/C_n}$$

Capacitors in Parallel
Rule 15–11

$$C_t = C_1 + C_2 + \cdots + C_n$$

Capacitive Reactance
Rule 15–12

$$X_C = \frac{1}{2\pi f C}$$

RC Time Constant
Rule 15–13

$$t = RC$$

Impedance of Series AC Circuit
Rule 15–14

$$Z = \sqrt{R^2 + (X_C - X_L)^2}$$

Phase Angle
Rule 15–15

$$\theta = \tan^{-1}\frac{V_L}{V_R} \quad \text{or} \quad \theta = \tan^{-1}\frac{X_L}{R}$$

Resonant Frequency
Rule 15–16

$$f = \frac{1}{2\pi\sqrt{LC}}$$

Power Relationships in AC Circuit
Rule 15–17

Apparent power: $P = V_t I$; $\quad P = \dfrac{V_t^2}{Z}$; $\quad P = I^2 Z$

Reactive power: $P = V_X I$; $\quad P = \dfrac{V_X^2}{X}$; $\quad P = I^2 X$

True power: $P = V_R I$; $\quad P = \dfrac{V_R^2}{R}$; $\quad P = I^2 R$

Rule 15–18

$$\text{True power} = P = VI \cos\theta$$

Power Factor
Rule 15–19

$$PF = \cos \theta$$

CHAPTER 16

Adding or Subtracting Complex Numbers
Rule 16–1

Add (subtract) the real parts and the imaginary parts separately.

Multiplying Complex Numbers
Rule 16–2

Treat the complex numbers as binomials, remembering $j^2 = -1$.

Dividing Complex Numbers
Rule 16–3

Treat them as ordinary binomials and rationalize the denominator using the conjugate of the denominator.

Relationship Between Complex Numbers in Rectangular and Polar Form
Rule 16–4

1. Given a number in rectangular form, $R + jX$, we can convert to the polar form $Z \angle \theta$ using the following Pythagorean and trigonometric relationships:

$$Z = \sqrt{R^2 + X^2} \quad \text{and} \quad \theta = \tan^{-1} \frac{X}{R}$$

2. Given a number in polar form, $Z \angle \theta$, we can convert it to rectangular form using the trigonometric relationships

$$Z \sin \theta = X \quad \text{and} \quad Z \cos \theta = R$$

Multiplying and Dividing Numbers in Polar Form
Rule 16–5

1. To multiply two numbers in polar form, simply multiply the magnitudes and add the angles. For example,

$$M \angle A° \times N \angle B° = MN \angle A + B°$$

2. To divide two numbers in polar form, simply *divide* the magnitudes, and *subtract* the angles:

$$\frac{M \angle A°}{N \angle B°} = \frac{M}{N} \angle A - B°$$

The General Solution of a Parallel AC Circuit
Rule 16–6

1. Convert each branch impedance to polar form $Z \angle \theta$.
2. Find the admittance (Y) of each branch by taking reciprocals: $Y = (1 \angle 0°)/(Z \angle \theta°)$.
3. Convert each polar admittance into rectangular form: $Y \angle \theta = G \pm jB$.
4. Find the total admittance Y_t by adding the individual admittances in step 3.
5. Convert the total admittance to polar form: $Y_t \angle \theta°$.
6. Take the reciprocal of the total admittance in order to find the total impedance: $Z_t = (1 \angle 0°)/(Y_t \angle \theta°)$.
7. Change Z_t to rectangular form to reveal the equivalent series circuit: $Z_t = R \pm jX \, \Omega$.

J Answers to Selected Problems

CHAPTER 1

Exercise Set 1–1

1.

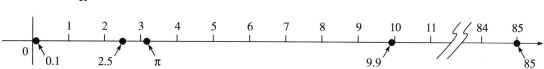

3.

a.

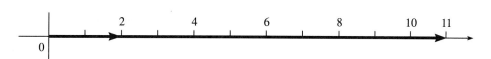

b.

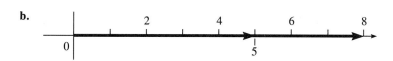

c.

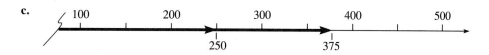

d.

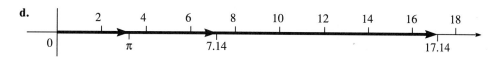

5. No difference, since addition is commutative.

Appendix J

6.
```
10 REM   This program computes the sum of
20 REM   three positive real numbers A, B, and C.
30 LET S=0
40 PRINT"Enter the value of A"
50 INPUT A
60 PRINT"Enter the value of B"
70 INPUT B
80 PRINT"Enter the value of C"
90 INPUT C
100 LET S=A+B+C
110 PRINT"The sum of A, B, and C is:";S
120 END
```

Exercise Set 1–2

1.
a.

b.

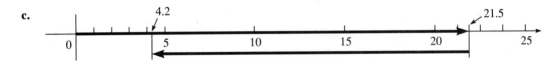

c.

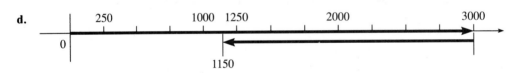

d.

3. 3
4. 1.5 h
5.
```
10 REM   This program computes the difference (D) between
20 REM   two real, positive numbers P and Q, where the
30 REM   number P is larger than the number Q, and adds the
40 REM   difference to 25.
50 LET S=0
60 PRINT"Enter the values P and Q in the proper order"
70 INPUT P, Q
80 LET S=25+(P-Q)
90 PRINT"The difference P-Q added to 25 is:";S
100 END
```

Exercise Set 1–3

1. a. 258.244
   b. 750
   c. 1.39533×10^6
   d. 1.8628×10^{12}
   e. 3.867×10^{18}
   f. 279.936×10^3

Answers to Selected Problems

    g. 1024
    h. 1.4058×10^9
    i. 30.36×10^{12}
    j. $70.5 \times 10^9 \, \Omega$ or $70{,}500 \, M\Omega$
    k. 8.7241×10^3
    l. 4.234×10^{15}
    m. 186×10^6 mi
    n. 1.2×10^9 m/s
    o. 31.25×10^{18}
    p. 6.975×10^{36}
    q. 10.247×10^3
    r. 5.198
3. 3×10^{10} cm/s
5. 6.9 in.
7.
```
10 REM   This program extracts the square root of the sum
20 REM   of R squared and X squared.
30 LET Z=0
40 PRINT"Enter the value of R."
50 INPUT R
60 PRINT"Enter the value of X."
70 INPUT X
80 LET Z=SQR(R^2+X^2)
90 PRINT"The square root of the sum is:";Z
100 END
```

Exercise Set 1–4

1. a. 321.57×10^3 **d.** 50
  b. 111.11×10^3 **e.** 64
  c. 3.559   **f.** $2 \, k\Omega$

Exercise Set 1–5

1. $\dfrac{29}{40}$ **11.** 0

3. $\dfrac{1}{6}$ **13.** $\dfrac{85}{32}$

5. $\dfrac{129}{256}$ **15.** $\dfrac{19}{16}$

7. $\dfrac{15}{4096}$ **17.** 3

9. $\dfrac{6}{5}$ **19.** $\dfrac{137}{60}$

Exercise Set 1–6

1. 21
3. a. 9   **c.** 1
  b. 360   **d.** 4

Exercise Set 1–7

1. a. 60%   **c.** 30.3%
  b. 33.3%   **d.** 400%
3. Yes.
5. 26%
6. 20%
7. 119.46, or slightly fewer than 120, employees
9. 19.65, or about 19, employees

CHAPTER 2

Exercise Set 2–1

1. **a.** $+100°C$
   b. Approximately $+370°F$ (for 60% tin, 40% lead)
   c. -12 V
   d. -20 V
   e. $-\$7.17$

Exercise Set 2–2

1. 56.8
3. -35
5. 4.32×10^{-3}
7. 1.625
9. 0
11. 11.71

Exercise Set 2–3

1. -3
3. 114
5. -4
7. -0.136
9. -0.70857

Exercise Set 2–4

1. **a.** $3VA, -12I^2R, 4ZLC^2$
   b. $\quad 3VA: \; 3, V, A, 3V, 3A, VA$
   $\quad -12I^2R: \; -12, I^2, R, -12I^2, -12R, I^2R$
   $\quad 4ZLC^2: \; 4, Z, L, C^2, 4Z, 4L, 4C^2, 4ZL, 4ZC^2,$
   $\quad\quad\quad\quad 4LC^2, ZL, ZC^2, LC^2, ZLC^2$

2. **a.** $V - 7i$
   c. No like terms
   e. $7Z/3 + R$
   g. No like terms
   i. $6000V$

3. **a.** 0   **f.** 9
   b. -6 **g.** 3750
   c. 19   **h.** 2.965
   d. -6 **i.** $\dfrac{13}{5}$
   e. 0   **j.** 0

Exercise Set 2–5

1. **a.** $-a$ **f.** $-2R^2 - 4X^2$
   b. $5V^2 - 4R$ **g.** $2\phi + 6\lambda + 3$
   c. $5R^3 - 4$ **h.** $V - 3I$
   d. $-a$ **i.** $-X - 6$
   e. $9 - 6Z$ **j.** $P - 5Q$

3. **a.** $4k - (-3V + 4k)$
   b. $2x - (4y - 5z)$
   c. $3R - (-2Xa + 4Xc)$
   d. $-a - 2a - (3a - 4a)$
   e. $2\lambda + 8f - (-\lambda + f)$

Answers to Selected Problems

CHAPTER 3

Exercise Set 3–1
1. +
3. a. -18
   b. 6
   c. $-\dfrac{1}{8}$
   d. -10
   e. 0.0000338445
   f. $-a$

Exercise Set 3–2
1. a. x^8
   b. 512
   c. ϵ^9
   d. -512
   e. $-a^4$
   f. 64
   g. $\dfrac{1}{4}$
   h. V^{10}
   i. I^7
   j. $\dfrac{1}{x^{10}}$
   k. $9x^2$
   l. $(-1)^n(3^n)k^{2n}$
   m. $\dfrac{1}{4}L^2C^2$
   n. $4\pi^2 f^2$
   o. $-2^{mr}x^{2r}$

Exercise Set 3–3
1. a^2x^3
3. $9V^4$
5. $6\pi^2 f^3 L$
7. $-3\pi^2 r^5$
9. $P^3 y^3$
11. $-c^4 m^6$
13. $\dfrac{3}{8}Z^4$
15. a^{11}
17. $27b^4 x^7$
19. a^{22}

Exercise Set 3–4
1. $-R^2 + R^3$
3. $-a^4 x + a^4$
5. $-R^2 X^2 - X^4$
7. $-18b^4 + 9b^5$
9. $4a^3 c^2 + 8a^2 c^3 - 8ac^4$
11. $-6i^5 - 2i^4 - i^5 = -7i^5 - 2i^4$
13. $a^6 k^2 - a^5 k^3 + a^4 k^4 - a^3 k^5 + a^2 k^6$
15. $-x^2 + \dfrac{2}{9}x^3 + 3x - 4$
17. $5V^3 - 8V^2$
19. 0

Exercise Set 3-5

1. $a^2 - b^2$
3. $V^2 - 4V + 4$
5. $2P^2 - PW^2 + 2PW - W^3$
7. $2Mk - M^4 + 2k^3 - M^3k^2$
9. $x^2 - \dfrac{x}{3}$
11. $2R^3 + 3R^2X - 2RX^2 - 3X^3$
13. $0.0927Vv^2 + 1.03V^4 - 0.03v^3 - \dfrac{1}{3}V^3v$
15. $101VI - 100V + 1.01V^2I^2 - V^2I$
17. $-\dfrac{1}{20}Y^2 + 0.35YZ - Y^2Z^3 - 0.3Z^2 + YZ^4$
19. $a^2 + ab^3 - 2a^2b - 2ab - 2b^4 + 4ab^2 - a^3b - a^2b^4 + 2a^3b^2$
21. ```
    10 PRINT"Enter the values of a and b."
    20 INPUT A,B
    30 LET N=(a-(2*b)-a^2*b)*(a+(b^3)-2*a*b)
    40 PRINT"The product of the two multinomial terms is:";N
    50 END
    ```

Exercise Set 3-6

1. a^4
3. $\dfrac{W}{4}$
5. $\dfrac{-V}{3}$
7. $-7X$
9. $7M^{x-k}$
11. ```
    10 PRINT"Enter the value of the base M."
    20 INPUT M
    30 PRINT"Enter the values of x and k."
    40 INPUT X,K
    50 LET Y=(M^X)/(M^K)
    60 PRINT"The value of the exponential quotient is";Y
    70 END
    ```

Exercise Set 3-7

1. $\dfrac{-a^2}{4}$
3. $-V^2I^2$
5. $-8.2 \times 10^{-12}P$
7. $3abc^6$
9. $\dfrac{-13Q^4}{P}$
11. c
13. $\dfrac{-0.424}{k}$
15. $0.033\tfrac{1}{3}f^2$

Answers to Selected Problems

17. $\dfrac{2\lambda^2}{m^3}$

19. $\dfrac{-1}{20v^3M}$

21.
```
10 REM  In this program, the Greek letter Lambda
20 REM  is replaced by the letter L.
30 PRINT"Enter the values of L and m."
40 INPUT L,M
50 LET Y=(-22*L^5*M^-2)/(-11*L^3*m)
60 PRINT"The value of the exponential quotient is";Y
70 END
```

Exercise Set 3–8

1. $\dfrac{-27t^2}{a} + 4t$

3. $\dfrac{-V^2}{2} + \dfrac{3}{2}V - \dfrac{1}{2}$

5. $-25P^2 + \dfrac{7}{P} - \dfrac{9}{P^2}$

7. $0.05V^6R^2 + \dfrac{0.8R^3}{V} - 7V^2R^2$

9. $\dfrac{4}{y} - \dfrac{8}{x} - \dfrac{2}{y} - \dfrac{2}{x}$

Exercise Set 3–9

1. $Z + 4$

3. $P - 1 + \dfrac{2 + 2R}{P + 1}$

5. $a - b$

7. $R - X$

9. $2C^2 + C + 2 + \dfrac{8C + 8}{C^2 - 5}$

11. $x^4 + x^3y + x^2y^2 + xy^3 + y^4$

13. $2W^2 - W + \dfrac{-12W}{2W - 3}$

14. $3x + 5 + \dfrac{x - 1}{x^2 + 2}$

15. $x^2 - 5x + 21 + \dfrac{-81x + 43}{x^2 + 3x - 2}$

17. $f^2 + 10f + \dfrac{-4}{f - 2}$

19. $x^2 + 5 + \dfrac{16}{x^2 - 2}$

21.
```
10 PRINT"Enter the value of Q."
20 INPUT Q
30 LET N=(Q^4+3*Q-7)/(Q^2+2*Q-3)
40 PRINT"The quotient is";N
50 END
```

CHAPTER 4

Exercise Set 4–1

1. 2
2. -2.5
3. 7
4. 4
5. 1
6. -26
7. 2
8. -1
9. 0
10. $\dfrac{5}{89.6} \approx 0.05580$
11. $\dfrac{5 + 6a}{-7}$
12. $\dfrac{6k}{5}$
13. 4
14. $\dfrac{-26g}{19}$
15. 1
16. -25
17. $-20q - 31a$
18. 10
19. $-\dfrac{32}{13}(1 + K)$
20. 1

Exercise Set 4–2

1. $V < 4$
3. $I \geq -7$
5. $R < 6$
7. $Z < 3$
9. $L < -2$
11. $Q \geq -3$
13. $H \leq 1$
15. $B < -5/2$
17. $d > 11$
19. $X > 12$

Exercise Set 4–3

1. 15, 16
3. The capacitors cost $0.445 each, and the inductor costs $1.495.
4. The rectifiers cost $0.55 each, and the zener costs $0.77.
5. 27 ft
6. 20, 25, 40
7. 4.3 ft, 15.7 ft
9. 11, 13, 15
10. 15 ft by 30 ft
11. 0.0239..., or about 2.4%

Answers to Selected Problems

13. $V_1 = 15$ V, $V_2 = 3$ V
15. 108°, 132°

CHAPTER 5

Exercise Set 5–1
1. Prime: 3, 13, 17, 53, 83, 101; composite: 15, 27, 48, 49, 51, 81, 99, 105

Exercise Set 5–2
1. $17cvx^2\sqrt{c}$
3. $9I\sqrt{9R}$
5. $4\sqrt{2V}$
7. $2\sqrt[6]{Z^m}$
9. $3Q^2\sqrt[4]{3}\sqrt{P}$
11. $5y^3\sqrt{5x}$
13. $\sqrt{6x}$
15. $13y^2\sqrt[4]{x}$

Exercise Set 5–3
1. $17Z(3Z - 2)$
3. $100k^2g(1 + 10g)$
5. $\frac{1}{4}PQ\left(PQ - \frac{1}{2}P + \frac{1}{8}Q\right)$
7. $\frac{1}{5}ax^2\left(\frac{17}{16}x - \frac{1}{4}a\right)$
9. $0.005I^2R^2(1 - 0.1IR)$
11. $25^2k^2M^2(k - 25M + 80k^2)$
13. $\frac{1}{2a^5}\left(\frac{1}{3} - \frac{1}{5a^2}\right)$
15. $\frac{X^2}{2}\left(X^3 - \frac{X}{2} + \frac{1}{8}\right)$

Exercise Set 5–4
1. $(V - e)^2$
3. $(1 + k)^2$
5. $\left(\frac{V^2}{2} - P\right)^2$
7. $\left(0.01 + \frac{b}{5}\right)^2$
9. $\left(\frac{1}{3} + 2x^2\right)^2$
11. $(\sqrt{4} - v)^2$
13. $\frac{1}{2}\left(\frac{a}{4} - \frac{b}{8}\right)^2$
15. $6y(2x + z)^2$

Appendix J

Exercise Set 5–5

1. $(1 - w)(1 + w)$
3. $(0.1Z - 1)(0.1Z + 1)$
5. $(X - 1)(X + 1)(X^2 + 1)$
7. $\left(\dfrac{1}{R} - \dfrac{\sqrt{2}}{r}\right)\left(\dfrac{1}{R} + \dfrac{\sqrt{2}}{r}\right)$
9. $\left(V - \dfrac{1}{10}\right)\left(V + \dfrac{1}{10}\right)$
11. $\left(k^2 - \dfrac{1}{2}\right)\left(k^2 + \dfrac{1}{2}\right)$
13. $\dfrac{1}{3}\left(t - \dfrac{1}{3}\right)\left(t + \dfrac{1}{3}\right)$
15. $\left(\dfrac{X}{2} - 4\right)\left(\dfrac{X}{2} + 4\right)\left(\dfrac{X^2}{4} + 16\right)$
16. Since division by zero is not a permissible operation, dividing by $V - k = 0$ in step 6 is incorrect, even though the axioms were applied correctly and the factoring method was appropriate. The lesson learned here is that rote, mechanical efficiency is not enough. The student must think at each step in any mathematical process.

Exercise Set 5–6

1. $(V - 2)(V - 7)$
3. $(I + 2e)(I - 3e)$
5. $(w - 2)(w + 11)$
7. $(i - 20)(i + 3)$
9. $(k^2 + 2t)(k^2 - 6t)$
11. $\left(m^2 - \dfrac{1}{2}\right)\left(m^2 + \dfrac{1}{4}\right)$
13. $\left(r - 4\dfrac{R}{x}\right)\left(r + 6\dfrac{R}{x}\right)$
15. $\left(\dfrac{1}{P} - 2\right)\left(\dfrac{1}{P} + 5\right)$
17. $\left(Z - \dfrac{1}{2}\right)\left(Z + \dfrac{1}{2}\right)\left(Z^2 + \dfrac{1}{8}\right)$
19. $(ab - 3c)(ab - c)$

Exercise Set 5–7

1. $(2t - 7u)(2t + 3u)$
3. $(2e + 4)(4e - 6)$
5. $(2Z + 3)(9Z + 2)$
7. $(4x^2 - 3y)(6x^2 - 3y)$
9. $(3Ve + 7k)(6Ve + 5k)$
11. $(3Q + 2P)(9Q - P)$
13. $(2mn - 3t)(5mn - t)$
15. $(6e + 5t)(7e - 4t)$
17. $\left(6ab - \dfrac{1}{4}\right)\left(8ab - \dfrac{1}{2}\right)$
19. $\left(XZ - \dfrac{1}{3}\right)\left(XZ - \dfrac{1}{2}\right)$

Answers to Selected Problems

Exercise Set 5-8

1. $(Z - X)(Z + X)(Z^2 + X^2)$
3. $(P - 3)(5P + 3)$
5. $3r(r + 3g)(r - 4g)$
7. $3(V + 4)(V - 4)$
9. $4(ir + xe)$
11. $3(2ab - 3cd + 5ad)$
13. $3(z - 3)(Z + 3)$
15. $6(x - 3)(x + 4)$
17. $(C - V - Q)(C - V + Q)$
19. $2(3Z^2 + a)(Z^2 - a)$

CHAPTER 6

Exercise Set 6-1

1. $\dfrac{-1}{-8} = -\dfrac{-1}{8} = -\dfrac{1}{-8}$
3. $-\dfrac{-R_2 R_t}{R_t - R_2}$
5. $\dfrac{V + e}{i - I} = -\dfrac{V + e}{I - i} = -\dfrac{-V - e}{i - I}$
7. $-\dfrac{-k}{p - 2} = +\dfrac{-k}{2 - p} = +\dfrac{k}{p - 2}$
9. $\dfrac{10}{3}$
11. $\dfrac{24R^2}{V}$
13. $\dfrac{V + e}{V - e}$
15. $\dfrac{x - 2}{5x - 3}$
17. $3I + i$
19. -1

Exercise Set 6-2

1. $\dfrac{1}{2}$
3. $\dfrac{-8P^2 + 30}{R}$
5. $\dfrac{V(8VI + 15)}{60I^2}$
7. $\dfrac{25n^2 - 6mn + 25m^2}{200m^3 n^3}$
9. $\dfrac{28Y^2 + 84Z^3 - 9Z^4}{42Z^5 Y^3}$

11. $\dfrac{-1}{Ve}$

13. $\dfrac{Z^2 + Z - 18}{Z^3(Z - 9)}$

15. $-\dfrac{R^2 + 4Rr - r^2}{R^2 - r^2}$

17. $\dfrac{2x^3 + 6x}{x^3 + x^2 - x - 1}$

19. $\dfrac{7}{6}$

Exercise Set 6–3

1. $\dfrac{3V^3 e^4}{5Ii^2}$

3. $\dfrac{R^2 + 2Rr + r^2}{8}$

5. $\dfrac{4P}{t}$

7. $\dfrac{n - m}{m - n}$

9. $\dfrac{5C - 4}{2}$

11. $(3y - 1)(y - 1)$

13. $\dfrac{1}{3}$

15. $\dfrac{147ac}{112xy}$

17. $\dfrac{2 - V}{2 + V}$

19. $\dfrac{n - m}{n + m}$

Exercise Set 6–4

1. $\dfrac{\sqrt{5e}}{2e}$

3. $\dfrac{\sqrt{6}}{2a}$

5. $6Z\sqrt{5}$
7. $A^2\sqrt{2A}$
9. $2(12 + 5\sqrt{2e} + e)$
11. $9I + 6\sqrt{Ii} + i$
13. $8 + 3\sqrt{6}$
15. $\dfrac{x^2 + 4x\sqrt{y} + 4y}{x^2 - 4y}$

CHAPTER 7

Exercise Set 7–1

1. About 357 mA
3. 1.85 A
4. 62 V
5. A voltmeter may be regarded as having infinite impedance, since it draws an insignificant current. This is especially true of VTVMs and DMMs. Therefore, for all practical purposes, there are no voltage drops across any of the individual resistors, so the voltmeter reads the supply voltage. Shorting out any resistor has no effect on the reading, since there is (practically)

Answers to Selected Problems

no current flowing in the circuit. However, shorting out the voltmeter is equivalent to shorting out the supply voltage, so the fuse blows.
6. -66.6 V
7. a. -4 V   b. 15 mA
8. 25 Ω
9. Approximately 10.9%
11. $R \propto T$, so $R = kT$
13. $R = \dfrac{k}{T}$
15. About 1070 Ω
17. $R \propto n$

Exercise Set 7–2
1. $3.80
3. a. Short across R_1, R_2, R_3, or any combination
   b. F_1 opens; V_s shorted
   c. R_2 open
   d. R_1 intermittently opens
   e. R_2 and R_3 not grounded
   f. F_1 blown
5. a. 24.75 Ω   b. 3.182 Ω
7. I. $I_1 = I_t \dfrac{G_1}{G_t} = 0.33 \times \dfrac{1 \times 10^{-3}}{3.884 \times 10^{-3}} = 84.964$ mA

   $I_2 = I_t \dfrac{G_2}{G_t} = I_t \times \dfrac{2.128 \times 10^{-3}}{G_t} = 180.77$ mA

   $I_3 = I_t \dfrac{G_3}{G_t} = I_t \times \dfrac{256.4 \times 10^{-6}}{G_t} = 21.786$ mA

   $I_4 = I_t \dfrac{G_4}{G_t} = I_t \times \dfrac{500 \times 10^{-6}}{G_t} = 42.482$ mA

   $\Sigma I = 330$ mA

   II. $R_t = \dfrac{1}{G_t} = 257.466$ Ω

   $I = \dfrac{V}{R} = \dfrac{85}{R_t} = 330$ mA

9.
```
10 REM                      filename: PARA-L-2
20 PRINT
30 REM   The following program computes the combined
40 REM   total value of resistors in parallel.
50 PRINT
60 INPUT"Enter the number of parallel branches";B
70 FOR X=1 TO B
80 PRINT"Enter the value of R";X;"in ohms."
90 INPUT R(X)
100 NEXT X
110 LET RT=R(1)
120 FOR X=2 TO B
130 LET RT=1/(1/RT+1/R(X))
140 NEXT X
150 PRINT"The total parallel combination resistance is";RT;"ohms"
160 PRINT
170 PRINT"Enter the value of the applied circuit voltage."
180 INPUT V
190 LET I=V/RT
200 LET P=V*I
210 PRINT"The total line current is";I"amperes."
220 PRINT"The power dissipated in the circuit is";P"watts."
```

Exercise Set 7-3

1. a. 7.82 mi
   b. 3.54×10^{12} ft/h
   c. 22,364 mi
   d. 88.5 km/h
   e. 41.6 mi/h
   f. 5.57 rad/s
   g. 9.8 m/s^2
3. 40.5×10^6 m
5. 162.6 cm
6. 155.3×10^3 lb (about 78 tons)
7. 0.265 lb
9. 20,000 kHz

CHAPTER 8

Exercise Set 8-1

1. and 3. *Measured* distance between points:
   a. 11¼ squares
   b. 6¾ squares
   c. 18½ squares
   d. 8 squares
   e. 7 squares
   f. 15¾ squares

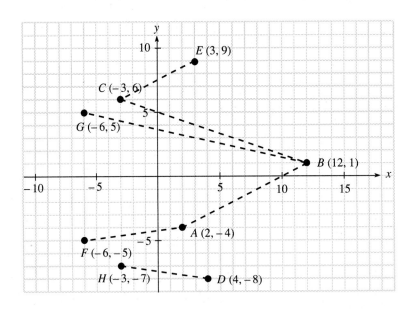

Exercise Set 8-2

1. 11.4 units

2.
Distance	Measured	Calculated	% Error
A to B	11.25	11.18	0.63%
C to E	6.75	6.71	0.6%
G to B	18.5	18.44	0.33%
F to A	8.0	8.06	0.74%
H to D	7.0	7.07	0.99%
C to B	15.75	15.81	0.38%

Answers to Selected Problems

3. a. $a = \pm\sqrt{c^2 - b^2}$ **b.** $b = \pm\sqrt{c^2 - a^2}$
4. a. 9
  b. 5.385
  c. 15.811
  d. 13.038
  e. 13.787
  f 17.661
  g. 1252.971
  h. 1.582
5. 7.607

7.

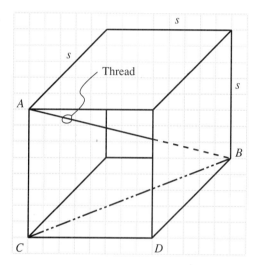

$\triangle ABC$ is a right triangle.

$\triangle CDB$ is also a right triangle.

Side $CB = \sqrt{s^2 + s^2} = s\sqrt{2}$

In $\triangle ABC$,

Side $AB = \sqrt{(s\sqrt{2})^2 + s^2}$
$= \sqrt{2s^2 + s^2} = \sqrt{3s^2}$
$= s\sqrt{3}$

Exercise Set 8–3

1. 209.6 in.^{2}
3. 6.46 ft
5. 51.52×10^3 m^2

Exercise Set 8–4

1.

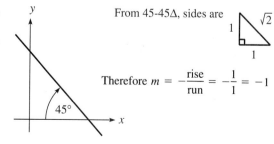

From 45-45$\triangle$, sides are

Therefore $m = -\dfrac{\text{rise}}{\text{run}} = -\dfrac{1}{1} = -1$

3. $y = \dfrac{x}{8}$

446 Appendix J

5.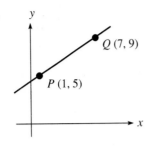

$m = \dfrac{9-5}{7-1} = \dfrac{2}{3}$

Therefore, $y = m(x - x_1) + y_1$

$y = \dfrac{2x}{3} + \dfrac{13}{3}$

Substitute coordinates of $R(3, 6)$

$y = \dfrac{2(3)}{3} + \dfrac{13}{3} = \dfrac{19}{3} \neq 6$

Therefore, R is not on the line connecting P and Q

7. $y = -3.5x - 7$

Exercise Set 8–5

1.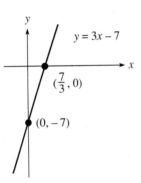

3.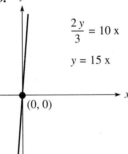

5.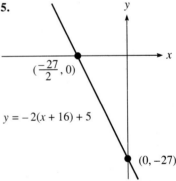

7.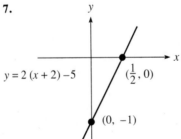

9.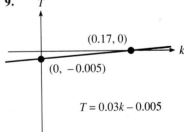

Answers to Selected Problems

11.
```
10 REM                    filename: ST-LINE
20 PRINT
30 REM      This program gives the coordinates of the
40 REM      X and Y intercepts of a linear equation in
50 REM      point-slope form Y = MX + B.
60 PRINT
70 PRINT"Enter the value of the slope."
80 INPUT M
90 PRINT"Enter the value of B."
100 INPUT B
110 LET X=-B/M
120 PRINT"The Y-intercept is (0,";B")"
130 PRINT"The X-intercept is (";X",0)"
```

Exercise Set 8–6

1. **a.** 25.62 V/ms
   b. Q
   c. 3.88 ms
   d. 0
3. 0.25 µs
5. 100 V
11. -1

CHAPTER 9

Exercise Set 9–1

1. $y = 3.5, x = 2$
3. $I = 8, E = 10$
5. No unique solution
7. $x = -5, y = -2$
9. $y = 0, x = 0$
11. $x = 40, y = 5$
13. $y = 3, x = \dfrac{1}{6}$
15. No unique solution
17. $y = \dfrac{S - Q}{PS - QR}, x = \dfrac{P - R}{PS - QR}$
19. $c = -2, b = 3, a = \dfrac{1}{2}$

Exercise Set 9–2

1. $V = 3, I = -4$
3. $b = -\dfrac{1}{14}, a = \dfrac{12}{7}$
5. $x = 2, y = 0$

448   Appendix J

7. No unique solution
9. $y = 1, x = 4$

Exercise Set 9–3

1. $a = 3, b = -2$
3. $x = 16, y = -10$
5. $x = \dfrac{3}{17}, y = \dfrac{49}{17}$
7. $m = 4, n = 4$
9. $P = 30, Q = 20$
10. $s = 1, t = 4$

CHAPTER 10

Exercise Set 10–1

1.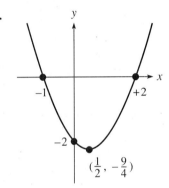

Vertex: $(\tfrac{1}{2}, -\tfrac{9}{4})$
y-intercept: $(0, -2)$
x-intercepts: $(-1, 0), (2, 0)$

Axis of symmetry: $x = \dfrac{1}{2}$

3.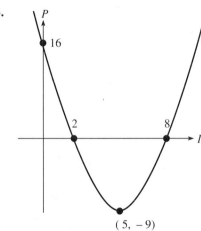

Vertex: $(5, -9)$
y-intercept: $(0, 16)$
x-intercepts: $(2, 0), (8, 0)$
Axis of symmetry: $x = 5$

Answers to Selected Problems

4.

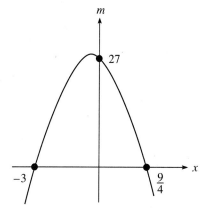

Vertex: $(-3/8, \; 441/16)$
y-intercept: $(0, 27)$

x-intercepts: $(-3, 0), \; (9/4, 0)$

Axis of symmetry: $x = -\dfrac{3}{8}$

5.

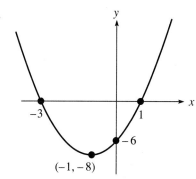

Vertex: $(-1, -8)$
y-intercept: $(0, -6)$
x-intercepts: $(-3, 0), \; (1, 0)$
Axis of symmetry: $x = -1$

7.

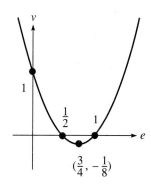

Vertex: $(3/4, \; -1/8)$
y-intercept: $(0, 1)$

x-intercept: $(1/2, 0), \; (1, 0)$

Axis of symmetry: $x = \dfrac{3}{4}$

9.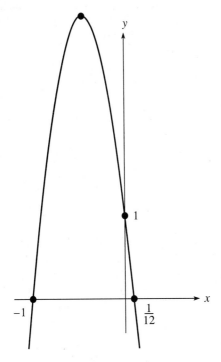

Vertex: $(-11/24, 507/144)$
y-intercept: $(0, 1)$

x-intercepts: $(-1, 0)$, $(1/12, 0)$

Axis of symmetry: $x = -11/24$

Exercise Set 10-2

1. Roots: $(3, 0)$ and $(2, 0)$; vertex: $(2.5, -1/4)$; y-intercept: $(0, 6)$
3. Roots: $(-3, 0)$ and $(2.25, 0)$; vertex: $(-0.375, 27.5625)$; y-intercept: $(0, 27)$
5. Roots: $(5, 0)$ and $(1, 0)$; vertex: $(3, -4)$; y-intercept: $(0, 5)$
7. Roots: $(0, 0)$ and $(0, 0)$; vertex: $(0, 0)$; y-intercept: $(0, 0)$
9. Roots: $(-1, 0)$ and $(3, 0)$; vertex: $(1, 8)$; y-intercept: $(0, 6)$
11.
```
10 REM                    filename: QUAD-2
20 PRINT
30 REM     The following program computes the REAL roots
40 REM     of the quadratic equation ax^2 + bx + c = 0.
50 REM
60 PRINT"Enter the value of A"
70 INPUT A
80 PRINT"Enter the value of B"
90 INPUT B
100 PRINT"Enter the value of C"
110 INPUT C
120 LET D=(B^2-4*A*C)
130 IF D<0 THEN 250
140 LET X1=(-B+(SQR((B^2)-(4*A*C))))/(2*A)
150 PRINT"One of the real roots is (";X1",0)"
160 LET X2=(-B-(SQR((B^2)-(4*A*C))))/(2*A)
170 PRINT"The other real root is (";X2",0)"
180 PRINT
190 LET VX=-B/(2*A)
200 LET VY=A*VX^2+B*VX+C
210 PRINT"The vertex is at the point (";VX",";VY")"
220 PRINT
230 PRINT"The y-intercept is at the point (0,";C")"
240 END
250 PRINT"There are no REAL roots for this quadratic equation."
```

Answers to Selected Problems

Exercise Set 10-3
1. Discriminant is 49; roots are real and unequal.
3. Discriminant is 0; roots are real and equal.
5. Discriminant is -4; roots are imaginary.
7. Discriminant is 8; roots are real and unequal.
9. Discriminant is 321; roots are real and unequal.
11.
```
10 REM                   filename: DISCRIM
20 PRINT
30 REM    The following program computes the discriminant
40 REM    of the quadratic equation ax^2 + bx + c = 0
50 REM    whose roots are real.
60 PRINT
70 PRINT"Enter the value of A"
80 INPUT A
90 PRINT"Enter the value of B"
100 INPUT B
110 PRINT"Enter the value of C"
120 INPUT C
130 LET D=B^2-4*A*C
140 PRINT "The value of the discriminant is";D
```

Exercise Set 10-4
1. $y = x^2 - 5x - 6$
3. $y = 5x^2 - \frac{61}{2}x + 3$
5. $y = 3x^2 + 2x - 5$

CHAPTER 11

Exercise Set 11-1
1. 2.17 V
2. $V_{R_1} \approx 1.5$ V; $V_{R_2} \approx 1.52$ V; $V_{R_3} \approx -0.49$ V (*Actual* polarity is shown in the figure)

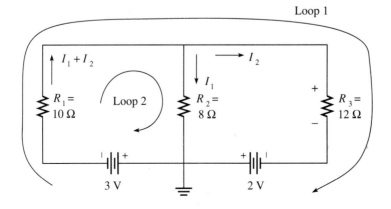

3. $V_{R_1} \approx 3.55$ V; $V_{R_2} \approx 0.59$ V; $V_{R_3} \approx -0.673$ V (Actual polarity and current flow are shown in the figure); $V_{R_4} \approx 2.96$ V; $V_{R_5} \approx 1.73$ V; $V_{R_6} \approx 3.55$ V; $V_{R_7} \approx 0.59$ V; $V_{R_8} \approx -0.673$ V (Actual polarity and current flow are shown in the figure).

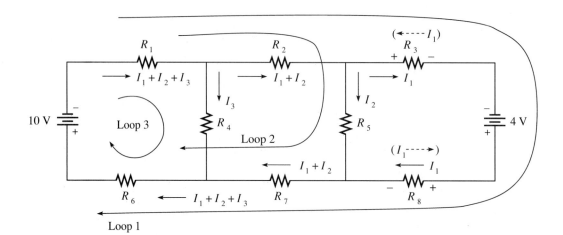

5. 229 mV
7. 10.3 V
8. -3.76 V
9. 30 Ω

CHAPTER 12

Exercise Set 12–1

1. a.

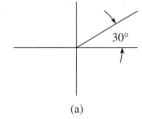

(a)

b.

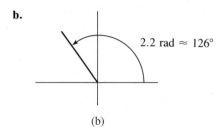

(b)

Answers to Selected Problems

c.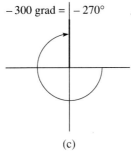

(c)

d.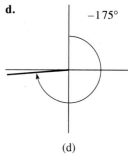

(d)

e.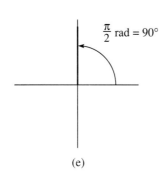

(e)

f.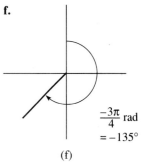

(f)

g.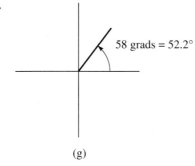

(g)

h.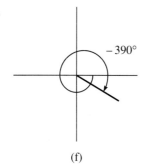

(f)

Appendix J

Exercise Set 12–2

1. -0.5789 (Actual value is -0.57735)
3.
```
10 REM                    filename: TRIG-INV
20 PRINT
30 PRINT
40 PRINT"Enter the angle D in DEGREES."
50 INPUT D
60 LET CSC=1/(SIN(D*.017453292#))
70 LET SEC=1/(COS(D*.017453292#))
80 LET COT=1/(TAN(D*.017453292#))
90 PRINT"The cosecant of"; D"degrees is";CSC
100 PRINT"The secant of"; D"degrees is";SEC
110 PRINT "The cotangent of"; D"degrees is";COT
```

Exercise Set 12–3

1. 53.34 m (about 175 ft)
2. Approximately 734 ft
3. 1291.5 ft
4. 3.76 in.
5. 7.8°
6. 57.5°

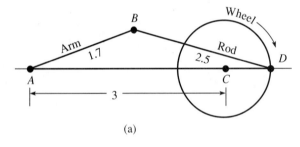

(a)

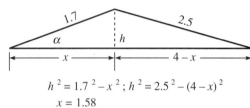

$h^2 = 1.7^2 - x^2$; $h^2 = 2.5^2 - (4-x)^2$

$x = 1.58$

Therefore, $\alpha = \cos^{-1}\dfrac{1.58}{1.7}$; $\alpha = 21.657°$

(c)

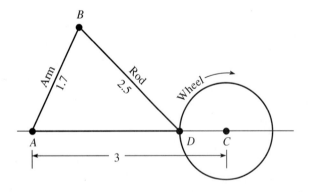

(b)

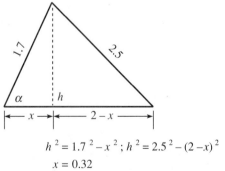

$h^2 = 1.7^2 - x^2$; $h^2 = 2.5^2 - (2-x)^2$

$x = 0.32$

Therefore, $\alpha = \cos^{-1}\dfrac{0.32}{1.7}$; $\alpha = 79.15°$

(d)

$79.15° - 21.657° \approx 57.5°$

7. 59.65 ft
9. 263.5 yd

Answers to Selected Problems

Exercise Set 12–4

1.

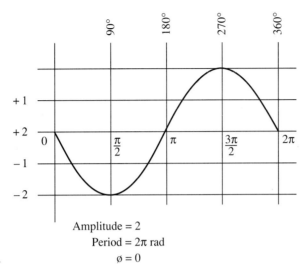

Amplitude = 2
Period = 2π rad
ø = 0

3.

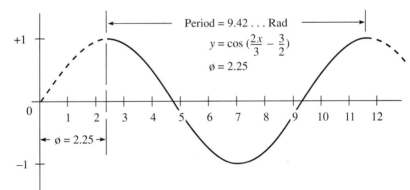

Period = 9.42 ... Rad
$y = \cos\left(\frac{2x}{3} - \frac{3}{2}\right)$
ø = 2.25

4.

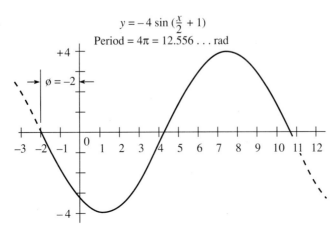

$y = -4 \sin\left(\frac{x}{2} + 1\right)$
Period = 4π = 12.556 ... rad
ø = −2

Appendix J

5.

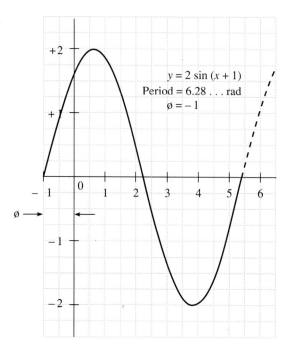

7.

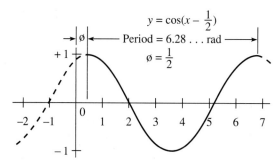

9.

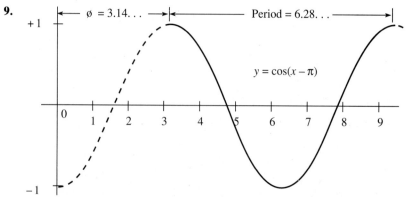

Answers to Selected Problems

10.

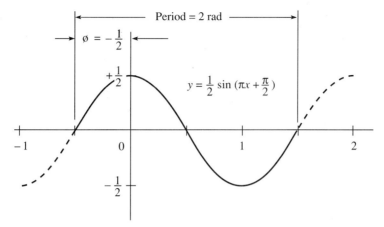

14.

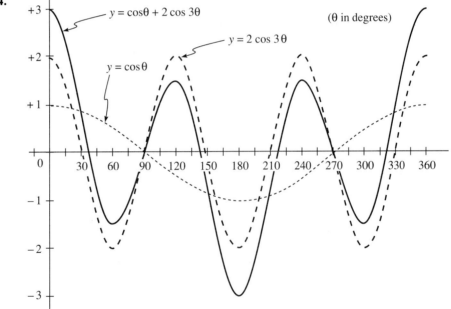

CHAPTER 13

Exercise Set 13–1

1. $1 + \log_3 5$
3. $2 + \log_2 x$
5. $\log_8 5 - \log_8 x$
7. $2 - \log_4 h$
9. $\log_7 4 + 2 \log_7 x - \log_7 3$
11. $\log_5 3$
13. $\log_b xy$

Appendix J

15. $2 \log_4 9$

17. $\log_b \left(\dfrac{\sqrt{x}}{49} \right)$

19. $\log_b M\sqrt{M}$

21. LIST
```
10 REM                    filename: 10 EXP X
20 PRINT
30 PRINT TAB(1)"X" TAB(6)"10^X"
40 FOR X=-20 TO 20 STEP 1
50 LET Y=10^X
60 PRINT TAB(1)X TAB(6)Y
70 NEXT X
```

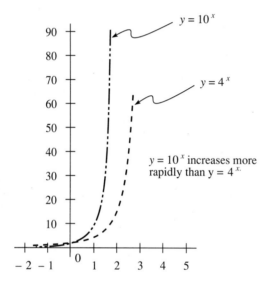

$y = 10^x$ increases more rapidly than $y = 4^x$.

Exercise Set 13–2

1. 1.915
3. −5.43
5. 0.46
7. −33.44
9. 5.42
11. 63.2
13. 0.25
15. 3162
17. 0.001
19. 11.6
21. a. $3.322 \log_{10} \dfrac{F}{f}$
   b. 0.638
23. 17 dB

Answers to Selected Problems

25.
```
10 REM                    filename: LOG ALL
20 PRINT"Enter the base (b) of the logarithm:"
30 INPUT B
40 PRINT"Enter the number (N) of the log you wish to find:"
50 INPUT N
60 LET L=LOG(N)/LOG(B)
70 PRINT
80 PRINT L;"is the log of";N;"to the base";B
90 PRINT
100 PRINT"CHECK: b^L = N. Therefore, N =";B^L
Ok
RUN
Enter the base (b) of the logarithm:
? 3
Enter the number (N) of the log you wish to find:
? 16

 2.523719 is the log of 16 to the base 3

CHECK: b^L = N. Therefore, N = 16
Ok
```

Exercise Set 13–3

1. 0.8614
3. 0.6784
5. 11.293
7. 1.430677
9. 1.442
11. 8.241
13. 0.104
15. −2.8614
17. 1.492
19. 0.6259
21. 0.25
23. 13.6495
25. 32.037
27. 0.125
29. 120.513
31. 16.2
33. 3
35. 0.625
37. 2.8708
39. 2
41. 0.0022 μF
42. Approximately 5.44 billion
43. $\log_2 512 > \log_5 625$
44. 0.073 in.

Exercise Set 13–4

1. 126 mW
2. **a.** 52.3 dB
   b. 27.9 dB
   c. $G' = 27.9 \text{ dB} + 10 \log \dfrac{2200 \ \Omega}{8 \ \Omega} = 52.3 \text{ dB}$
3. 8.85 V
5. 0.00022 dB/ft
7. The output may be as low as 79 W/channel and still be within spec.
9. 85 mW

CHAPTER 14

Exercise Set 14-1

1.

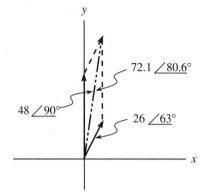

3.

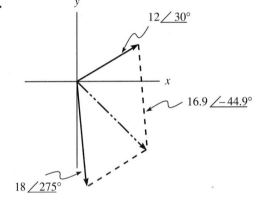

5.

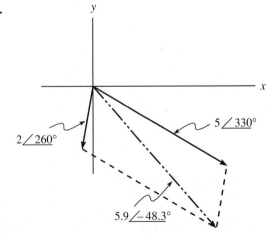

Answers to Selected Problems

7.

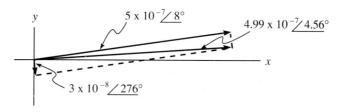

9.

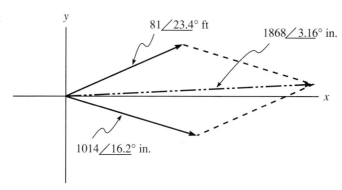

13. 4.74 h
14. a. $A = 3.2 \angle 0°$ V
$B = 0.32 \angle 150°$ V
B leads A by 150°.

b.

CHAPTER 15

Exercise Set 15–1

1. 8.3 mA
3. 76°C
5. 2.14 A_P

Exercise Set 15–2

1. 180 μH
3. 3⅓ H
5. a. $L_2 = \dfrac{L_1 L_t}{L_1 - L_t}$ **b.** $L_1 = \dfrac{L_t L_2}{L_2 - L_t}$

Appendix J

7. 300 mA
8. 100 turns
9. 400 Hz

Exercise Set 15-3

1. 66 pF
3. 11 pF
5. Half of the original value
7.
```
30 REM       This program calculates the capacitive
40 REM       reactance for any values of FREQUENCY
50 REM       in hertz, and CAPACITANCE in farads.
60 REM
70 PRINT"Enter the capacitance in farads."
80 INPUT C
90 PRINT"Enter the frequency in hertz."
100 INPUT F
110 LET XC=1/(2*3.141592654#*F*C)
120 PRINT"The capacitive reactance is";XC"ohms."
```

Exercise Set 15-4

1. 4700 Ω
3. 45.14 V
5.
```
30 REM       This program computes the voltage drop across the
40 REM       resistor in series with a capacitor as C discharges
50 REM       through R.
60 PRINT
70 PRINT"Enter the value of the resistor in ohms."
80 INPUT R
90 PRINT"Enter the value of the capacitor in farads."
100 INPUT C
110 PRINT"Enter the value of the applied DC voltage."
120 INPUT V
130 PRINT"Enter the time in seconds."
140 INPUT T
150 LET X=-T/(R*C)
160 LET VR=V*2.718281828#^X
170 PRINT"The voltage drop is";VR "volts."
```

Exercise Set 15-5

1. 55.5 kΩ
3. 3.6%
5. 128 mH
7. 81.21°
9.
```
20 REM       This program calculates the impedance of a series circuit
30 REM       consisting of resistance, inductance, and capacitance.
40 PRINT
50 PRINT"Enter the frequency of the source voltage."
60 INPUT F
70 PRINT"Enter the value of the resistance in ohms."
80 INPUT R
90 PRINT"Enter the value of the capacitance in farads."
100 INPUT C
110 IF C=0 THEN GOTO 130
120 LET XC=1/(2*3.141592654#*F*C)
130 PRINT"Enter the value of the inductance in henrys."
140 INPUT L
150 LET XL=2*3.141592654#*F*L
160 LET Z=SQR((R^2)+(XL-XC)^2)
170 LET A=ATN(XL/R)
180 LET AA=A*57.29577951#
190 PRINT"The circuit impedance is";Z "ohms."
200 PRINT"The phase angle is";AA
```

Answers to Selected Problems

Exercise Set 15–6

1. 16.63 kHz
2. $L = 2$ H; $R_L = R_i = 150$ kΩ
3. 55 μH
5. 0.038 μF

Exercise Set 15–7

1. $P = 0.225$ μW; $P_{app} = 0.512$ μW; $P_{reactive} = 0.461$ μW
3. 6 W
5. 7.4 μF
7.
```
30 REM    This program computes the capacitance required to raise
40 REM    the power factor of an inductive load by a specified amount.
50 PRINT
60 PRINT"Enter the supply voltage."
70 INPUT V
80 PRINT"Enter the frequency."
90 INPUT F
100 PRINT"Enter the power dissipated by the load."
110 INPUT P
120 PRINT"Enter the PRESENT power factor."
130 INPUT PF
140 PRINT"Enter the DESIRED power factor."
150 INPUT DF
160 LET T=SQR((1-PF^2)/PF^2)
170 LET TT=SQR((1-DF^2)/DF^2)
180 LET A=P*T
190 LET B=P*TT
200 LET C=A-B
210 LET XC=V^2/C
220 LET CX=1/(2*3.141592654#*F*XC)
230 PRINT"The required capacitance is"; CX "farads."
```

CHAPTER 16

Exercise Set 16–1

1.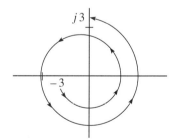

3. $+j3$
5. $\dfrac{1}{j} = \dfrac{1}{\sqrt{-1}} \cdot \dfrac{\sqrt{-1}}{\sqrt{-1}} = \dfrac{j}{j^2} = \dfrac{j}{-1} = -j$
7. $-jR$
9. $\dfrac{-j12}{\sqrt{\pi}}$

Appendix J

Exercise Set 16–2

1.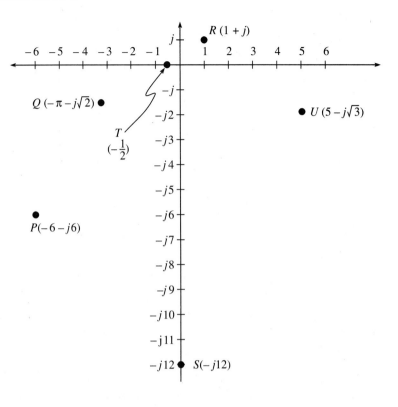

3. 5.831
5. **a.** 45° **b.** 24.24° **c.** 45° **d.** 90° **e.** 0° **f.** −19.1°
   See art, page 465.

Exercise Set 16–3

1. $Z_t = 6.6k - j0.155 \text{ k}\Omega$
3. $4.56 - j13$
5. $\dfrac{7}{3}$
7. $-685 - j660$
9. $-39 - j2$
11. $-0.2 - j0.1$
13. $\dfrac{-12.5 - j9.5}{4.25}$
15. $3.625 - j7.125$

Exercise Set 16–4

1, 3. See art, page 465.
5. **(1)** $57.36 - j81.91$ **(2)** $58.52 \angle -19.98°$ **(3)** $58 + j33.5$ **(4)** $150 \angle 36.87°$

Answers to Selected Problems

Exercise Set 16–2, Problem 5

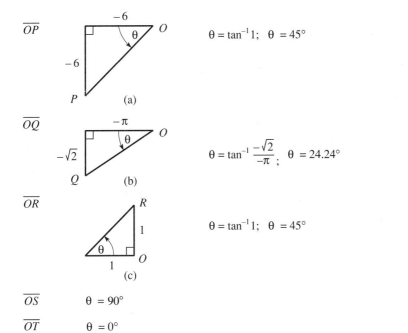

$\overline{OS}$ $\theta = 90°$

$\overline{OT}$ $\theta = 0°$

$\overline{OU}$

Exercise Set 16–4, Problems 1 and 3

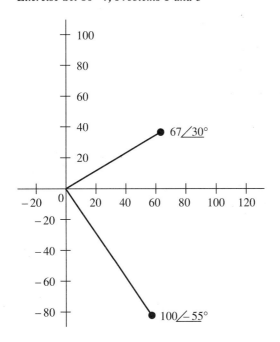

Appendix J

Exercise Set 16-5 (All answers are approximate.)

1. $88 \angle -92°$
3. $374 \angle 47°$
5. $106 \angle 100°$
7. $12{,}409 \angle 0.4°$
9. $7.7 - j29$
11. $12.7 \angle -45°$

Exercise Set 16-6

1. a. $44.55 + j58.16 \ \Omega$
   b. $Z_t = 73.26 \angle 52.55° \ \Omega$
   c. $52.55°$
   d. 164 mA
   e. 1.2 W
3. $30 \ \mu\text{W}$
5. $Z_t = (Z_1 \parallel Z_2) + Z_s = \dfrac{Z_1 Z_2}{Z_1 + Z_2} + Z_s$, where the impedances are expressed as complex numbers
6. $40 + j55 \ \Omega$
7. $53.63 - j6.81 \ \Omega$
9. $10.11 \angle -46.2° \ \Omega$
11.
```
 30 PRINT"Enter the resistance R1 and reactance X1 of branch Z1,"
 40 PRINT"paying careful attention to the sign (+ or -) of X1."
 50 INPUT R1,X1
 60 PRINT"Enter the resistance R2 and reactance X2 of branch Z2,"
 70 PRINT"paying careful attention to the sign (+ or -) of X2."
 80 INPUT R2,X2
 90 LET K1=57.295779 51#
100 LET K2=.017453292#
110 LET Z1=SQR((R1^2)+(X1^2))
120 LET A1=ATN(X1/R1)*K1
130 LET Z2=SQR((R2^2)+(X2^2))
140 LET A2=ATN(X2/R2)*K1
150 LET Y1M=1/Z1
160 LET Y1A=(-1)*A1
170 LET Y2M=1/Z2
180 LET Y2A=(-1)*A2
190 LET P1=Y1M*SIN(Y1A*K2)
200 LET Q1=Y1M*COS(Y1A*K2)
210 LET P2=Y2M*SIN(Y2A*K2)
220 LET Q2=Y2M*COS(Y2A*K2)
230
240 LET YTM=SQR((P1+P2)^2+(Q1+Q2)^2)
250 LET YTA=ATN((P1+P2)/(Q1+Q2))*K1
260 LET ZTM=1/YTM
270 LET ZTA=(-1)*YTA
280 PRINT"The polar impedance is";ZTM"/___";ZTA
290 LET ZTMY=ZTM*SIN(ZTA*K2)
300 LET ZTMX=ZTM*COS(ZTA*K2)
310 PRINT"The equivalent series impedance is (";ZTMX") + j(";ZTMY")"
```
13. $146.9 - j128.7 \ \Omega$

Glossary

Abscissa The horizontal or x-axis of a Cartesian coordinate system. Used to plot the independent variable in a two-variable relationship.

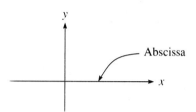

Absolute value The magnitude of any number N without regard to sign, indicated by $|N|$. For example, $|-3| = 3$.

Algebraic expression Any mathematical idea written using literal numbers along with the signs and symbols of algebra.

Algorithm A recipe, or step-by-step process, for solving a given type of mathematical problem.

Amplitude In the trigonometric relationship given by $y = A \sin(Bx - C)$ or $y = A \cos(Bx - C)$, A is the amplitude. It is the maximum y-value attained by the graphs of sine or cosine functions.

Analysis of units A method of making conversions between sets of units by treating the units as if they were literal factors. For example, the speed of light in meters per second may be converted to feet per second as follows:

$$\frac{300 \times 10^6 \, \text{m}}{\text{s}} \times \frac{3.28 \, \text{ft}}{\text{m}} = 984 \times 10^6 \, \text{ft/s}$$

Angle An amount of rotation. An angle may be measured in degrees, radians, grads, or a variety of other units.

Antilog An abbreviation meaning "the inverse log of." If y is the logarithm of x to the base b ($y = \log_b x$), then antilog $y = x$. Also called inverse log and commonly symbolized by $\log^{-1}$.

Apparent power In an AC circuit, the product of the applied voltage and the total line current. Apparent power is related to true and reactive power as shown in the diagram.

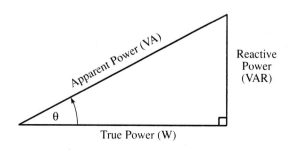

467

Associative property In addition (multiplication), the terms (factors) may be grouped together in any combination without affecting the sum (product). For example, $2 + (3 + 4) = (2 + 3) + 4 = 9$. Also, $2 \times (3 \times 4) = (2 \times 3) \times 4 = 24$. (Note: Subtraction and division are *not* associative.)

Axiom A statement whose truth is so obvious that no formal proof is required. Axioms are used in solving many types of equations.

Axis of symmetry The vertical axis dividing a parabola into two identical halves.

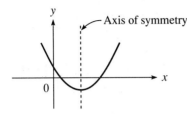

Base The number to be raised to a given power, as indicated by an exponent. In the expression $2^3 = 8$, 2 is the base.

BASIC An acronym for <u>B</u>eginners' <u>A</u>ll-purpose <u>S</u>ymbolic <u>I</u>nstruction <u>C</u>ode. A type of computer program language closely associated with ordinary English sentence rules and common mathematical expression.

Capacitive reactance (X_C) The opposition measured in ohms (Ω) that a capacitor presents to an AC voltage.

$$X_C = \frac{1}{2\pi f C} \text{ ohms}$$

Cartesian coordinate system Two mutually perpendicular straight lines dividing a plane into four quadrants, used for the mathematical analysis of equations and geometric relationships.

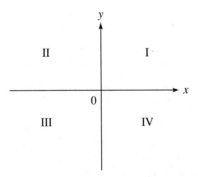

Coefficient Any factor of a product is said to be the coefficient of the remaining factors. In the product $-8xyz$, -8 is said to be the numerical coefficient. A literal coefficient may be any single literal factor or any combination of the literal factors. In $-8xyz$, for example, y is the literal coefficient of $-8xz$.

Common denominator A denominator that is evenly divisible by all other denominators in a group of fractions. For example, in the addition of $-3/12a + 2/6ab + 1/4b$, the common denominator is $12ab$.

Common logarithm A system of logarithms having 10 as the base, such that if $y = 10^x$, then $x = \log_{10} y$.

Glossary

Commutative property In addition (multiplication), the order in which the terms (factors) are added (multiplied) does not affect the result. For example, 2 + 3 = 3 + 2. Also, 2 × 3 = 3 × 2. (Note: Subtraction and division are *not* commutative.)

Complex fraction A fractional expression in which both terms (numerator and denominator) are themselves fractions:

$$\frac{a - \dfrac{b+c}{d}}{b + \dfrac{a-d}{c}}$$

Complex number A number of the form $a \pm jb$, where a is the real part and $\pm jb$ is the imaginary part.

Composite number A number that is not itself prime but rather consists of the product of prime factors. For example, the number 112 consists of the prime factors 2 × 2 × 2 × 2 × 7.

Constant of proportionality If x and y are proportional, the constant of proportionality (k) indicates the exact manner in which y varies as x changes. For example, Ohm's law states that I is proportional to V ($I \propto V$). To make this proportion an equation, we need to add the constant k telling exactly how I changes with variations in V. In Ohm's law, $k = 1/R$, and we write $I = (1/R)V$, or $I = V/R$.

Coordinates The numbers in a Cartesian plane used in locating a point in two-dimensional space. Complete coordinates consist of the ordered pair (x, y). For example, point P has the coordinates $P(2, 5)$.

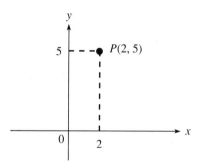

Cosine See trigonometric relationships.

Decibel (dB) A logarithmic unit used to express the ratio of one signal level to another.

$$\text{For voltage:} \quad A \text{ (dB)} = 20 \log_{10} V_2/V_1$$
$$\text{For power:} \quad G \text{ (dB)} = 10 \log_{10} P_2/P_1$$

Degree One 360th part of a complete revolution. One degree equals 0.017453292 rad. Degrees are used as one measure of the magnitude of an angle.

Delta method A method of finding the slope of a curve between two points on the curve by dividing the y increment (Δy) by the x increment (Δx). The slope is then given by

$$m = \frac{\Delta y}{\Delta x} = \frac{\text{rise}}{\text{run}}$$

Denominator In the fraction P/Q, Q is the denominator. Also called the divisor in long division. For example, 3 is the denominator in the fraction $12/3$.

Glossary

Direct variation Two mathematical quantities that either increase or decrease together are said to be directly related. Direct variation is often expressed as direct proportionality using the symbol $\propto$, as in $y \propto x$, which means y is directly proportional to x.
   A statement of a direct proportion may be changed to an equation by adding the constant of proportionality k, as in $y = kx$.

Discriminant In the quadratic equation, the portion under the radical sign, $b^2 - 4ac$. Used to reveal the nature of quadratic roots.

$$b^2 - 4ac = 0 \quad \text{Roots are real and equal}$$
$$b^2 - 4ac > 0 \quad \text{Roots are real and unequal}$$
$$b^2 - 4ac < 0 \quad \text{Roots are imaginary}$$

Dividend When P is to be divided by Q, P is called the dividend. For example, 12 is the dividend in the division problem $12 \div 3 = 4$. In a fraction, the dividend is the numerator.

Divisor When P is to be divided by Q, Q is called the divisor. For example, 3 is the divisor in the division problem $12 \div 3 = 4$. In a fraction, the divisor is the denominator.

Effective value Also called the RMS (root mean square) value. Refers to that value of an alternating current that will produce the same heating effect as a corresponding direct current. The RMS voltage is given by $V = 0.707 V_p$, where V_p is the peak AC value.

Engineering notation A method of expressing powers of ten such that the power is always some multiple of 3. For example, the number 2.09×10^{-7} is expressed in engineering notation as 209×10^{-9}. On a digital hand-held calculator, the key marked ENG converts the displayed number to engineering notation form.

Equal sign The symbol (=) used to indicate that two quantities in an equation are equivalent in value. The quantity to the left of the equal sign is called the left member; that to the right is called the right member.

Equation A mathematical expression stating that two quantities are equal.

Exponent A number written to the upper right of another number (the base) that indicates how many times the base is to be taken as a factor. For example, in the expression $2^3 = 8$, the number 3 is the exponent. In BASIC, exponentiation may be written using the symbol $\wedge$, as in $2 \wedge 3 = 8$.

Exponential equation An equation in which the unknown appears as an exponent.

Extrapolation Any of several techniques for approximately determining the value of some variable lying beyond the actual data points of a curve or table.

Factor When two or more numbers are multiplied together, each is said to be a factor of the product.

Factoring Various methods used to reduce an algebraic expression to two or more factors whose product is the given expression.

Formula A scientific law or relationship expressed as an equation. For example, Ohm's law may be expressed by the formula $I = V/R$.

Fraction An indicated division of the form P/Q. Usually, the actual division of P by Q is *not* to be carried out.

Fractional exponent An exponent of the form p/q, such that

$$b^{p/q} = \sqrt[q]{b^p}$$

Greatest common factor (GCF) The largest factor that will evenly divide two or more terms. For example, the GCF of $12x^3y - 16x^2y^4$ is $4x^2y$. The GCF is also referred to as the highest common factor (HCF) or the greatest common divisor (GCD).

Hypotenuse In a right triangle, the side opposite the right angle. The hypotenuse is always longer than either of the other sides.

Glossary

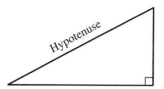

Imaginary number   A number of the form $\pm jN$ expressing the solution to $\sqrt{X}$, where X is a negative number. In particular, $\sqrt{-9} = j3$.

Imaginary root   In a quadratic equation, a solution (root) that is itself an imaginary number. An imaginary root occurs when $b^2 - 4ac < 0$.

Impedance (Z)   The total combined opposition offered to the flow of an alternating current by a circuit containing both resistance and reactance. Impedance is measured in ohms (Ω).

$$Z = \sqrt{R^2 + (X_L - X_C)^2} \quad \text{ohms}$$

Index of the root   The number indicating the root to be extracted from the radicand. For example, 4 is the index of the root in $\sqrt[4]{272}$.

Inductive reactance (X_L)   The opposition measured in ohms (Ω) that an inductor offers to the flow of alternating current. Inductive reactance is given by

$$X_L = 2\pi fL \quad \text{ohms}$$

Inequality   A statement that one quantity is greater than ($>$), less than ($<$), greater than or equal ($\geq$), less than or equal ($\leq$), or not equal ($\neq$) another. For example, $3 < x + 2$ is an inequality for which the solution is $x > 1$.

Infinity (∞)   A quantity is said to become infinite when it increases or decreases without limit.

Juxtaposition   In algebra, a method of indicating multiplication by writing the factors adjacent to one another without any other symbols of operation. For example, the product $15 \times A \times C$ may be written using juxtaposition as $15AC$.

Left member   In an equation or inequality, the expression to the left of the equal or inequality sign.

Like terms   Algebraic terms having the same literal parts. For example, $-152ax^2$, ax^2, and $0.33ax^2$ are all like terms.

Linear equation   An equation of the form $ax + b = 0$.

Literal number   A letter used to represent some numeral or other quantity.

Logarithm   An exponent. In the exponential equation $y = B^x$, x is the logarithm of y to the base B ($x = \log_B y$).

Logarithmic equation   An equation containing one or more logarithmic expressions, such as $3 = \log_5 y$. Since this may be written in exponential form as $y = 5^3$, then $y = 125$ is the solution to the logarithmic equation.

Lowest common denominator (LCD)   The smallest quantity that is evenly divisible by two or more denominators. In the fractional sum $a/b^2c + d/bc^2$, the LCD is b^2c^2.

Minuend   In a subtraction problem, the number from which another (the subtrahend) is to be subtracted. For example, 5 is the minuend in the subtraction problem $5 - 2 = 3$.

Monomial   An algebraic expression having only one term. For example, I^2R is a monomial.

Multinomial   An algebraic expression having more than one term. For example, $ax^{-3}b^2 - cd^4$ is a multinomial (see polynomial).

Natural logarithms   A system of logarithmic relationships whose base (e) is approximately 2.71828. Also called the Naperian logarithm.

Negative number   A number less than zero. That is, a negative number is any number to the left of zero on the number line.

Number line A straight line used to represent real numbers graphically as points on the line. The middle of the number line is zero, with negative numbers to the left and positive numbers to the right.

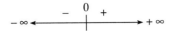

Numerator In a fraction, the term above the vinculum. N is the numerator in the fraction N/D.

Ordinate The vertical axis, or y-axis, of a Cartesian coordinate system.

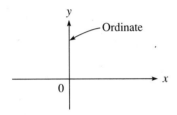

Origin In a Cartesian coordinate system, the point where the two axes intersect. The origin has the coordinates (0, 0).

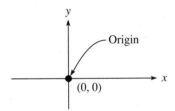

Parabola The shape of a curve expressed by a quadratic equation. For example, the following is a parabola.

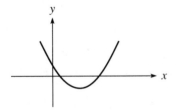

Percent Literally, "by the hundreds." A numerical strategy in which the whole of something is considered to be made up of 100 parts (100%). If, for example, we are considering 3 parts out of 15, we see from the proportion

$$\frac{3}{x\%} = \frac{15}{100\%}$$

that 3 corresponds to 20%.

Glossary

Period The time required for a periodically changing event to complete one cycle. In the case of the sinusoidal voltage given by $y = A \sin(Bx - C)$, the period is $2\pi/B$.

Phase angle In an AC circuit containing resistance and reactance, the angle between the applied voltage and the total line current.

Phase shift The angular displacement between two similarly varying voltages or currents. In the case of the sinusoidal voltage given by $y = A \sin(Bx - C)$, the phase shift is C/B.

Phasor An electrical quantity treated like a vector.

Plane A two-dimensional surface, such as the surface partitioned into four quadrants by a Cartesian coordinate system.

Point-slope form The equation of a straight line having the form

$$y = m(x - x_1) + y_1$$

where $m = $ slope and (x_1, y_1) is a point on the line.

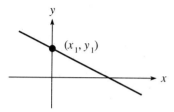

Polar form A complex number in the form of $A \angle \theta$, where A is the magnitude and θ is the angle.

Polarity The sign of a number ($+$ or $-$) indicating whether it is greater than zero (positive) or less than zero (negative).

Polynomial An algebraic expression of the form ax^n, where n is a positive whole number. A polynomial may have only one term (see multinomial).

Power factor An indication of the amount of true power being dissipated in a sinusoidal AC circuit. Power factor values may range from 0 to 1, with 1 indicating that 100% of the power is true power. Numerically, power factor is equal to the cosine of the phase angle.

$$\text{P.F.} = \cos \theta$$

Powers of ten A number such as 10^n.

Prime number A number that is not evenly divisible by any number other than itself and 1. For example, 2, 3, 5, and 17 are prime numbers.

Product The result of multiplying two or more numbers together.

Proportion A statement that two fractions are equal.

Proportionality symbol The symbol ($\propto$) indicating the manner in which one quantity varies with another. For example, $X \propto Y$ means that X varies directly with Y. To indicate that X varies inversely with Y, we may write $X \propto 1/Y$ or $X \overline{\propto} Y$.

Pythagorean theorem An algebraic expression relating the three sides of a right triangle.

$$A = \sqrt{B^2 + C^2}$$

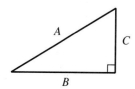

474 Glossary

Quadrant One of four areas in a plane partitioned by a Cartesian coordinate system.

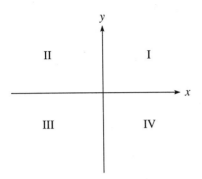

Quadratic equation An equation of the form $y = ax^2 + bx + c$ $(a \neq 0)$.

Quadratic formula A formula for finding the roots of a quadratic equation.

$$x = \frac{-b \pm \sqrt{b^2 - 4ac}}{2a}$$

Quotient The result obtained by dividing one number by another.

Radian An angle with its vertex at the center of a circle, and whose sides intercept an arc whose length is equal to that of the radius of the circle. Angle θ has a magnitude of 1 rad. In degrees, 1 rad ≈ 57.29577951°.

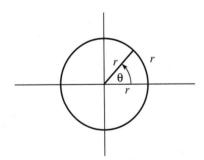

Radical sign A symbol ($\sqrt{}$) used to represent that a root is to be extracted from a given number, called the radicand. The radical sign is actually a composite symbol using a vinculum to group the terms or factors of the radicand.

Radicand The number under the radical sign whose root is to be extracted. For example, 25 is the radicand in $\sqrt{25}$.

Ratio A fraction used for the purpose of comparing or relating two similar quantities.

Reactive power The power that appears to be dissipated by a pure reactance. Reactive power is given by I^2X, V_XI, or V_X^2/X, and is measured in units of VAR (volt-amperes reactive).

Real number Any number, positive, negative, or zero, that can be represented as a point on the number line.

Glossary

Rectangular form   A complex number in the form of $a \pm jb$, where a is the real part and $\pm jb$ is the imaginary part.

Resonant frequency   That frequency at which all reactances cancel each other out, leaving only pure resistance. The resonant frequency may be calculated by

$$f = \frac{1}{2\pi\sqrt{LC}}$$

Resultant   The net result of two or more plane vectors acting at a point.

RMS   The effective R(oot) M(ean) S(quare) value of a sinusoidal voltage or current. The **RMS** value of a sinusoidal AC voltage, for example, is given by

$$V = 0.707 \times V_p$$

where V_p is the peak value of the voltage.

Root   A value of a literal variable that satisfies an equation is said to be a root of that equation.

Satisfy   A number is said to satisfy an equation (and hence be a solution) when replacement of the variable by its numerical value produces identical results on both sides of the equal sign. For example, $x = 3$ is said to satisfy the equation $8 = 11 - x$, since $8 = 11 - 3$, or $8 = 8$.

Scalar   A quantity having magnitude only. For example, the pound is a scalar quantity.

Scientific notation   A method of expressing very large or small quantities as numbers between 1 and 10 times an appropriate power of ten. For example, the quantity 0.0001345 may be expressed in scientific notation as 1.345×10^{-4}.

Secant line   A straight line that crosses a curve in two points.

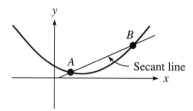

Signed numbers   Positive (+) or negative (−) numbers. The sign (polarity) of the number indicates whether it is to the right or left of zero on the number line.

Simultaneous equations   A system of two or more equations whose roots simultaneously satisfy all the given equations. In the simplest case, the two equations may have one unique solution, no solution, or an infinite number of solutions.

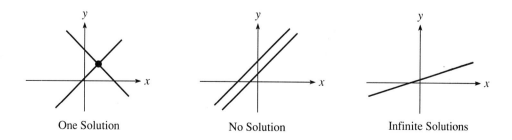

One Solution   No Solution   Infinite Solutions

Sinewave A signal whose amplitude varies with time according to the sinusoidal function $y = A \sin(Bx - C)$.

Sinusoidal Varying in accordance with the sine of an angle.

Slope A measure of the steepness of a straight line. Numerically, slope (m) is the ratio of rise over run.

$$m = \frac{\text{rise}}{\text{run}} = \frac{\Delta y}{\Delta x}$$

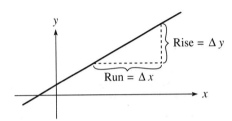

Slope-intercept form The equation of a straight line having the form

$$y = mx + b$$

where $m =$ slope and $b = y$-intercept.

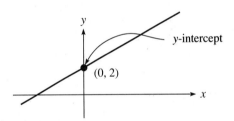

Solidus fraction A means of writing a fractional quantity on a single line, as in $y = (ab)/(a - b)$.

Solve To solve an equation or inequality means to find the values of the literal factors that satisfy the equation.

Standard position An angle is said to be in standard position when its initial side extends along the positive x-axis and its vertex is at the origin of a rectangular coordinate system.

Subtrahend In a subtraction problem, the number that is to be subtracted. For example, in the subtraction $8 - 3 = 5$, 3 is the subtrahend.

Summation notation A shorthand way of indicating the sum of several terms. For example, the sum of two or more resistors may be written as

$$R_t = \sum_{k=1}^{n} R_k = R_1 + R_2 + \cdots + R_n$$

Symbols of grouping Symbols used to indicate that an expression is to be regarded as a single quantity. The conventional symbols are the parentheses ( ); the brackets [ ]; the braces { }; and the vinculum _____, which is most commonly used with fractions and radicals.

Glossary

Tangent line A line that touches a curve in one and only one place.

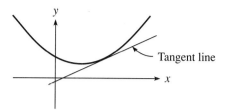

Terminal side A straight line used to represent the extent of rotation of an angle about its vertex.

Terms Those parts of an algebraic expression whose individual pieces are not separated by a plus or minus sign. For example, in the algebraic expression $13a^2bx - 5a^2$, the terms are $13a^2bx$ and $-5a^2$.

Transpose An informal expression meaning to transfer a term from one side of an equation to the other by means of addition or subtraction. For example, in the equation $y - 3 = x$, we may transpose the -3 from the left member to the right member and write $y = x + 3$.

Trigonometric relationships In a right triangle, the following definitions are made:

$$\sin \theta = \frac{\text{opposite side}}{\text{hypotenuse}}$$

$$\cos \theta = \frac{\text{adjacent side}}{\text{hypotenuse}}$$

$$\tan \theta = \frac{\text{opposite side}}{\text{adjacent side}}$$

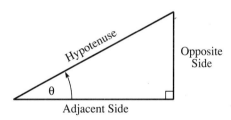

Vector quantity A quantity having both magnitude and direction. A vector may be represented by a straight line whose length indicates its magnitude, whereas its direction may be indicated by the angle it makes with some reference point.

Vertex (1) The point where the sides of an angle intersect. (2) The point where the axis of symmetry crosses the graph of a parabola.

Vinculum A straight line used as a symbol of grouping. The vinculum is commonly used to group terms in fractions and in the radicand, as in $\frac{a-b}{a+x}$ and $\sqrt{25ax - 5ac}$.

Word problem A problem in which verbal statements are used to express the conditions or circumstances of a situation involving numerical quantities.

x-intercepts The point or points where the graph of an equation crosses the x-axis.

y-intercepts The point or points where the graph of an equation crosses the y-axis.

Z The symbol for impedance. The impedance offered to the flow of an AC sinusoidal current is given by

$$Z = \sqrt{R^2 + (X_L - X_C)^2} \quad \text{ohms}$$

Index

Abscissa, 164, 251
Absolute gain, 308
Absolute value, 35
Addition
   algebraic fractions, 112–114
   arithmetic fractions, 18
   decimals and whole numbers, 6
   signed numbers, 34, 45
Algebraic expression, 38
Algebraic terms, 39
Amplitude, 276
Analysis of units, 154–155
Angle, 248
Antilog. *See* Inverse logarithm
Apparent power, 358
Arithmetic, operations of, 5
Array. *See* Determinants
Associative property, 8, 21
Axioms
   algebraic equations, 72
   algebraic inequalities, 78
Axis of symmetry 220

BASIC computer language, 9
Base, 12, 49, 295, 297
Braces, 41
Brackets, 41
Brigg's system. *See* Common logarithm

Calculator, 7
Calculator conversions (P–R and R–P), 380–381
Capacitive reactance, 342, 375
Cartesian coordinate system, 164
Coefficient, 40
Coefficient of coupling, 332
Common denominator, 18
Common factor, 90–91
Common logarithm, 295–296
Commutative property, 8, 21
Completing the square, 222
Complex numbers, 371–381
Composite number, 86
Conjugate of complex numbers, 376
Coordinates of a point, 165
Cosine of an angle, 253

Cosine curve, 274
Counterelectromotive force (Cemf), 337
Current divider, 150–151

Decibel (dB), 305–308
Degree of an angle, 249
Delta method, 185
Denominator, 18, 108
Determinants
   in computer-assisted analysis, 208, 213
   second order, 205–210
   third order, 210–214
Digit, 5
Direct variation, 135
Discriminant, 226
Dissimilar terms. *See* Unlike terms
Distance, 168–170
Dividend, 16
Division
   algebraic fractions, 116–118
   arithmetic, 16
   arithmetic fractions, 20
   exponents, 56, 68
   monomials, 64, 69
   multinomial by monomial, 62–63, 69
   multinomials, 60, 69
   signed numbers, 60, 69
Divisor, 16

Effective value, 328–329
Electrical vector. *See* Phasor
End (BASIC statement), 9
Eng key (calculator), 14
Engineering notation, 14
Equal sign, 72
Equations
   defined, 72
   exponential, 299–300
   straight line, 178
   linear, 72
   solving, 74–77
Exp key (calculator), 14
Exponent 12, 49, 288
   fractional, 88
   negative, 57, 68

Exponential relationship, 289
Extrapolation of data, 183

Factoring
   $acx^2 + (ad + bc)x + bd$, 98–101
   common factor, 90–91
   differences of squares, 93–94
   perfect trinomial squares, 92–93
   $x^2 + (a + b)x + ab$, 95–98
Factors, 39
   product of factors, 11
   in radicand, 88–89
Formula, defined, 128
Fractions
   algebraic, 108
   arithmetic, 18
   complex, 118
   containing radicals, 119–122
   properties, 108–110
   solidus, 130
Frequency, 276

GCF. *See* Greatest common factor
Grad, 250
Graphing linear equations, 180
Greatest common factor, 90
Ground, electrical, 32

HCF. *See* Greatest common factor
Hypotenuse, 168, 256, 258

Imaginary number, 4, 367
Imaginary operator. *See* Operator j
Imaginary roots, 221, 226
Impedance, 350
   complex representation, 374
   geometrical representation, 374
Impedance matching, 335–337
Index of a root, defined, 88
Indicated division, 18
Inductive reactance, 338, 375
Infinity, 5
Initial side, 248
Input (BASIC statement), 9
Instantaneous rate of change, 183
Inverse logarithm, 296
Inverse relationship, 290
Inverse trigonometric relationships, 259–260
Inverse variation, 135

Juxtaposition, 48

Kirchhoff's current law (KCL), 235
Kirchhoff's voltage law (KVL), 234

LCD. *See* Lowest common denominator
Laws of exponents, 50, 56, 68

Let (BASIC statement), 9
Like terms, 39
Literal number, 38
Logarithmic equation, 300–302
Logarithmic relationship, 290
Loop equations, 237–239
Lowest common denominator, 18, 113

Mathophobia, 4
Matrix. *See* Determinants 206
Member of an equation, 72
Monomial, 51
Multiplication, 52
   algebraic fractions, 115–116
   arithmetic fractions, 20
   exponents, 49
   monomials, 51–52, 68
   multinomial by monomial, 53, 68
   multinomials, 54–55, 68
   signed numbers, 48, 68
   whole numbers and decimals, 11
Mutual inductance, 332

Naperian log. *See* Natural logarithm
Natural logarithm, 295–297
Negative number, 32
Network analysis, 238–240
Number line, 5
Numerator, 18, 108

Ohm's law, 128
Operator j, 368
Ordinate, 164
Origin of coordinate system, 165

Parabola, 218
Parallel circuit
   AC, 365–387
   DC, 146–151
Parentheses, 41
Percentage, 25
Percent of error, 58
Perfect trinomial square, 92–93
Phase angle, 352
Phase shift, 278
Phasor, 317
Pi, 250
Point-slope form, 178
Polar form, 379
Polarity, 32
Polynomial, 52
Power, 12
Power factor, 359
Power key (calculator), 12
Powers of ten, 13, 57–58, 69
Prime factorization, 86–87
Prime number, 86

Index

Principal diagonal, 206
Print (BASIC statement), 9
Proportion, 23
Proportionality constant, 135
Proportionality symbol, 135
Pythagorean theorem, 167, 266

Quadrant, 252
Quadratic formula, 221–223
Quadratic relationship, 218
Quotient, 16

RMS. *See* Root mean square
Radian, 249
Radical sign, 15
Radicand, 15
Ratio, 23
Rationalizing denominator, 119–120
Reactive power, 358
Real number, 5
Rectangular components of a vector, 314
Rectangular form of complex number, 378
Rem (BASIC statement), 9
Renum (BASIC statement), 11
Resistor color code, 14
Resonance, 355
Resultant, 313
Right triangle, 256
Root of an equation, 72, 221
Root mean square, 329

Scaler quantity, 312
Scientific notation, 7, 13
Secondary diagonal, 206
Self-inductance, 330
Series circuits
  AC, 325–364
  DC, 128–146
Signed numbers, 34
Similar terms. *See* Like terms
Simultaneous linear equations, 194–216
Simultaneous solution, 194
Sine, 253
Sine wave, 272
Sinusoidal, 272
Slope of a line, 176
Slope-intercept form, 178
Solution
  by addition and subtraction, 196–202
  by comparison, 204
  by substitution, 202
Special triangles. *See* Triangle
Square root, 15
Standard position of an angle, 248
Subtraction
  arithmetic, 9
  signed numbers, 36, 45
Summation notation, 133
Superposition theorem, 240
Symbols of grouping, 41, 45
  nested, 42
  removing, 43

Tangent, 257–258
Terminal side of an angle, 248
Terms of a fraction, 108
Time constant, 344–348
Transformers, 333–337
Transposition, defined, 75
Triangle
  30-60, 173
  45-45, 172
Trigonometry, 248
Trinomial, 51

Unlike terms, 39

Vector, 264, 312
Vector diagram, 312
Vector summation, 313–320, 326
Vertex
  of angle, 248
  of parabola, 220
Vinculum, 41
  of fraction, 18, 108
Voltage divider, 137–142

Whole number, 5
Word problems, 79–82

X-intercepts, 220

Y-intercept, 218

Zero exponent, 59, 69
Zero
  division by, 21
  problems with, 21–22